Encyclopedia of Herbicides: Research and Reviews

Volume VI

Encyclopedia of Herbicides: Research and Reviews

Volume VI

Edited by **Molly Ismay**

New York

Published by Callisto Reference,
106 Park Avenue, Suite 200,
New York, NY 10016, USA
www.callistoreference.com

Encyclopedia of Herbicides: Research and Reviews
Volume VI
Edited by Molly Ismay

International Standard Book Number: 978-1-63239-260-2 (Hardback)

Contents

Preface

The researches and case studies detailed in this book provide a great deal of information concerning herbicide use in a number of settings. Both in agricultural and non-crop settings, herbicides have found utility in weed management strategies universally. However, applications and methodologies have been continually undergoing advancements. The aim of this book is to facilitate a better understanding of perennially evolving weed compositions while maintaining the effectiveness of current options. This book, therefore, provides detailed research review on the prevalent trends in the utilization of herbicides, toxicity of herbicides and its impact on aquatic, soil biota and human health.

This book is a result of research of several months to collate the most relevant data in the field.

When I was approached with the idea of this book and the proposal to edit it, I was overwhelmed. It gave me an opportunity to reach out to all those who share a common interest with me in this field. I had 3 main parameters for editing this text:

1. Accuracy – The data and information provided in this book should be up-to-date and valuable to the readers.
2. Structure – The data must be presented in a structured format for easy understanding and better grasping of the readers.
3. Universal Approach – This book not only targets students but also experts and innovators in the field, thus my aim was to present topics which are of use to all.

Thus, it took me a couple of months to finish the editing of this book.

I would like to make a special mention of my publisher who considered me worthy of this opportunity and also supported me throughout the editing process. I would also like to thank the editing team at the back-end who extended their help whenever required.

Editor

Research Reviews

Herbicide Resistant Weeds:
The Technology and Weed Management

Jamal R. Qasem

Additional information is available at the end of the chapter

1. Introduction

Pest resistance to control methods in general is not an isolated phenomenon but usually expected and well demonstrated when any method is repeatedly applied over a long period of time without being changed or modified in nature, structure, principals of application or formulation. All pests that growers must control in agricultural land have the capacity to become resistant to whatever tactic is used to control them [11]. It is usually expressed as a gradual adaptation or "fitness" of some individuals or populations of the targeted pest or organism to the frequently applied control methods and available conditions. This adaptation may be physical, morphological or phenological, physiological, anatomical or biochemical or could result from the interaction between any two or more of these. It may also be due to some genetic changes as mutations occur on the key site at which a specific method operates. These mutations are at least partially dominant and inherited. Traits are conferred by modifications to single nuclear genes. This indicates that the rate of resistance evolution will be driven by mutation, the intensity of selection, the dominance and relative fitness of mutations in presence or absence of the herbicide and by dispersal of resistance alleles within and between weed populations [28]. However, no proof that the herbicides cause the mutations leads to resistance [37]. However, most often resistance is controlled by a single, dominant or semi-dominant gene [38] although recessive genes control of herbicide resistant trait in natural weed populations has been also implicated in resistance to dintroanaline, while wild populations exposed to herbicide stresses for the first time may efficiently express herbicide-resistant genes.

Most weed modifications and adaptations, if not all, are advantageous to the pest, since allow its escape on time and/or place and thus avoid external hazard or threat to its existence and genetic line. Resistance therefore should not be confused with natural tolerance or low susceptibility due to a normal physiological or behavioristic property of an unselected population [23].

Organisms are varied in sensitivity, responses and thus adaptability to such conditions and in responses to any treatment or imposed external factors. Tolerance and then gradual resistance of agricultural pests to any control method or environmental stress is thus a strategy through which organisms/ or pests encounter hazards and maintain life and therefore may be applied to any method of pest or weed control including prevention, mechanical, cultural, physical, biological and chemical [30]. For example, weeds resisting soil mulch cover show some morphological and/or physical characteristics that allow penetration of the mulch layer; also, flooding of resistant species possess water impermeable seed coat or generate O_2 and reduce CO_2 penetration. Firing or flaming is resisted through presence of a hard seed coat or deeply buried regenerative propagules; certain weed species show feedback mechanisms or luxury accumulation of mineral nutrients and thus avoid toxicity; high temperature and low soil moisture harmful effects are avoided by adoption of secondary or enforced seed dormancy, while harmful effects of excessive light is avoided by some morpho-physiological alterations. Soil acidity may be encountered in the microhabitat by root exudates or selective mineral absorption and salinity by excretion of salt through different mechanisms and formation of salt glands or vacuoles or shedding salt saturated organs; microbes attack is avoided by production of repellent allelochemicals, and pests through some morpho-chemical adaptations. However, the mechanism behind tolerance or resistance is different and based on the type of target pest or the hazard imposed.

Herbicides represent one of the external factors and form a group of synthetic- plus some biochemicals used to suppress or kill unwanted vegetation and are a major component of pesticides. They assist in management and restoration of areas invaded by invasive species. Herbicides are a major technological tool and responsible, in part, for an agricultural revolution and increase in food production in the last few decades. However, at present this technology faces radical changes in effectiveness under field conditions that lead in different cases to failure of weed control operation due to continued development of weed tolerance/resistance and evolution and limitations in the herbicide industry and development.

2. Agriculture practices and weed evolution

General weed control methods (tillage, hoeing, hand weeding, flooding, cuttings or mowing, flaming, use of general herbicides) are all nonselective and usually applied to a composite weed species or vegetation of inter and intra-specific variations in richness, morphology, growth habit and responses. Each species may adapt, or not, to any of these methods. Since weeds are widely different in mechanisms by which they encounter hazards they are exposed to, they are different in plasticity and responses. With continued use of a single control method for a long period of time, species migrate, flourish or die. Flourishing species gradually became better fit and adapted, and increase in number and population size in absence of others. The only surviving individuals are those possessing rare single gene mutations and evolved resistance will be monogenic, resulting in a large change in the resistance phenotype. However, when doses are lower and selection acts within the range of standing genetic variation, polygenic responses will be possible and resistance will evolve by a gradual change in the mean susceptibility of the population [28]. On the other hand, population of not or less adapted individuals, decline in growth and number until greatly suppressed, limited and may become extinct. Therefore, with continuous dependence on a single method of weed control, a weed population is usually shifting toward better adapted species or individuals that cope well with existing control measures and new conditions. Self-thinning of a weed population is continued toward complete tolerance to employed control measures. Therefore, weeds adapted to mowing tend to grow short, in a rosette form, creeping above the soil surface or show high plasticity and softness of aerial parts and stems and become difficult to mow and also escape hand weeding. Deep rooted weed species are difficult to pull out even by soil tillers. Seasonal dormancy and shifts in the weed population in the growing season is well recognized for certain weed species such as *Senecio vulgaris* [29; 37], while physiological adaptation of *Echinochloa crusi-galli* and *Cyperus rotundus* to flooding conditions and the role of Alcohol dehydroginase enzyme (Adh) in *E. crusi-galli* is well documented [5; 14]. Similar adaptations of *Cirsium arvense* ecotypes to temperature variations [43] and *Typha anguistifolia* and *Typha latifolia* genetic and clonal variations [27; 40] have also been reported. In this regard, it is important to differentiate between tolerance and resistance of weeds to herbicides. Tolerance is the inherited ability of a species to survive and reproduce after herbicide treatment; it refers to the natural variability to herbicides and exists within individuals of a species and quickly evolves. It usually refers to relatively minor or gradual differences in intraspecific variability. Resistance is the inherited ability of a plant or a biotype to survive and reproduce following exposure to a dose of herbicide that is normally lethal to wild type [16; 23; 30; 37]. Therefore, it is a decreased response of a population of weed to herbicides as a result of their application. However, both terms sometimes are misused or used interchangeably.

Tolerant weed species are less harmed by herbicides; they exhibit a certain degree of avoidance or adaptation strategy that allows recovery and thus escape control measures. They may respond by timing stomata closure or having sunken pores or stomata, thick waxy cutical on upper leaf surface, encased growing points or some biochemical, physiological or anatomical properties better developed by time until they become best fit and adapted to applied herbicides and become thereafter resistant. This, however, leads to gradual but radical changes in the weed population composition and distribution spectrum at which resistant individuals or certain weed species increased and dominate and susceptible ones are reduced and replaced. Adaptation or exclusion of the less tolerant species depends on performance of these by time. Generally a weed population becomes rich in individuals and poor in species with the continuous use of the same herbicide or different herbicides of similar mode/mechanism of action. This shift does not however, reflect better competitiveness or higher regenerative ability but most likely due to absence of sensitive highly competing species or forms that allow resistant individuals to utilize more resources [9; 22].

In cultivated fields, associating weeds bear more resemblance to crop plants in morphology, physiology and responses to control measures and other agricultural practices in general. They mimic crops from sowing and germination until harvest. Since herbicides used on crop plants are selective, weeds respond by exhibiting similar morphology, physiology and biochemistry as crop plants to avoid hazards. However, weeds derived from crop plants as hybrids, crop relatives or wild-weedy forms are better fit to such conditions than others. Weed-crop associations also exist between weed species of different taxa from crop plants. In this case, the longer the use of the same herbicide/s, the greater the close association between crops and certain well performed weed species that later transfer into adapted weed races. Crop relative weeds however, are of great potential to intra- and inter- gene exchange and efficient mating system among themselves and with crops, thus become best adapted and more difficult to control.

3. Selection pressure and weed races

With continuous use of the same agricultural practice/s, interspecies selection occurs and plant species are gradually purified (intraspecific selection) by time until they become best adapted. Since all control measures including herbicides aim to eliminate weeds without causing injury to crop plants, weeds respond by developing mechanism/s allowing escape of chemical hazards. Under such conditions, sensitive individuals are first limited or disappear. Tolerant individuals increase in number and accumulate tolerance until they become resistant. Therefore, a resistant population of any weed species is exposed to long-term selection pressure through which it is purified and performs well under prevailing conditions in absence of sensitive weed species. With continuous exposure to herbicide pressure, a population of resistance is usually developed.

Weeds tend to avoid herbicide toxicity by changing normal growth habits, or exhibiting some phenological (such as changes in germination patterns), physical and/or physiological changes through which they adjust emergence time, external appearance or physiology. These however, are inherited traits that allow plants to survive herbicide treatments. One best adaptation is that of weeds similar to crop plants in most or all growth aspects. These form weed races similar to crop plants and well adapted to their habitats. Among reported weed races are *Camelina sativa* to flax crop, *Echinochloa crus-galli var. Oryzicola* that associate with rice and the weedy wild rice or red rice in India and east-south Africa [8; 20]. All are genetically irrelevant to crop plants. However, in some cases weed races are of the same botanical family or belong to the same crop species. This kind of association leads to development of "crop-races" that possess weedy characters very well adapted to cultural practices; they are similar to crop plants in most growth aspects and difficult to control by herbicides or other control methods including hand weeding. They take an advantage from conditions under which crop plants are growing until they become difficult to leave their habitats or even become dependent on crop plants in their growth and environment. These weeds are specialized to certain crop plants or cultivars. Moreover, many genetically related species can exchange genes with crop individuals and mimic crops. It can be concluded that any agricultural practice exerts selection pressure and may become troublesome to farmers when repeatedly applied for a long period. Its positive impact on crop growth and productivity is usually negated with time until it becomes a real trouble. Its residual negative effects may not possible to overcome for a long period after abandonment.

4. Field evidence of weed resistance and herbicide resistance protocol

In the field all growth patterns and distribution of weed species may be observed. Some species grow in colonies, in certain growth patterns, forming an ecological niche, sporadically distributed, or randomly scattered within crop plants. Certain species are dominant while others show moderate growth or are suppressed while some grow vigorous or have limited growth and short stature. This however, depends on the microhabitat and place they occupy in the field and their performance. Under intense cultivation and thick crop stands, individuals of certain weed species express phenotypic plasticity (phenotypes) at which they change/ modify their appearance, reduce or drop lower branches and thus lateral growth, elongate and increase cell divisions, overtopping crop plants and trapping light, although some shade tolerant species perform better under such conditions. Phenotypic plasticity modifying the mode of growth and energy allocation in response to environmental changes is considered to be important adaptive mechanism. These phenological variations can be easily observed among different weed species. Uniform application of herbicides in the field should equally affect all individuals of a single weed species. When herbicides are best timed and properly applied they should yield similar mode of action on species individuals. While differences in influence of a herbicide on different weed species is expected, hence differences in taxonomy, morphology, physiology and biochemistry, but such differences among individuals of a single species should have resulted from some morphogenetic or other variations within the same or different populations of that species. Certain individuals are totally killed, others less injured and some escape control unharmed. When the same herbicide or herbicides of the same mechanism of action are used, it becomes clearer that previously less or unaffected individuals should exhibit similar responses as were first shown. Gradually these individuals increase in number and growth until they dominate the site with continuous use of the same herbicide or its analogues while sensitive individuals are suppressed or removed. This however, takes a relatively long time for the population to shift from susceptible to complete resistant and depends on herbicide, environment and plant factors. These are positive signs on possible herbicide-resistance development in the field. If less affected or unharmed individuals in the first herbicide application are killed or severely injured in repeated treatments then there should be another cause of escape or partial control at first application and herbicide resistance should be then excluded. On the other hand, unharmed individuals may also tolerate higher application rates. Therefore, farmers should keep observing changes in the weed population as long as the herbicides are in use. They must get familiarized with weed species, populations and densities at pre- and post- herbicide treatments, comparing weed growth, performance and densities and recording any changes in populations thereafter. Less or unharmed individuals of any species should be followed up throughout subsequent applications of the same herbicide or herbicides of similar mode of action.

Sometimes partial effect or failure of the applied herbicide to control certain weed species or individual weeds in the first application may be thought as due to wrong calibration, misapplication, incomplete coverage treatment by a general herbicide or unsprayed gaps resulting from low sprayer boom during spray, unfavorable weather conditions, improper timing of herbicide application, and weed flushes after application of a non-resisted herbicide [16]. This could be easily judged in the repeated application to these species or individuals. When the herbicide failed to control these for the second time or at higher rates then resistance may be underway. With continued use of the same herbicide for different times, resistant individuals aggregate forming irregular patches while other weeds are controlled. A patch of uncontrolled weeds starts spreading and healthy weeds are mixed with uncontrolled weeds of the same species (Fig. 1).

Therefore irregularly shaped patches of a single weed species in the field are an indicator of herbicide resistance, especially when:

• There are no other apparent application problems.

• Other weed species on the herbicide label are effectively controlled.

• Field history indicates extensive use of the same herbicide or herbicides of the same mechanism of action.

• No or minimal herbicide symptoms appear on the single uncontrolled weed species.

• There has been a previous failure to control the same species or population in the same field with the same herbicide or with herbicides of the same site of action.

However, the rate at which a resistant weed population is selected depends on the number and frequency of herbicide applications it receives, the size of the population and its genetic diversity, and characteristics of the herbicide target site. Resistance buildup is accelerated when the management of crops does not include different weed control methods that limit herbicide use. In addition, this may be greatly enhanced in conservation or zero tillage because weeds are not killed by mechanical disturbance and general herbicides.

5. Interaction between environment and genetics

Growth and productivity of any plant species are mainly influenced by genetics, ecology and their interactions. Weeds are different from crops in their responses to both factors. They are more flexible and thus better responsive and adapted to extremes in environmental conditions such as high temperature, freezing, excessive light, salinity, drought, etc. Tolerance of weeds and better responses are mainly due to better and rapid interaction between environment and genetics compared to crop plants. In addition, the long term breeding and selection pressure imposed on crop plants has lead to selection of less adapted species or cultivars that are highly sensitive to ecological stresses and deficient in certain characteristics that offer protection or defense mechanisms against unfavorable environment. Weed fitness in natural habitats and their rapid responses to the changing environment allow evolution of weed ecotypes, genotypes, biotypes or phenotypes. Some of the basic differences in the definitions of pest resistance depend on these terms. The basic unit of plant classifications is the "species" that is defined as a group of individuals displaying common characteristics and having the ability to mate and produce fully viable progeny. A species usually consists of several to many populations. A population is a group of organisms within a species that co-exist in time and space [35; 36] and share a distinct range of genetic variations. While a genotype is the sum of the genetic coding or the genome of an individual, a biotype may not be coincident with genotype as an individual has many genes. Certain genes may be expressed or unexpressed and not pertain to the phenotype associated with the biotype. A biotype is a phenotype that consistently expresses or exhibits a specific trait or set of traits; it represents a group of individuals or a population within a species with a distinctive genetic variation of biochemical or morphological traits. Phenotype refers to the physiological and morphological profile of the expressed gene in an individual [42]. A single genotype can produce different phenotypes in response to environmental conditions and the fundamental properties of organisms are known as phenotypic plasticity. The epigenetic change is thus reflecting the alteration of phenotype (morphological or biochemical) without change in either the coding sequence of a gene or the upstream promoter region. Therefore biotypes within the same species may be developed due to this interaction. On the other hand, ecotype is a population within a species that has developed distinctive morphological or physiological characters (herbicide resistance) in response to a specific environment and persists when individuals are moved to a different environment. Ecotypes are of different germination and growth optima for the same environmental factor and phenotypes may be emerged and observed in weed populations. These alter their morphological features in response to certain prevailing environmental conditions which aim at protection of their individuals against unfavorable ecological stresses. Somatic polymorphism of certain weed species is well recognized and expressed as seed polymorphism of different morphological or physiological requirements for germination on different parts of the same weed individual. These however, are somatic rather than genetically based differences.

6. Herbicide resistance and crop relative weeds

Crop relative weeds are usually derived from the same species of crop plants and thus are genetically related. Most crop species have wild relatives and can interact with them under field conditions. Examples are radish, carrots, vetch, celery, lettuce, fennel, eggplants, wheat, barley, oat, etc. In addition, crop plants which are domesticated from wild forms possess a high degree of compatibility with crops. These are referred to as wild and weedy relatives, in spite of the fact that all species are related because their cells can read a common genetic code [15]. Crop weedy relatives are genetically compatible with crop plants and easily exchange genes. The emerged hybrids may become noxious weeds with certain weedy characteristics derived from both crop plants and wild forms. They could exhibit a certain degree of dormancy that is usually weak or absent in its parents and possess other weed traits making them difficult to control. These new generations have the ability to resist environmental hazards much better than parents and can exist and dominate in both productive and unproductive habitats. These

Figure 1. Three resistant weed species (a, b, c) to glyphosate herbicide at different growth stages and spray times. (a). *Conyza canadensis* resistant to glyphosate until harvest stage of wheat. Source http://www.sciencephoto.com/ media/ courtesy of the Montana State University (b). A field infested by suspected glyphosate- resistant *Kochia*, after the field was sprayed with three applications of glyphosate. Photo181407/enlarge Southern Agricultural Research Center. By Dillon Tabish, 08-11-12.Available at: http://www.flatheadbeacon.com/articles/article/scientists_discover_possible_herbicide_resistant_weed_in_montana/29184 (c). Palmir Amaranth (*Amaranthus palmeri*) resistance to glyphosate in corn at early growth. Source: E. Larson, April 21st, 2011.Availableat:http:// www.mississippi-crops.com/ 2011/04/21/how -to-deal-with-glyphosate-resistance-and- weed-issues-in-corn/.

are of a high genetic plasticity allowing their individuals to adapt to extensive herbicide applications and thus resist chemical treatments. Crop-weed crossed forms can easily exchange genes with crop plants as well as with weedy relatives and therefore are becoming troublesome weeds in fields with genetically modified crops.

7. Gene flow potential with wild/weedy relatives of world crops

In nature, genetic information is transferred between different individuals, populations, and generations (to progeny) and across spatial dimensions [2; 15]. This phenomenon, known as gene flow, serves as a mechanism to maintain the biological diversity that helps to ensure long-term survival of populations and species in various environments.

Gene flow is a critical determinant of population genetic structure, playing an important role in both evolutionary and applied plant population genetics [12]. It is also known as 'migration' [13] or admixture [1] and can be defined as the movement of genes between populations of a species and between these populations and inter-fertile relatives [39; 41], conferring new traits, the biophysical characteristics of the organism to individuals of the recipient population [34].

Gene flow could occur through dispersal of pollen (via outcrossing between sexually compatible individuals within or among populations) or seeds (via seed dispersal), or vegetative parts capable of clonal propagation [34; 41]. Pollen dispersal is the typical method for such exchange of genetic information [15] and pollinating visitors or other agents including wind, animal, water current and other factors could play a significant role in this issue. This happens by cross-pollination (hybridization), that is, the pollination of members of one population or genetic pool with that of another [34]. These are natural and ordinary phenomena that occur in conventional as well as genetically modified crops.

Movement of pollen away from its site of production can result in true gene flow only if (1) the pollen first effects fertilization to form seeds, and (2) seeds germinate, produce plants that express the gene (i.e., are not silen8ced), and are able to reproduce [15]. Gene flow can be from crop to crop or landrace, from crop to wild relative, and even from wild relative to crop plant [34]. Spread of this phenomenon would lead to radical changes in vegetation composition and weed ecological distribution and their economic significance.

However, two types of gene flow are known; horizontal and vertical. Stewart [39] showed that 'horizontal' gene flow is the movement of genes between disparate, unrelated species, such as between plants and microbes while horizontal gene flow is more theoretic.

Among the world's 180 most damaging weeds, however, cause 90% of all crop losses, only five groups (related weeds of rice, sorghum, rape seed, sugarcane, and oats) are sexually compatible with the most important crops (Table 1). This fact emphasizes that the number of weed-crop crosses likely to lead to extremely troublesome or unmanageable problems is small.

Weed crosses with herbicide-tolerant biotech crops are likely to be favored in some agricultural fields where the herbicide is used. In areas where little or no herbicide is applied (e.g., native lands), the weed–biotech crop crosses will not be favored [15]. Self-pollinating crops are considered of low risk in terms of gene flow to weeds. Roundup Ready, Clearfield, or Liberty Link canola, in contrast, could pollinate nearby herbicide-susceptible canola as well as weedy canola relatives, resulting in volunteer canola plants and weeds that may be resistant to several herbicide families [38]. However, several pieces of evidence clearly show an escape of weedy transgene from fields via seed flow and this escape occurs via man-mediated long-distance dispersal events [4]. Other results revealed that development of weed resistance via selection pressure from repeated herbicide applications in herbicide resistant crops (in the absence of gene flow), often poses greater risks than that from gene flow to related weed species [15].

Rank	Crop	Scientific Name	Related weeds: sexually compatible with crops
1	Wheat	Triticum aestivum	T.aestivum
		Triticum durum	Aegilops cylindrical
			A. tauschii
			A. triumcialis
			Agropyron spp
2	Rice	Oryza sativa	O. sativa
		Oryza glaberrima	O. glaberrima
			O. barthii
			O. longistaminata
			O. rufipogon
			O. punctata
3	Maize	Zea mays	Z. mays ssp Mexicana
4	Soybean	Glycine max	G. soya
5	Barley	Hordeum vulgare	H. spontaneum
6	Sorghum	Sorghum bicolor	S. bicolor
			S. almum
			S. halepense
			S. propinguum
			S. sudanense
7	Canola	Brassica napus, B. rapa, B. juncea	B. napus, B. rapa, B. nigra
8	Sunflower	Helianthus annus	Helianthus annus

Source: Different references

Table 1. Examples of some important food crops and their sexually compatible weed species

In this regard, biotech crops conferring stress tolerance (e.g., to water deficits, diseases, insects, salt stress, or nutritional deficiencies) may need more scrutiny because their crosses with weedy relatives may impart selective advantages in both agricultural and nonagricultural areas. Thus, some traits obtained from biotech crops could theoretically facilitate development into problematic weedy or wild species [15].

The economic consequences due to gene flow from biotech crops will primarily impact the agricultural fields in which those crops are grown, but potentially could impact natural areas given the proper rare combination of sexually compatible relatives, favorable environment, and reproductive/fitness advantages. As an example, rice grown in tropical countries may be relatively more prone to such processes because of the substantial populations of its wild/weedy relatives that grow naturally in or adjacent to the rice-producing areas [8; 26].

Crop-wild hybridization may also create genotypes with the potential to displace parental taxa in new environments [7]. However, the most important variable affecting gene flow is the degree of relatedness and distance between the crop and the weed, because gene flow is only possible if close relatives are growing near the crop. As a result the possibility of gene flow depends mainly on presence of wild or weedy relatives [11]. Transgene (s) transfer may have unpredictable and out of control ecological impacts under intensive cultivation of biotech crops [25]. While different crops can exchange genes with wild relatives, gene escape to wild or weedy relatives and its ecological impacts are outrated. The ecological consequences of gene flow however, depends on the amount of transgenes moved out to a wild population and the genetically modified traits and whether they have an evolutionary advantage under natural selection pressure or not and if enhanced fitness of wild and weedy relatives then the transgene followed by gene flow would persist and spread rapidly in the population of wild relatives through introgression, invade a new area and outcompete other individuals under natural conditions [24]. Weeds receiving transgenes will continue to evolve when exposed to selection pressure and it becomes nearly impossible to move them out from the environments if they can persist and spread in the populations.

8. Transgenic crops and weed evolution

The development of crops that are resistant to herbicides is a relatively new technology aimed to improve weed control in agricultural land. Herbicide-resistant crops can be created by standard methods of plant breeding, but the use of genetic engineering techniques is more usual. Herbicide-resistant crops are made resistant by either transgene technology or by selection in cell or tissue culture for mutations that confer herbicide resistance [10]. Glyphosate and glufosinate are herbicides most used in this regard. For example, soybean, corn, cotton, sugar beet, and canola are available as glyphosate- resistant cultivars and some are now widely planted in different countries. Importance of genetically engineered crops is to:

- Develop crops more tolerant/resistant to herbicides and thus increase herbicides uses and selectivity.

- Eliminate possible injury effects of soil persistent herbicides to crop plants.

- Increase options for weed control when the number of herbicides is limited, such as in minor crops.

- Effective control of certain difficult weed species and widening of weed control spectrum

- Achieve more effective weed control

- Increase bio-safety and enhance better eco-friendly use of new and less toxic herbicides

- May be more cost- effective weed control method

However, public concern about the impact of genetically modified crops on the natural environment encouraged more studies on this aspect in the last few years. Among the possible impacts, the 'escape' of the transgene, either through dispersal of the crop plant outside the agricultural area or through hybridization with wild relatives and thus increase the possibility of "weediness" [41].

In the majority of instances, there is a very low probability that an approved biotech crop introduction could create an environmental risk different from that of a nonbiotech version of the same crop. This however, does not lessen the serious concerns about possible consequences of the escape of transgenes into the environment [41]. Examples of the risks mentioned in the context of gene flow from genetically modified plants are: i) new emerged weeds resulting from an escape by the crop itself; ii) super weeds resulted by hybridization of a (wild/weedy) species with the transgenic crop; iii) genetic erosion (loss of original diversity of wild relatives). To date, all instances of weeds becoming resistant have resulted from the weed evolving its own biochemical mechanism and not by acquiring genes for resistance from the crop. However, in some cases it would be possible for the herbicide resistance gene to flow from the crop to the weed [11].

Possible consequences of hybridization and introgression depend on the plant, gene, trait, and ecological factors [39]. In the case where transgenes might be introgressed into "weedy wild relatives", there are concerns about exacerbating "weediness" traits or even the disruption of natural ecosystems. Therefore, to assess the risk of gene flow it needs to be examined not only the probability of genes moving between plants, but how possible is it for the new plants to survive [39].

In general, people ideally would like to minimize or prevent gene flow from transgenic organisms to weedy wild relatives or to places where extensive crop breeding takes place [39]. Three approaches to gene flow mitigation are possible [3].The first is by keeping the genetic modification out of the pollen, preventing the formation of pollen, and keeping the pollen inside the flower. It requires transplastomic plants hence the modified DNA is not situated in the cell's nucleus but is present in plastids, which are cellular compartments outside the nucleus. The second approach relies on male sterile plants unable to produce functioning flowers and therefore cannot release viable pollen. Cytoplasmic male sterile plants are known to produce higher yields. The third approach works by preventing the flowers from opening "cleistogamy" that occurs naturally in some plants. Cleistogamous plants produce flowers which either open only partly or not at all.

However, herbicide-resistant genes have no ecological significance in places where the corresponding herbicide is not used. When paired with a gene that might have an effect in a natural ecosystem, there is a potential problem with gene flow. Repeated application of the herbicide (especially general herbicides) would select for and protect crosses and backcrosses, increasing the possibility of successful gene flow to wild, related species [10].

9. Weed control spectrum of selective herbicides and population shifts

Some plants are genetically tolerant to certain herbicides while others have evolved resistance after repeated exposure to an herbicide. Tolerant and resistant plants usually degrade or metabolize the chemical to nonphytotoxic substances. In some cases of resistance, such as with triazine herbicides, the herbicide does not reach the key site in treated plants. Although tolerance and resistance are common, herbicide selectivity among plants is often conditional; thus it depends on plant, herbicide and environment factors.

Some of the factors that influence herbicide selectivity are as follows:

- Physiological or biochemical tolerance to the herbicide

- Herbicide application rate

- Time of application

- Herbicide formulations and surfactants used.

- Growth stage of weed and crop or other plant development

- Weather patterns (temperature, light, wind, rain, etc.)

- Variation in microenvironment or micro- topography

- Variation in resource level

- Soil type and pH

Many of the principles and practices of how herbicides used or applied to attain selective chemical and effective weed control are important. These involve the role of plant morphology and physiology, chemical properties, and environmental factors [31]. Herbicide selectivity in one way or another is in direct link with herbicide resistance. Crops are resistant to herbicides selectively used to kill weeds. Even with repeated treatment, crop plants can resist or tolerate higher rates of selective applied herbicide or repeated treatments. This depends on some level of tolerance/resistance higher in crop plants compared with weeds for that specific herbicide or herbicide group. For example, Syrian marjoram (*Origanum syriacum*) was found to withstand up to 4 times higher rates of oxadiazon and oxyfluorfen herbicides either applied on foliage parts or through the soil [32; 33]. Certainly many factors have an important role in giving a resistant value for crop plants. Some of these are listed below:

9.1. Plant factors and herbicide selectivity

Plant factors that influence the way weeds and crops respond to herbicides are genetic inheritance, age, growth rate, morphology, growth form and anatomy, and physiological and biochemical processes. The most effective use of herbicides results from considering these factors when selecting an herbicide or application method.

9.2. Plant age and growth rate

Weed seedlings or young plants are usually killed more easily than large or mature vegetation. In addition, some preemergence herbicides that suppress seed germination are often not effective when used to control larger, better established plants. Plants that are growing rapidly or in shaded places generally are more susceptible to herbicides than are plants of slow growth or unshaded.

9.3. Morphology

The morphology or growth habit of plants can determine the degree of sensitivity to some herbicides. Morphological differences in root structure, location of growing points, and leaf

properties between crops or other desirable plants and weeds can determine the selectivity pattern of some herbicides. Annual weeds in a perennial crop, meadow, or pasture
usually can be controlled by herbicides because of their different root distribution and
structure compared to those of perennial plants. For example, perennial crops such as
alfalfa can recover from moderate contact herbicide injury to foliage whereas annual weeds,
because of their small size and shallow root system, will be killed by the same herbicide
application.

The meristematic regions of most grasses, such as cereal crops and grassy weeds, are located
at the base of the plant or even below the soil surface. The growing points are protected from
herbicide exposure by the foliage or soil that surrounds them. Thus, herbicide that contacts
only foliage may injure some leaves but will not typically impair the ability of the plant to
grow. In contrast, most dicot plants have their meristems exposed at shoot tips and leaf axils.
For this reason, these plants are more susceptible than grasses to foliage-applied herbicides,
especially of contact action.

Leaf properties of some plants can impart selectivity to certain herbicides, while other plants
are effectively controlled. Spray droplets do not adhere well to the surfaces of narrow, upright,
waxy leaves that characterize many monocot plants like cereals, onion, and most grasses. Thus,
spray droplets do not adequately cover such leaves following herbicide application and the
effect of the herbicide is reduced. In contrast, dicot plants have relatively wide leaves that are
usually horizontal to the main stem. Leaves of dicot plants, therefore, intercept more spray
solution than leaves of grasses and spray droplets spread more evenly over dicot foliage.
Herbicide effectiveness is best when spray interception and coverage are greatest and with use
of surfactants. However, ecological factors and geographical regions under which weeds are
growing have significant influence on herbicide selectivity and rates of applications since they
affect or modify weeds morphology and internal anatomy.

9.4. Physiological and biochemical processes

Plant physiology influences herbicide passage after its application. This process is called
"absorption". The extent of herbicide movement in a plant- "translocation"- after it has been
absorbed is also a physiological process. Both absorption and translocation are important
processes governing herbicide activity and vary markedly among plant species. Generally,
plant species that readily absorb and translocate herbicides are most easily killed.

Biochemical and biophysical processes are also important plant factors determining herbicide
selectivity. Herbicide adsorption can be responsible for differential herbicide susceptibility
among plant species. During this process an herbicide is bound so tightly by cellular constituents (usually cell walls) that it cannot be translocated readily and thus is inactivated.
Membrane stability is another biochemical/biophysical process that results in herbicide
selectivity among plants. In this case, the cell membranes of tolerant plants can withstand the
disruptive action of the herbicide. The ability of carrot to withstand the toxicity of certain oils
is an example of this form of herbicide selectivity.

9.5. Genetic inheritance

Plant species within a genus usually respond to herbicides in a similar manner, while responses to herbicides by plants in different genera often vary. The reason is that plants with similar taxonomic traits often have similar morphogenetic and enzymatic components. Thus, crops and weeds that belong to the same genera are usually susceptible to the same herbicides and are similarly affected since they have similar biochemistry. This rule is not absolute, however, because varieties of many crops are known to respond differently to the same herbicide and weeds usually adopt different mechanisms of herbicide resistance while crop plants have lost many of their traits in breeding programs that present in wild relatives.

10. Herbicides and edaphic factors

Soil factors affect herbicide performance and their effectiveness. These including soil-organic matter content, microorganism populations, soil water table and moisture content and soil pH. Organic matter acts through adsorption and release of chemical molecules. Certain herbicides are tightly adsorbed on soil particles and thus become unavailable to weeds. These molecules may be totally inactivated upon their release. Therefore weed control may be complete or not based on the amount of the herbicide adsorbed and whether the held amount on soil colloids is compensated or not before applied. The higher the percentage of organic matter and clay particles, the greater the adsorption in amount and time of herbicide molecules and the lower the herbicide activity and *vice versa*. This requires that some operations should be well managed when soil applied herbicides are used including their incorporation or placement in/on the soil.

Activity of soil microorganisms is another factor affecting activity of soil- applied herbicides and persistence. Microorganisms may degrade herbicide molecules and feed on organic herbicides. In general, favorable soil factors to microorganism populations stimulate their activity and thus rapid herbicide degradation. Therefore, soil-microbe population is an important factor in increasing or decreasing herbicide persistence and weed control duration.

Soil water also affects herbicide activity and performance. When high amounts of soil water are available or at high soil water levels, herbicide molecules may by hydrated. On the other hand, moisture is necessary to transfer herbicide molecules into the root system and then translocate these upward to vegetative parts through the xylem.

Soil pH affects cation exchange capacity of soil particles. Salt or mineral forms of certain herbicides may interact with soil particles under these conditions by exchanging cations or anions and thus lead to breakdown of herbicide molecules and inactivation.

All above soil factors and others such as soil- root temperature and soil mechanical properties can affect herbicide activity and performance and their effectiveness in controlling weed species and herbicide selectivity. Weeds may become adapted to certain soil conditions, escape control operations and lead to dominance of well adapted species or populations.

11. Weed resistance and dormancy, avoidance and weed density

Dormancy is the state at which seeds in the soil or buds are not germinating or growing due to external conditions exert influences on physiological and biochemical internal processes including enzymes activities, food transport to embryo and metabolism. This state is keeping seeds or buds safe until the cause of dormancy is over. This behavior is important to maintain genetic line and continuity of the species in changeable environment. Under conditions of herbicide application, some of these chemicals are absorbed by seeds or dormant buds while others are not. These result differences in germination, emergence and growth patterns of different weed species. However, some herbicides may stimulate seed germination while others inhibit this process or even kill seed embryo. Differences also exist in hardness and permeability of seed coat of different weed species at which species of Chenopodiaceae and Fabaceae are good examples. These characters cause differences in germination and growth of seedlings and may confer another cause of herbicide resistance. Avoidance of herbicide toxicity may result from seed interring into dormancy and not further responding to the applied herbicide with no absorption or translocation of the herbicide into the embryo. In addition, herbicide molecules may be deactivated or degraded inside the seed itself by some oxidative enzymes or may bound into certain constituent inside the seed.

On the other hand, stimulation of weed seeds to germinate using certain herbicides also exist and allows higher seedlings emergence and partitioning of herbicide molecules among individuals of weed species. Division of herbicide molecules among high number of emerged seedlings would further diluted herbicide inside weed plants.

All above mentioned factors should be considered when herbicide-resistance is discussed. These may cause great differences in weed growth patterns and distribution in the field.

12. Weed resistance updates and resistance mechanisms

With continued dependence on herbicides for weed control and with the absence of other methods and herbicide rotation, the resistance problem is extenuated and the number of resistant weed species and biotypes is dramatically increased. At present, the reported herbicide resistant weeds are approaching 393 (species and their biotypes). These represent 211 species (124 dicots and 87 monocots) and detected from over 680,000 fields [21; 44] reported from 61 countries from all over the globe. However, the highest number of resistant species was reported from the advanced countries indicating efficient and rapid detection with available technology to diagnose, discover and deal with this issue. However, the highest number of weeds reported resist the main three groups of herbicides based on site of action including; the ALS (127 weeds), Photosystem II (69) and the ACCase (42) inhibitors. The highest number of weed resistant species and biotypes came from the USA (141), Australia (61) and Canada (58). Most numbers of resistant species belong to the families Poaceae, Asteraceae and Amaranthaceae and most frequently mentioned are genera of *Amaranthus* (30 times and 11 species), *Echinochloa* (23 times and 6 species), *Lolium* (20 times and 4 species), *Alopecurus* (12 times and 3 species), *Avena* (11 times and 3 species), *Bromus* (11 times and 5 species), *Conyza* (10 times and 3 species), *Setaria* (9 times and 5 species), *Poa* (8 times and one species), *Ambrosia* (7 times and 2 species), *Digitaria* (6 times and 4 species), *Phalaris* (6 times and 3 species), *Hordeum* (5 times and 2 species) and *Sorghum* (6 times and 3 species). Most are of the grass family usually exhibiting distinct morphological features allowing wide dispersal and escape of herbicide treatment such as encased growing points, vertical leaf arrangement and thick waxy cuticle that reduce herbicide penetration and lead to herbicide droplets bouncing off leaves. Other genera reported are characterized by their prolific seed production and/or seed polymorphism. All above mentioned genera however, showed multiple resistance to different herbicides groups. Most resisted are herbicides widely and repeatedly used including: glyphosate, paraquat, atrazine and 2,4-D and others used in fields cultivated by genetically modified crops. Some recently developed herbicides are also resisted including chlorsulfuron and sufonylurea group. This phenomenon demonstrates that the herbicide industry and development is far behind weed evolution. On the other hand, weed species and biotypes showing multiple resistance are most common and some are among the world's worst weeds [19] including: *Amaranthus* spp., *Echinochloa* spp., *Avena* spp. and *Chenopodium album* characterized by their polymorphic seed production and phenotypic plasticity. This reflects a great ability to maintain and exhibit high plasticity and possess various mechanisms of herbicide resistance.

The precise molecular mechanism of resistance varies with different plants, but in general plants resist herbicides in one of the following ways:

- Avoiding the herbicide by not absorbing it or, if absorbed, the weed compartmentalizing it away from its target site.

- Reducing the uptake or herbicide uptake is not enough to injure the weed or reach lethal level.

- Changing the structure of the target site of the herbicide so the plant is no longer sensitive

- Reduce herbicide translocation to the key site or binding it into certain plant constituent

- Sequestration by complete physical removal of the herbicide from the key site

- Target site mutation and changes in structure lead to insensitive plants and failure herbicide binding.

- Deactivating the herbicide by chemical alteration or herbicide metabolism before reaching target site

However, resistance mechanisms through which different weed species resist herbicide treatments are many and varied but most are physio-chemically based (Table 2).

Herbicide Group	Site of Action	HRAC Group
ALS inhibitors	Inhibition of acetolactate synthase ALS (acetohydroxyacid synthase AHAS)	B
Photosystem II inhibitors	Inhibition of photosynthesis at photosystem II	C1
ACCase inhibitors	Inhibition of acetyl CoA carboxylase (ACCase)	A
Synthetic Auxins	Synthetic auxins (action like indoleacetic acid)	O
Bipyridiliums	Photosystem-I-electron diversion	D
Glycines	Inhibition of EPSP synthase	G
Ureas and amides	Inhibition of photosynthesis at photosystem II	C2
Dinitroanilines and others	Microtubule assembly inhibition	K1
Thiocarbamates and others	Inhibition of lipid synthesis - not ACCase inhibition	N
PPO inhibitors	Inhibition of protoporphyrinogen oxidase (PPO)	E
Triazoles, ureas, isoxazolidiones	Bleaching: Inhibition of carotenoid biosynthesis (unknown target)	F3
Nitriles and others	Inhibition of photosynthesis at photosystem II	C3
Chloroacetamides and others	Inhibition of cell division (Inhibition of very long chain fatty acids)	K3
Carotenoid biosynthesis inhibitors	Bleaching: Inhibition of carotenoid biosynthesis at the phytoene desaturase step (PDS)	F1
Glutamine synthase inhibitors	Inhibition of glutamine synthetase	H
Arylaminopropionic acids	Unknown	Z
Unknown	Unknown	Z
4-HPPD inhibitors	Bleaching: Inhibition of 4-hydroxyphenyl-pyruvate-dioxygenase (4-HPPD)	F2
Mitosis inhibitors	Inhibition of mitosis / microtubule polymerization inhibitor	K2
Cellulose inhibitors	Inhibition of cell wall (cellulose) synthesis	L

Source: 21; Updated: November, 2012

Table 2. Herbicide resistant weeds summary table (Thursday, November 08, 2012)

13. Factors enhancing herbicide resistance

All natural weed populations, regardless of the application of any herbicide, probably contain biotypes that resist herbicides. Repeated application of an herbicide exposes the weed population to a selection pressure which may lead to an increase in the number of surviving resistant individuals in the population. As a consequence, the resistant weed population may increase to a level that adequate weed control cannot be achieved by the application of that herbicide [18]. Factors enhancing herbicide resistance include: the use of a single herbicide or herbicides of same mechanism of action, same formulation, same method of application, time of application, weather conditions during spraying, weed-density and application rate, surfactants, herbicide family and mechanism of action, crop rotation, and employed control methods.

Because weeds contain a tremendous amount of genetic variation that allows them to survive under a variety of environmental conditions, the development of a resistant species is brought about through selection pressure imposed by the continuous use of an herbicide or herbicides of similar mechanism of action. Long residual pre-emergence herbicides or repeated application of post-emergence herbicides will further increase selection pressure.

Factors in general that can lead to or accelerate the development of herbicide resistance include weed characteristics, chemical properties and cultural practices.

Weed characteristics conducive to rapid development of resistance to a particular herbicide include:

- Weeds having short life cycles (annuals).
- High seed production.
- Level of selection pressure imposed by the herbicide
- Relatively rapid turnover of the seed bank due to high percentage of seed germination each year (i.e., little seed dormancy).
- Several reproductive generations per growing season.
- Extreme susceptibility to a particular herbicide.
- One weed which would normally be controlled but not controlled while others were removed.
- High frequency of resistant gene (s).

Herbicide characteristics which lead to rapid development of herbicide resistance in weed biotypes include:

- A single site of action of the same herbicide continuously is used.
- Broad spectrum of weed control.
- Long residual activity in the soil.

Cultural practices can also increase the selection pressure for the development of herbicide-resistant biotypes. In general, complete reliance on herbicides for weed control can greatly enhance the occurrence of herbicide-resistant weeds. Other factors include:

- Shift from crop rotations towards mono cropping.
- Little cultivation or zero tillage for weed control or no elimination of weeds that escape herbicide control.
- Continuous or repeated use of a single herbicide or several herbicides that have the same mechanism of action.
- High herbicide use rate relative to the amount needed for weed control.
- Complete weed control

- Orchard and vineyard weeds.

- Roadside weeds.

14. Management of herbicide resistance

Herbicide-resistant weed populations can be managed following an integrated weed control program. The following practices are important for an effective management strategy:

- Herbicide rotation. Adopting this method, it should be known that herbicides of different chemical families may have the same site of action.

- Using mixtures of herbicides with different modes of action and overlapping weed spectrums. This would help in managing evolution of weed resistance.

- Crop rotation. Crops differ in their competitiveness against weeds. Plant crops having a different season of growth, different registered herbicides and crops for which there are alternate methods of weed control. Rotation breaks down weed population and prevents the build up of resistance to herbicides. In addition, different crops may require different types of herbicides and thus herbicides may be rotated as well. However, some herbicide groups include different chemicals that can be used in different crops; therefore crop rotation alone may not be enough to avoid resistance development in this case.

- Herbicides with the same site of action should not be applied or used in both fallow years and in the crop(s) planted within 3 years.

- Growers should keep rotating methods of weed control. Non-chemical control techniques including tillage, hand-weeding before flowering, mulching, soil solarization, prevention methods of weed dispersal (certified seed, clean equipments, use a power washer or compressed air to remove seeds).

- Herbicide-resistant weeds should be controlled before flowering and seed setting.

- Farmers should only use non- or short-residual herbicides and avoid using persistent chemicals and not applying them repeatedly within a growing season. This method would reduce the selection of herbicide-resistant weed biotypes. However, repeated applications within a single growing season of certain herbicides (paraquat, glyphosate) also lead to development of resistant weed populations.

- Where possible mechanical weed control such as rotary hoeing and cultivation is recommended to be combined with herbicide treatments.

- Weed escapes of resistant biotypes may be eliminated by cultivation in row crops. Fallow tillage can control herbicide-resistant and susceptible weed populations when they emerge at about the same time.

- Accurate record keeping. Farmers should be familiar with the history of herbicides use in their fields. Also keep tracking the weed species that have been present in a given field and of how well particular herbicides have controlled them. Farmers should check for weedy patches in patterns consistent with application problems and hand-weeding these patches.

- Always weed free crop seeds should be used that greatly minimize introduction seeds of herbicide-resistant biotypes.

- Implementation of integrated weed management. This is important for effective control of all weeds including herbicide-resistance.

- Monitoring fields for weed escapes for resistant and susceptible biotypes. A resistance problem may not become visible until 30 percent or more of the weed population is no longer controlled. Check to see if the escapes are of one species or a mixture of species. If a mixture, the problem is more likely related to the environment or the herbicide application. If only one species was not controlled, the problem is likely to be resistance, especially if the species was controlled by the herbicide in the past and if the same herbicide has been used repeatedly in the field.

- Implementation of prevention methods of weed control. All measures aimed at prevention of weed introduction to fields and their dispersal should be strictly followed including governmental quarantine regulations.

- Alternating spring and winter crops, thus tillage and herbicides are used at different times in the different crops. Weed biotypes that survive in one crop could be killed in the other.

- Changing herbicide program, if weed resistance occurs, herbicides with other sites of action and other weed management practices must be used in an integrated management strategy. However, weed management strategies that discourage the evolution of herbicide resistance should include the following:

  - Use herbicide only when necessary and where possible herbicide application should be based on economic threshold.

  - Apply herbicides in tank mixed, pre-packed, or sequential mixtures of multiple site of action.

  - Never use unregistered mixtures, follow label recommendation at all times

  - Regularly monitor your crops so that resistant patches can be observed in time to be controlled with, for instance, spot spraying.

  - Apply the herbicide at the correct leaf stage of the weed and the crop.

  - Calibrate sprayer correctly before using herbicides

  - Planting new herbicide-resistant crop varieties should not result in more than two consecutive applications of herbicides with the same site of action against the same weed unless other effective control practices are also included in the management system.

  - Respond quickly to changes in weed populations to restrict spread of weeds that may have been selected for resistance.

○ Encourage railroads, public utilities, highway departments and similar organizations that use total vegetation control programs and vegetation management systems that do not lead to selection of herbicide resistant weeds. Resistant weeds from total vegetation control areas frequently spread to cropland. Chemical companies, governmental agencies, and farm organizations can all help in this effort.

• To keep herbicide-resistant weeds under control, the following strategies should be also incorporated into a weed management plan:

○ Clean tillage and harvest equipments before moved from infested to clean fields from weed resistant species.

○ Total weed control in uncultivated places or sites

○ Close cultivation

○ Monitor hand weeding to insure more than 90% removal of weeds in the crop row.

○ Prevention of weed seed spread through:

– Use of clean equipment.

– Enter the field with resistant plants last.

– Use a power washer or compressed air to remove seeds.

– Recognizing patterns of weed escapes typical of resistant plants

– Watch for small weed patches that appear in the same place in the next crop.

– Watch for weed patches that do not have a regular shape that would indicate an herbicide application problem.

Herbicide resistance however, provides a basic understanding of the genetic basis of weediness, while the development of weed genomics would provide three predictable and useful outcomes. The first is the identification of genes that could improve crop yields. The second is to improve our understanding of the evolution of herbicide resistance and the to aid in the identification of novel herbicide targets. Currently, there is little (if any) solid predictive capability of why some weeds develop resistance and others do not. Third, our understanding of weed biology would be exponentially expanded [6].

Research has recently been performed to assess the ability to cripple the effect of transgenes. The goal here is for the transgenic effect to not be as strong if it went to a wild relative. In one case, the genetic background of the crop weakened the weedy relative. In another case, the weakness was built into the genetic construct, called *transgenic mitigation*, in which an herbicide-resistant gene was paired with a dwarfing gene. In either case, transgenic weeds were less competitive than their non-transgenic parent weeds [39].

15. Conclusion

Weeds either leave (disappear), adapt, tolerate or resist any unfavorable environmental conditions that influence their normal growth and life strategies. Herbicide resistance is a complex phenomenon resulting from altered herbicide target enzyme, enhanced herbicide metabolism or reduced herbicide absorption and/or translocation. It is a survival strategy through which many successful weed species and/or biotypes counteract or escape chemical hazards. Weeds expressing this phenomenon have developed some morpho- (behaviorist), physio-, and/ or biochemical mechanism/s allowing existence. However, two theories are mainly considered: the mutation and the natural selection [17]. Colonizers, as well as some specialist weeds of high seed production and polymorphic characteristics, have rapid responses to prevailing environmental conditions and high ability to express herbicide-resistant genes and exhibit wide ecological variations [28]. This phenomenon is well documented in agricultural as well as other disturbed habitats while the list of weed resistant species gets longer with continued dependence on herbicides for weed control. From the information presented in this chapter, it is clearly demonstrated that herbicide resistance in weeds is far exceeding herbicide technology and industry. Most problematic weed species are genetically related to major food crops including wheat, rice and maize. This may pose another danger for the genetic industry and genetically engineered crops of wild relatives. Away from weed biology and resistance control, methods of weed control must be integrated and continuously rotated for effective weed control and prevention of weed resistance. This however, may not be achieved in absence of information and field data and well managed weed control strategies, considering all the factors that influence weed life and development.

Author details

Jamal R. Qasem

Address all correspondence to: jrqasem@ju.edu.jo

Department of Plant Protection, Faculty of Agriculture, University of Jordan, Amman, Jordan

References

[1] Anonymous-I. (2010) Evolution. Retrieved December, 2010. Available at: *http://www.biologydaily.com/biology/Evolution*.

[2] Anonymous-II. (2010). Gene flow. Retrieved December, 2010. Available at: *http://en.wikipedia.org/wiki/Gene_flow*.

[3] Anonymous-III. (2010) Gene flow mitigation. Retrieved December, 2010. Available at: *http://en.wikipedia.org/wiki/Gene_flow*.

[4] Arnaud, J.F.; Viard, F.; Delescluse, M. and Cuguen, J. (2003) Evidence for gene flow via seed dispersal from crop to wild relatives in *Beta vulgaris* (Chenopodiaceae): consequences for the release of genetically modified crop species with weedy lineages. *The Royal Society, Published online, 19 March, http://www.ncbi.nlm.nih.gov/pmc/articles*.

[5] Barrett, S.C. (1988). Genetics and evolution of agricultural weeds. In: *Weed management in Agroecosystems: Ecological Approaches*. Altieri, M.A. and Liebman, M. (eds.). CRC Press, Inc, Baco Raton, Florida.

[6] Basu, C.; Halfhill, M.D.; Mueller, T.C and Stewart Jr, C. N. (2004). Weed genomics: new tools to understand weed biology. *Tends in Plant Science* 9(8), 391-398.

[7] Campbell, L.G.; Snow, A.A and Ridley, C.E. (2006). Weed evolution after crop gene introgression: greater survival and fecundity of hybrids in a new environment. *Ecology Letters* 9, 1198–1209.

[8] Chen, L.I.J., Lee, D.S., Song, Z.P., Shu, H.S. and Lu, B.R. (2004). Gene flow from cultivated rice (*Oryza sativa*) to its weedy and wild relatives. *Annals of Botany* 93, 67-73.

[9] Conard, S.G. and Radosevich, S.R. (1979). Ecological fitness of *Senecio vulgaris* and *Amaranthus retroflexus* biotypes susceptible or resistant to atrazine. *Journal of Applied Ecology* 16, 171-177.

[10] Duke, S.O.and Cerdeira, A.L. (2005). Transgenic herbicide-resistant crops. *Outlooks of Pest management*. DOI10.1564/16 oct06, pages 2008-211. Research Information Ltd.

[11] DuPont (2008). Herbicide resistant crops and weed management: scientific summary and the DuPont perspective. Available at: http://www2.dupont.com/Biotechnology/ en_US/science_knowledge/herbicide_resstance.

[12] Ellstrand, N.C. (1992). Gene flow among seed plant populations. *New Forests Journal* 6(1-4), 241-256.

[13] Ellstrand, N.C. (2003).Current knowledge of gene flow in plants: implications for transgene flow. *The Royal Society, published online 12 May at: http:// www.ncbi.nlm.nih.gov/pmc/articles*.

[14] Fuentes, R.G., Baltazar, A.M., Merca, F.E., Ismail, A.M. and Johnson, D.E. (2010). Morphological and physiological responses of lowland purple nutsedge (*Cyperus rotundus* L.) to flooding. *AoB Plants* Vol. 2010.plq010, doi: 10.1093/oobpla/plq010.

[15] Gealy D.R.; Bradford K.J.; Hall L.; Hellmich R.; Raybould A. and Wolt J. (2007). Implications of gene flow in the scale-up and commercial use of biotechnology-derived crops: Economic and policy considerations. *Issue Paper 37. Council for Agricultural Science and Technology (CAST), Ames, Iowa*.

[16] Gunsolus, J.L.(2008). Herbicide resistant weeds. 2002 Regents of the University of Minnesota.. University of Minnesota Extension. Available at:http://www.extension.umn.edu/distribution/cropsystems/DC6077.html.

[17] Hager, A.G. and Refsell, D. (2008). Weed resistance to herbicides. In: *2008 Illinois Agricultural Pest Management Handbook*. Department of Crop Science, University of Illinois.

[18] Herbicide Resistance Action Committee (HRAC). (2009). Guidelines to the management of herbicide resistance. Available at http://www.hracglobal.com/Publications/ManagementofHerbicideResistance/tabid/225/Default.asps(1-16)1/18/2009.

[19] Holm, L.G., Plucknett, D.L., Pancho, J.V. and Herberger, J.P. (1977). *The World's Worst Weeds: Distribution and Biology*. University of Hawaii, Honolulu. USA.

[20] Holzner, W. and Numata, M. (eds.). (1982). *Biology and Ecology of Weeds*. Dr. W. Junk Publishers. The Hague-Boston-London.

[21] http://www.weedscience.org/summary/UspeciesMOA.asp?1stMOAID=19.

[22] Kending, A. (2009). Herbicide resistance in weeds. University of Missouri Extension. Available at: http://extension.missouri.edu/publications/DisplayPub.aspx?P=G4907.

[23] LeBaron, H.M. and Gressel, J.(eds.). (1982). *Herbicide Resistance in Plants*. John Wiley & Sons, New York.

[24] Lu, B-Rong and Yang, C. (2009). Gene flow from genetically modified rice to its wild relatives: assessing potential ecological consequences. *Biotechnology Advances* 27, 1083-1091.

[25] Lu, B-Rong. (2008). Transgene escape from GM crops and potential biosafety consequences: an environmental perspective. International Center for Genetic Engineering and Biotechnology (ICGEB), vol. 4. Collection of Biosafety Reviews 66-141.

[26] Lu, B-Rong. and Snow, A. (2005). Gene flow from genetically modified rice and its environmental consequences. *BioScience*, 55, 669–678.

[27] Mashburn S. J. Sharitz, R. R. and Smith, M. H. (1978). Genetic variation among *Typha* populations of the southeastern United States. *Evolution* 32: 681-685.

[28] Neve, P. (2007). Challenges for herbicide resistance evolution management: 50 years after Harper. *Weed Research* 47, 365-369.

[29] Putwain, P.D., Scott, K.R. and Holliday, R.J. (1982). The nature of resistance to traizine herbicides: case histories of phenology and population studies. In: *Herbicide Resistance in Plants*; LeBaron, H.M. and Gressel, J. (eds.), John Wiley & Sons, New York pp. 99-115.

[30] Qasem, J. R. (2003). *Weeds and their Control*. University of Jordan Publications. Amman, Jordan. 628 pp.

[31] Qasem, J. R. (2011). Herbicides applications: Problems and considerations, In, *Herbicides and Environment*, Andreas Kortekamp (Ed.), *ISBN*: 978-953-307-476-4, InTech, Available from: http://www.intechopen.com/articles/show/title/herbicides-applications-problems-and-considerations.

[32] Qasem, J. R. and Al-Jebury, I. S. (2001). Weed control in marjoram (*Origanum syriacum* L.) under field conditions. Dirasat 28, 194-207.

[33] Qasem, J.R. and Foy, C.L. (2006). Selective weed control in Syrian marjoram (*Origanum syriacumL.*) with oxadiazon and oxyfluorfen herbicides. *Weed Technology*, 20 (3), 670-676.

[34] Quist D. (2010). Vertical (Trans) gene Flow: Implications for crop diversity and wild relatives published. *Third World Network, Penang, Malaysia Jutaprint, ISBN*: 978-967-5412-26-4.

[35] Radosevich, S. R., Holt, J.S. and Ghersa, C.M. (2007). *Ecology of Weeds and Invasive Plants. Relationship to Agriculture and Natural Resource Management.* Hobokin, N.J.: Willey Interscience.

[36] Radosevich, S., Holt, J. and Ghersa, C. (1997). *Weed Ecology: Implication for Management* . 2nd Edition. John Wiley & Sons Inc. New York.

[37] Smith, C.M. and Namuth, D. (2005). Herbicide resistance: mechanisms, inheritance, and molecular genetics. eLearn & Grow Library, 7pp. available at: http://plantand-soil.unl.edu/croptechnology2005/pageincludes/printModule.jsp?inform.

[38] Smith, C.M.; Hulting, A.; Thill, D.; Morishita, D. and Krenz, J. (2007) Herbicide-resistant weeds and their management, weed control strategies. In: *Pacific Northwest Conservation Tillage Handbook, University of Idaho*.

[39] Stewart N.C. (2008). Gene flow and the risk of transgene spread. University of Tennessee. Retrieved December 2010 Available at: http://agribiotech.info/details/Stewart-GeneFlow%20Mar%208%20-%2003.pdf.

[40] Tsyusko, O. V., Smith, M. H., Sharitz, R. R. and Glenn, T. C. (2010). Genetic and clonal diversity of two cattail species, *Typha latifolia* and *T. angustifolia* (Typhaceae), from Ukraine. *American Journal of Botany*, 97 (12), 2061-2067.

[41] van de Wiel, C.; Groot, M.; and Den Nijs, H. (2005). Gene flow from crops to wild plants and its population ecological consequences in the context of GM-crop biosafety including some recent experiences from lettuce. *Wageningen UR Frontis Series, Volume 7 Environmental Costs and Benefits of Transgenic Crops*. Retrieved, December 2010 Available at: http://library.wur.nl/ojs/index.php/frontis/article/viewArticle/914.

[42] Vencill, W., Grey, T. and Culpepper, S. (2011). Resistance of weed to herbicides. In: *Herbicides and Environment* (Kortekamp, A. ed.). pp. 585-594. InTech, Available at: http://www.intechopen.com/articles/show/title/herbicides-applications-problems-and-considerations.

[43] White, D.E. (1979). *Physiological Adaptations in Two Ecotypes of Canada Thistle (Cirsium arvense* (L.) Scope. MSc. Thesis, University of California, Davis.

[44] WSSA, (2012). Data from the international survey of herbicide resistant weeds obtainable at WeedScience.com.website.

Chapter 2

Toxicity of Herbicides:
Impact on Aquatic and Soil Biota and Human Health

Maria Aparecida Marin-Morales,
Bruna de Campos Ventura-Camargo and
Márcia Miyuki Hoshina

Additional information is available at the end of the chapter

1. Introduction

During the last decades, the scientific community, including government and non-government organizations have increased their interest in detecting and controlling the environmental agents responsible for damages to the human health and sustainability of the ecosystems. This interest has been intensified by the frightening increase on the reports of the anthropogenic action on the environment responsible for damages to the ozone layer, accidental release of wastes and radioactive gases, as well as contamination by pesticides used in agriculture. However, the growth of the human population and of the activities associated with agriculture, industrialization and urbanization have contributed to the depredation of the biodiversity and genetic variability, resulting in the compromise of several species, including man [9].

After the industrial revolution, a great number of chemical substances have been released into the terrestrial and aquatic environments and in the atmosphere. These substances can be transported and transformed by different processes, whose transformation by-products can cause adverse effects on man, as well as damages to the terrestrial and aquatic ecosystems. Several studies have shown the presence of residues of several chemical substances in the air, water, soil, food and organisms in general [10].

Environmental pollution by genotoxic and mutagenic products affects the exposed organism and its future generations, this fact is observed both for animals, and in this case man is included, and for the other groups of organisms such as plants and microorganisms. In order to evaluate the consequences of the anthropogenic activities on the ecosystem it is necessary that the scientific community pays a special attention in the search for understanding the modes of action of xenobiotics present in the ecosystem in the biota exposed. For this, extensive, detailed and ordered studies of the contaminants must be developed with the purpose of preventing the biological impairment, such as inductions of alterations in the genetic materials of the organisms [11].

Some studies have been performed in the attempt to evaluate the behaviour, transforma- tions and effects of chemical agents, both in the environment and in the organisms. Toxicology establishes the limits of concentration or quantity of chemical substances acceptable in the environment by studies on the toxic effects of these substances in the organism and ecosystems [12].

Considering that the use of agrochemicals, such as herbicides, have caused a great environ- mental contamination, due to their widespread use, it has become indispensable to perform the assessment of the toxicity of these compounds.

1.1. The importance of herbicides

Living beings are exposed to the action of numerous agents that are potentially toxic. These agents can be physical, chemical or biological and can provoke in the organisms physiological, biochemical, pathological effects and, in some cases, genetic effects [13]. A great variety of chemical substances with mutagenic potential, both natural and synthetic, have been investi- gated. Many of these substances are found in food, pharmaceutical drugs, pesticides and in complexes of domestic and industrial effluents. It is known that these compounds can cause detrimental inheritable changes in the genetic material, without these changes being expressed immediately [14]. Thus, several compounds dispersed in the environment can represent danger to human health, since they present a potential to induce mutations [15].

The production of food can occur both by agricultural activities and by livestock. The yield of food production is directly related with the relationship established between the species of interest for production and the other plant, animal, microbial and parasitic biological systems that compete for resources available in the environment [16]. Among the species that jeopard- ize the agricultural production there are the weeds that, when invade crops, can cause significant loss in the yield and quality of the harvest [17]. Therefore, in order to enhance the productivity and the quality of crops, the removal of weeds from agriculture becomes important.

Before the introduction of selective herbicides as an agricultural practice, the removal of weeds was accomplished manually in an extremely laborious form. Thus, the farmers sought other forms to control weeds, such as, integrating other weed control practices such as crop rotation, tillage and fallow systems [17].

The introduction of selective herbicides in the late 40's and the constant production of new herbicides in the following decades gave farmers a new tool in the control of weeds [17]. Therefore, the process of modernization of agriculture introduced, in the 60's, the use of new biological varieties considered more productive, but dependent on chemical fertilizers and intensive use of pesticides, in order to increase productivity. The use of these chemical agents resulted in the increase of productivity, but, on the other hand, brought adverse consequences, since many are harmful substances for man and the environment. The world practice of using agrochemicals for long periods, often indiscriminate and abusive, has raised concerns among the public authorities and experts of public health and sustainability of natural resources [16].

Many agrochemicals are very toxic substances whose absorption in man are almost exclusively oral and can also occur by inhalation or dermally. As a consequence of the human exposure to pesticides, a series of disturbances can be observed, such as gastric, neurological and muscular [18].

Among the pesticides, the main agents of intoxication are the herbicides and insecticides. According to Vasilescu and Medvedovici [19], herbicides are defined as any substance, individually or in mixtures, whose function is to control, destroy, repel or mitigate the growth of weeds in a crop.

The use of herbicides, despite the fact that they are characterized as a highly effective tool in the control of weeds, has led to a change in the phytosociological composition of weeds and to a selection of biotypes resistant to herbicides, besides also causing impacts in the environment and human health. According to He et al. [20], herbicides are the most used chemical substances throughout the world. During the 90's, the global pesticide sales remained relatively constant, between 270 and 300 billions of US dollars, and 47% of this value corresponded to herbicides and 79% to insecticides. Since 2007, herbicides assumed the first place among the three major categories of pesticides (insecticides, fungicides/bactericides, herbicides) [21].

The use of herbicides to control weeds has been a common practice in global agriculture, mainly with the objective to increase agricultural production. However, when these chemicals are used in an uncontrolled manner, they can cause impacts on non-target organisms, especially on those that live in aquatic environments [22].

According to Chevreuil et al. [23], Kim and Feagley [24] and Abdel-Ramham et al. [25], most of the toxic effects of the herbicides on animals and plants were insufficiently investigated. As a consequence of the lack of information about the action of herbicides in the biological environment, these chemical agents can also represent a problem to human health [26, 27]. The impact of a pesticide in the environment depends on its dispersion mode and its concentration, as well as its own toxicity [28]. The mutagenic effects of the herbicides can result from several reactions with the organism, as a direct action of the compound on the nuclear DNA; incorporation in the DNA during cell replication; interference in the activity of the mitotic or meiotic division, resulting in incorrect division of the cell [29].

Some herbicides interfere directly in the cell division of plants, elongation and/or cell differentiation, causing disturbances in the functioning of the roots or vascular tissues [30]. In animals, herbicides can act in several tissues or organs and, sometimes, are associated with tumorigenic processes [31].

Jurado et al. [32] listed the general advantages and disadvantages of using herbicides. In this list, the authors cited as advantages: kill unwanted plants; help crops grow since it eliminates weeds that compete with crops for water, nutrients and sunlight; can be safely used in plantations, while the manual or mechanical removal processes of weeds can cause damages to crops; can be used in geographically close crops; in most cases, only one application of the herbicide is sufficient to control the weeds, while the other methods must be constantly used; are easy to use; have fast action; are relatively inexpensive and are economically more viable than manual removal; non-selective herbicides can be used to eliminate vegetation cover in areas intended for the construction of residences and/or roads; to eradicate plants bearing diseases; and since some herbicides are biodegradable, they can become relatively inert after some time. The disadvantages listed by the authors are: some herbicides are not biodegradable and, thus, can persist in the environment for a long period of time; all herbicides are, at least, mildly toxic; can cause diseases and even accidental death (case of paraquat); can be carried into rivers by rainwater or be leached to groundwater polluting these environments; some herbicides can accumulate in the food chain and are toxic for animals, including man.

1.2. Herbicides classification

According to Moreland [33], herbicides are designated by common names approved by the Weed Science Society of America (WSSA) or by the British Standards Institution. Organic herbicides are classified according to their application method, chemical affinity, structural similarity, and by their mode of action [34]. In relation to the application methods, herbicides can be classified into two groups: soil application and foliar application. According to Jurado et al. [32], all the herbicides applied in the pre-planting (surface or incorporation) and pre-emergence (in crops, weeds or both) are classified as herbicides of soil application and those applied in the post-emergence are classified as foliar application.

Moreover, herbicides can be classified according to their mode of action. Following, it will be presented the classes of herbicides, according to their mode of action, based in the classification of Moreland [33] :

1. *chloroplast-associated reactions:* photo-induced electron transport and reaction coupled to phosphorylation occur in the chloroplast, any interference in these reactions inhibit the photosynthetic activity. Herbicides that inhibit the photo-chemically induced reactions are divided into the following classes:

a. electron transport inhibitors: electron transport is inhibited when one or more intermediary electron carriers are removed or inactivated or even when there is interference in the phosphorylation. Example: diuron, atrazine.

b. uncouplers: uncouplers dissociate the electron transport of the ATP formation through the dissipation of the energetic state of the thylakoid membrane, before the energy can be used to perform the high endergonic reaction of ADP phosphorylation. Example: perfluidone.

c. energy transference inhibitors: inhibition of energy transference inhibitors acts directly in the phosphorylation, as well as inhibitors of the electron transport, which inhibit both the electron flow and the formation of ATP in coupled systems. Example: 1,2,3-thiadiazol-phenylurea, nitrofen.

d. inhibitory uncouplers: the term "inhibitory uncouplers" was used by Moreland [33] to indicate that the herbicides interfere in reactions affected by electron transport inhibitors and by uncouplers; These "inhibitory uncouplers" inhibit the basal transport, uncoupled and coupled of electrons. The herbicides classified in this group affect both the electron transport and the gradient of protons. Examples: acylanilides, dinitrophenols, imidazole, bromofenoxim.

e. electron acceptors: the compounds classified in this group are able to compete with some component of electron transport and consequently suffer reduction. Examples: diquat, paraquat.

f. inhibitors of the carotenoid synthesis: this class of herbicides acts to inhibit the synthesis of carotenoids, resulting in accumulation of precursors of carotenoid devoid colour (phytoene and phytofluene). The inhibition of carotenoid synthesis leads to the degradation of chlorophyll in the presence of light; degradation of 70s ribosomes; inhibition of the synthesis of proteins and loss of plastids. Examples: amitrole, dichlormate, SAN6706.

2. *mitochondrial electron transport and phosphorylation:* herbicides that interfere in the mitochondrial system are classified as:

a. electron transport inhibitors: defined as substances that have the ability to interrupt the electron flow in some point of the respiratory chain, acting in one of the complexes. Examples: diphenylether herbicides.

b. uncouplers: in appropriate concentrations, the classic uncouplers, that are weak lipophilic acids or bases, prevent the phosphorylation of ADP without interfering in the electron transport. Generally, any compound that promotes the dissipation of the energy generated by the electron transport, except for the production of ATP, can be considered as uncoupler. Example: isopropyl ester glyphosate.

c. energy transfer inhibitors: compounds of this group inhibit the phosphorylating electron transport, when the apparatus of energy conservation of the mitochondria is intact and the inhibition is circumvented by uncouplers. They combine with an intermediary in the coupling energy chain and, thus, block the phosphorylation sequence that leads to the ATP formation. No herbicide seems to act as an energy transfer inhibitor.

d. inhibitory uncouplers: most of the herbicides that interfere in the oxidative phosphorylation present a great variety of responses and are classified as uncoupling inhibitors. At low molar concentrations, herbicides fulfil almost all, if not all, of the requirements established for uncouplers, but at high concentrations they act as electron transport inhibitors. Herbicides that present this behaviour are the same classified as uncoupler inhibitors of the photoinduced reactions in the chloroplast. Example: perfluidone.

3. *interactions with membrane:* herbicides can affect the structure and function of membranes directly or indirectly. When the herbicides disaggregates a membrane, they can influence directly the transport processes by interacting with the protein compounds, such as, ATPases and by altering the permeability by physicochemical interactions, or indirectly by modulating the supply of ATP needed to energize the membrane. Interactions with the membrane can cause:

a. compositional alterations: can modify or alter the composition of lipids in the membrane and can also act in the metabolism and synthesis of lipids. Examples: dinoben, chlorambem, perfluidone.

b. effects in the permeability and integrity. Examples: paraquat, diquat, oryfluorfen, oryzalin.

4. *cell division:* herbicides may suppress cell division by interfering in the synthesis or active transport of precursors into the nucleus, which are necessary for the synthesis of DNA during interphase; modify the physical or chemical properties of the DNA or of their complexes; interfere in the formation and function of the spindle; and/or inhibits the formation of the cell wall. Several of the processes mentioned previously need energy and, therefore, interferences in the amount of energy caused by an herbicide could modulate the mitotic activity. The effects of the inhibitors of the cell division are dependent on the concentration and vary according to the species and the type of tissue. There is a relationship between cell division and cellular energy. In higher plants, cell division is prevented or suppressed in conditions in which the glycolysis or the oxidative phosphorylation is inhibited. Another form of the herbicide to alter cell division would be interacting with the microtubules, since these cellular structures are responsible for the orientation and movement of chromosomes during cell division. Examples of herbicides that interfere in cell division: N-phenylcarbamates, ioxynil, trifluralin.

5. *Synthesis of DNA, RNA and protein:* there are correlations between inhibition of RNA and protein synthesis and low concentration of ATP in tissues and these correlations suggest that interferences in the energy production, necessary to perform biosynthetic reactions, could be the mechanism by which the herbicides could express their effects. Moreover, they can inhibit the synthesis of DNA or RNA by altering the chromatin integrity and, in these cases, the synthesis of proteins is also affected. Examples: glyphosate, trifluralin.

The herbicides can still be classified according to the chemical affinity. Table 1 shows the chemical classes and examples of each class, according to Rao [34].

Class of the herbicide	Examples of herbicides
Acetamides	Acetochlor, alachlor, butachlor, dimethenamid, metolachlor, napropamide, pronamide, propachlor, propanil
Aliphatics	Chlorinated aliphatic acid (TCA), acrolein, dalapon
Arsenicals	Disodium methanearsonate (DSMA), monosodium methanearsonate (MSMA), cacodylic acid
Benzamides	Isoxaben
Benzoics	Dicamba
Benzothiadiazoles	Bentazon
Bipyridiliums	Diquat, paraquat
Carbamates	Asulam, desmedipham, phenmedipham

Class of the herbicide	Examples of herbicides
Cineoles	Cinmethylin
Cyclohexanediones (cyclohexenones)	Clethodin, cycloxidim, sethoxydim, tralkoxydim
Dinitroaniniles	Benefin, ethalfluralin, fluchloralin, pendimethalin, prodiamine, trifluralin
Diphenylethers	Acifluorfen, bifenox, fluoroglycofen, fomesafen, lactofen, oxyfluorfen
Imidazolidinones	Buthidazole
Imidazolinones	Imazapyr, imazaquin, imazethapyr, imazamethabenz
Imines	CGA-248757
Isoxazolidinones	Clomazone
Nitriles	Bromoxynil, dichlobenil, ioxynil
Oxadiazoles	Oxadiazon
Oxadiazolidines	Methazole
Phenols	Dinoseb
Phenoxyalkanoic acids Phenoxyacetics Phenoxybutyrics Arylophenoxy propionics	2,4-D, MCPA, 2,4,5-T 2,4-DB Dichlorprop, diclofop, fenoxaprop, fluazifop-P, quizalofop-P
N-phenylphthalimides	Flumiclorac
Phenylpyridazines	Pyridate
Phenyl Triazinones (Aryl Triazinones)	Sulfentrazone
Phthalamates	Naptalam
Pyrazoliums	Difenzoquat
Pyridazinones	Norflurazon, pyrazon
Pyridinecarboxylic Acids	Clopyralid, picloram, triclopyr
Pyridines	Dithiopyr, thiazopyr
Pyridinones	Fluridone
Pyrimidinythio-benzoates (Benzoates)	Pyrithiobac
Quinolinecaryoxylic acids	Quinclorac
Sulfonylureas	Bensulfuron, chlorimuron, chlorsulfuron, halosulfuron, metsulfuron, nicosulfuron, primisulfuron, prosulfuron, sulfometuron, thifensulfuron, triasulfuron, tribenuron
Tetrahydropyrimidinones	Yet to be commercialized
Thiocarbamates	Butylate, diallate, EPTC, molinate, pebulate, thiobencarb, triallate
Triazines	Ametryn, atrazine, cyanazine, hexazinone, prometryn, simazine
Triazinones	Metribuzin
Triazoles	Amitrole
Triazolopyrimidine Sulfonanilides	Flumetsulam
Uracils	Bromacil, terbacil, UCC-C4243
Ureas	Diuron, fluometuron, linuron, tebuthiuron,
Unclassified herbicides	Bensulide, ethofumesate, fosamine, glufosinate, glyphosate, tridiphane

Table 1. Classification of the herbicides according to the chemical affinity.

1.3. Aquatic and soil contamination due to the presence of herbicides

When a herbicide is used to control weeds, sometimes a majority of the compound ends up in the environment, whether it is in the soil, water, atmosphere or in the products harvested [17]. Due to the widespread use of these chemicals over the years, there has been an accumulation of these residues in the environment, which is causing alarming contaminations in the ecosystems [35] and negative damages to the biota. To Bolognesi and Merlo [3], the widespread use of herbicides has drawn the attention of researchers concerned with the risks that they can promote on the environment and human health, since they are chemicals considered contaminants commonly present in hydric resources and soils. According to the same authors, herbicides represent a high toxicity to target species but it can be also toxic, at different levels, to non-target species, such as human beings. Herbicides can cause deleterious effects on organisms and human health, both by their direct and indirect action [2]. Among the biological effects of these chemicals, it can be cited genetic damages, diverse physiological alterations and even death of the organisms exposed. Some herbicides, when at low concentrations, cannot cause immediate detectable effects in the organisms, but, in long term can reduce their lifespan longevity [4]. Herbicides can affect the organisms in different ways. As with other pesticides, the accumulation rate of these chemicals on biota depends on the type of the associated food chain, besides the physicochemical characteristics (chemical stability, solubility, photo-decomposition, sorption in the soil) of the herbicide [5-6]. Thus, despite the existence of several toxicological studies carried out with herbicides, in different organisms, to quantify the impacts of these pollutants and know their mechanisms of action [7, 8, 2], there is a great need to expand even more the knowledge about the effects of different herbicides in aquatic and terrestrial ecosystems. Data obtained from *in situ*, *ex situ*, *in vivo* and *in vitro* tests, derived from experiments of simulation, occupational exposure or environmental contaminations, need to enhance so that it is possible to obtain even more consistent information about the action of these compounds.

According to Jurado et al. [32], when herbicides are applied in agricultural areas they can have different destinations, since being degraded by microorganisms or by non- biological means or even be transported by water, to areas distant from the application site. Thus, according to the same authors, the organisms can be then exposed to a great number of these xenobiotics as well as their metabolites.

The fate of the compound in the soil depends on the characteristics of the compound and the soil. The hydrogenionic properties of a compound in the soil determines its sorption characteristics, such as, acid herbicides in soils with normal pH are negatively charged and consequently are movable in most of the soils [17]. Some groups of pesticides are neutral in soils with normal pH but due to electronic dislocations in the molecules, they can bind to soil colloids by several forms [36].

According to Kudsk and Streiberg [17], during the last two decades, several studies have been completed to predict the behaviour of pesticides in the soil. Despite the numerous efforts to assess the effects of herbicides in the soil, there are conflicting data in the literature on the subject, where some studies show that the residues of pesticides can be sources of carbon and energy to microorganisms, and then are degraded and assimilated by them, while other reports affirm that pesticides produce deleterious effects to the organisms and biochemical and enzymatic processes in the soil [37]. According to Hussain et al. [37], in general, the application of pesticides, and here it is also included herbicides, made long term, can cause a disturbance in the biochemical balance of the soil, which can reduce its fertility and productivity.

Once in the soil, herbicides can suffer alteration in their structure and composition, due to the action of physical, chemical and biological processes. This action on the herbicides is the one that will determine their activity and persistence in the soil. Some molecules, when incorporated into the soil, are reduced by volatilization and photo-decomposition. Once in the soil, herbicides can suffer the action of microorganisms, which, added to the high humidity and high temperature, can have their decomposition favoured [38]. If they are not absorbed by plants, they can become strongly adsorbed on the organic matter present in the colloidal fraction of the soil, be carried by rainwater and/or irrigation and even be leachate, thus reaching surface or groundwater [39].

The prediction of the availability of herbicides to plants has two purposes: 1. ensure that the herbicide reaches the roots in concentrations high enough to control weeds, without compromising the agricultural productivity; 2. predict if the compound is mobile in the soil to estimate how much of the herbicide can be leachate from the roots zone to groundwater [17].

The contamination of aquatic environments by herbicides has been characterized as a major world concern. This aquatic contamination is due to the use of these products in the control of aquatic plants, leachate and runoff of agricultural areas [40]. According to He et al. [20], it is a growing public concern about the amount of herbicides that have been introduced into the environment by leachate and runoff, not to mention that the contaminations of the aquatic environments generally occur by a mixture of these compounds and not by isolated substances.

Guzzella et al. [1] did a survey on the presence of herbicides in groundwater in a highly cultivated region of northern Italy. The researchers monitored for two years the presence of 5 active ingredients and 17 metabolites resulting from these compounds. The authors verified that atrazine, although banned in Italy since 1986, was the major contaminant of the groundwater of the sites studied, they also observed that the concentration of at least one of the compounds studied exceeded the maximum allowed concentration in 59% of the samples likely due in both cases to off-label herbicide use. This scenario could be, in long term, a serious problem for the quality of this water, which is used as drinking water.

Toccalino et al. [41] carried out a study to verify the potential of chemical mixtures existing in samples of groundwater used for public supply. In these samples, the most common organic contaminants were herbicides, disinfection by-products and solvents. The authors concluded that the combined concentrations of the contaminants can be a potential concern for more than half of the samples studied and that, even though the water destined to public supply pass through treatments to reduce contaminations and meet the legislations, it can still contain mixtures at worrying concentrations.

Saka [42] evaluated the toxicity of three herbicides (simetryn, mefenacet and thiobencarb) commonly used in rice planting in Japan, on the test organism *Silurana tropicalis* (tadpoles). The authors observed that the three herbicides, particularly thiobencarb, are toxic for tadpoles

(LD50 test), even for concentrations found in waters where the rice is cultivated. In a similar study carried out by Liu et al. [43], it was observed that the effect of the herbicide butachlor (most used herbicide in rice planting in Taiwan and Southeast Asia) on the organism *Fejervarya limnocharis* (alpine cricket frog) exposed to concentrations used in the field. In this study no effect on the growth of tadpoles of *F. limnocharis* was observed, but there was a negative action on survival, development and time of metamorphosis. The authors suggested that the herbicide butachlor can cause serious impacts on anurans that reproduce in rice fields, but this impact varies from species to species.

In a study conducted by Ventura et al. [8], it was observed that the herbicide atrazine has a genotoxic and mutagenic effect on the species *Oreochromis niloticus* (Nile tilapia). In this study, the authors observed that the herbicide can interfere in the genetic material of the organisms exposed, even at doses considered residual, which led the authors to suggest that residual doses of atrazine, resulting from leaching of soils of crops near water bodies, can interfere in a negative form in the stability of aquatic ecosystems.

Bouilly et al. [44] studied the impact of the herbicide diuron on *Crassostrea gigas* (Pacific oyster) and observed that the herbicide can cause irreversible damages to the genetic material of the organism studied. Moreover, the authors affirm that, due to the persistence of diuron in environments adjacent to its application site and that it is preferably used in spring, the pollution caused by its use causes negative impact in the aquatic organisms during the breeding season.

In general, when herbicides contaminate the aquatic ecosystem, they can cause deleterious effects on the organisms of this system. Thus, organisms that live in regions impacted by these substances, whose breeding period coincides with the application period of the herbicides, can suffer serious risks of development and survival of their offspring.

Hladik et al. [45] evaluated the presence of two herbicides (chloroacetamide and triazine), as well as their by-products, in drinking water samples of the Midwest region of the United States. The authors detected the presence of neutral chloroacetamide degradates in median concentrations (1 to 50 ng/L) of the water samples. Furthermore, they found that neither the original chloroacetamide herbicides nor their degradation products were efficiently removed by conventional water treatment processes (coagulation/flocculation, filtration, chlorination). According to Bannink [46], about 40% of the drinking water from Netherlands is derived from surface water. The Dutch water companies are facing problems with the water quality due to contamination by herbicides used to eliminate ruderal plants. These data serve as alerts for the presence of herbicides and their degradation products in drinking water, pointing out the need for the development of new treatment systems that could be more efficient to eliminate this class of contaminants.

According to Ying and Williams [40], organic herbicides, when in aquatic ecosystems, can be distributed in several compartments depending on their solubility in water. These compartments include water, aquatic organisms, suspended sediment and bottom sediment. The more hydrophilic the organic pesticide, the more it is transported to the aqueous phase, and the more hydrophobic a pesticide is, the more it will be associated to the organic carbon of the suspended and bottom sediment [47]. The sorption of the herbicides in sediments in suspension can reduce the degradation rate of the herbicides in water, and the movement of the sediment in suspension can transport the pesticides from one place to another, entering into the tissue of organisms or settling on the bottom [40].

A study conducted by Jacomini et al. [48] evaluated the contamination of three matrices (water, sediment and bivalve molluscs) collected in rivers influenced by crops of sugar cane in São Paulo State-Brazil. In this study, the authors observed that the highest concentrations of residues of the herbicide ametrin were present in the sediment, showing the persistence of this compound in the sediments of rivers and its potential to mobilize between the compartments of the aquatic system, such as water and biota.

When the herbicides are dispersed in the water or sediments in suspension of the rivers, they can end up in other ecosystems such as estuaries. Duke et al. [49], when studying the effect of herbicides on mangroves of the Mackay region, found out that diuron, and even other herbicides, are potentially responsible for the mangrove dieback. According to the authors, the consequences for this death would be the impoverishment of the quality of the coastal water with an increase of the turbidity, nutrients and sediment deposition, problems in the fixation of seedlings and consequent erosion of the estuaries.

In a review conducted by Jones [50], the author highlights the contamination of marine environments by herbicides (such as diuron), discussing that the contamination of these environments can occur by transport of these substances of agricultural or non cultivated areas (roadsides, sports fields, train tracks), runoff by storms and tailwater irrigation release), pulverizations and accidental spills. These contaminations mean that the photochemical efficiency of intracellular symbiotic algae of the coral, in long term, may be compromised, leading to a loss in the symbiotic relationship of the coral with the algae and a consequent bleaching of corals. Still considering the marine ecosystem, Lewis et al. [51] verified that the runoff of pesticides from agricultural areas influence the health of the Great Barrier Reef in Australia and can disturb this sensitive ecosystem.

Considering the prior literature, it is likely possible that the effects of herbicides do not occur only at the places that they are applied but also in places distant from their application. Moreover, herbicides can induce alterations in non-target organisms, altering the survival and the equilibrium of the ecosystems, whether they are aquatic or terrestrial. Thus, much care must be taken when introducing these substances into the environment and more studies should be conducted in order to thoroughly understand the environmental consequences that herbicides can cause.

2. The effects of herbicides using different bioassays and test-organisms

Many studies have evaluated the impact of different chemical classes of herbicides using different doses, organisms and bioassays, focusing on toxic, cytotoxic, genotoxic, mutagenic, embryotoxic, teratogenic, carcinogenic and estrogenic effects.

With respect to the toxicity, some herbicides pose major concerns when applied in regions close to water resources due to their highly toxic potential to many aquatic organisms [52].

Biological tests of toxicity and mutagenicity are, according to Moraes [53], indispensable for the evaluation of the reactions of living organisms to environmental pollution and also for the identification of the potential synergistic effects of several pollutants. The impact that toxic materials can promote in the integrity and function of DNA of several organisms has been investigated [54]. Several biomarkers have been used as tools for the detection of the toxic, genotoxic and mutagenic effects of pollution. Among them we can cite the presence of DNA adducts, chromosome aberrations, breaks in the DNA strands, micronuclei formation and other nuclear abnormalities, besides induction of cell death [55].

Most of the tests used to detect the mutagenic potential of chemical substances are based on the investigation of possible inductions of chromosome damages such as structural alterations, formation of micronuclei, sister chromatid exchanges, assessment of mutant genes or damages in the DNA, using different test organisms, such as bacteria, plants and animals, both *in vitro* and *in vivo* [56].

According to Veiga [57], it is possible to estimate the genotoxic, mutagenic, carcinogenic and teratogenic effects of agrochemicals by relatively simple methods. Several studies have been carried out by several researchers concerned with the harmful effects of pesticides in an attempt to verify their possible physiological [58, 59], mutagenic [7, 8, 60, 61, 62] and carcinogenic effects [63].

The interaction between different methods of evaluating the toxic, genotoxic and mutagenic potential provides a more global and comprehensive view of the effect of a chemical agent. For the monitoring of organisms exposed to chemical agents, the chromosome aberration test, micronucleus test and comet assay have been widely used [64]. A few studies also have shown the toxic effects of chemicals, by cell death processes, both necrotic and apoptotic [65].

According to Kristen [66], the dramatic expansion in the production of xenobiotic compounds by anthropogenic activities has compromised the environment by the introduction of millions of chemicals with toxic potential to biological systems.

Cytogenetic tests are adequate to identify the harmful effects of substances, in their several concentrations and different periods of exposure. These tests, generally performed with test organisms, are commonly applied in biomonitoring to the extent of pollution and in the evaluation of the combined effects of toxic and mutagenic substances on the organisms in the natural environment [53]. Micronuclei assays are efficient to assess the mutagenic activity of herbicides both in laboratorial and field assays [67]. The comet assay can be used to evaluate damages in proliferating cells or not, in *in vitro* or *in vivo* tests and can be applied with the purpose of genotoxicological analyses [68]. According to these same authors, these tests are considered one of the best tools to biomonitor several chemical compounds, including herbicides. According to Ribas et al. [69], the simplicity, reproducibility and rapidity of the comet test, associated to the ability of this assay in evaluating damages in the DNA, makes this technique highly applicable to environmental genotoxicology.

The toxic, cytotoxic, genotoxic, mutagenic, embryotoxic, teratogenic carcinogenic and estrogenic effects caused by herbicides on various organisms could be exemplified by studies as described below.

2.1. Atrazine

Atrazine is a triazinic herbicide, classified as moderately toxic of pre- and post-emergence, used for the control of weeds in crops of asparagus, corn, sorghum, sugarcane and pineapple [70]. According to Eldridge et al. [71], triazinic herbicides are among the most used pesticides in agriculture due to their ability to inhibit the photosynthesis of weeds in crops [16].

Triazine herbicides are extensively used in the United States to control grass, sedge and broadleaf weedsduring the cultivation of maize, wheat, sorghum, sugarcane and conifers [72]. In Brazil, these herbicides are widely used on crops of sugarcane and maize. Due to the widespread use of triazine herbicides in the agriculture and, therefore, its high exposure potential for humans, the United States Environmental Protection Agency (USEPA) has conducted a special review on the published and non published data of several triazine herbicides [73]. According to Nwani et al. [22], the herbicide atrazine is widely used in crops worldwide. The dangers, both toxic and genotoxic of this herbicide have been revised; however, there is an urgent need for more detailed studies on the mode of action of this compound. Atrazine has been tested in several systems, but there are shortcomings in relation to certain tests performed and some evidences of the genotoxic effects, *in vivo*, still need to be confirmed [74].

Several studies using the test system *Aspergillus* have shown that atrazine is not mutagenic to these organisms [75, 76, 77], although it is considered mutagenic for other test systems such as *Drosophila melanogaster* [78, 79]. According to Ribas et al. [74], atrazine was responsible for a significant frequency of aneuploidies in *Neurospora crassa*, given by the chromosomal non-disjunction in *Aspergillus nidulans*, and by the induction of loss of sexual chromosomes in *Drosophila melanogaster*.

Sorghum plants treated with atrazine presented an increase in the number of their chromosomes, multinucleated cells, aneuploidy and polyploidy, and abnormalities in the mother cells of the pollen grain, which suggests that this herbicide interferes in the stability and also in the meiosis [80].

Popa et al. [70] observed that atrazine, when applied in high concentrations in maize seedlings, can induce chromosome breaks, visualized by the presence of single and paired chromosome fragments; a high frequency of chromatids and chromosome bridges; lagging chromosomes and presence of heteropolyploid or polyploid cells. Grant and Owens [81] showed that atrazine induced chromosome breaks (in mitosis and meiosis) in the species *Pisum sativum* and *Allium cepa*.

Hayes et al. [82] investigated the effect of the herbicide atrazine on wild leopard frogs (*Rana pipens*), in different regions of the United States. The authors observed that a great percentage of males exposed to the herbicide presented abnormalities in the gonads, such as development retardation and hermaphroditism. This effect can, in long term, lead to a decline in the amphibian population of the sites contaminated with this herbicide.

According to Gammon et al. [83], some publications have reported a possible feminization of frogs, both in laboratorial assays and field studies. This effect is mainly due to the action of the enzyme aromatase; however, published research not shown the measures of this enzyme. Thus, there are doubts about the feminization theory, except for the studies that presented a great number of frogs with morphological alterations related to very high levels of atrazine.

Nwani et al. [22] evaluated the genotoxic and mutagenic effects of the herbicide Rasayanzine, whose active ingredient is atrazine, using the comet assay and micronucleus test, in erythrocytes and gill cells of the fish *Channa punctatus*. By the data analysis of the two cell types, significant effects for all the concentrations (4.24, 5.30 and 8.48 mg/L) and exposure periods tested (1, 3, 5, 7, 14, 21, 28 and 35 days) were observed. The highest damages were observed for the highest concentrations and exposure times, showing the genotoxic and mutagenic dose-response potential of atrazine for the aquatic organism. Furthermore, it was found that gills were more sensitive to the action of the herbicide, when compared to erythrocytes. From the results obtained, the authors suggested a careful and judicious use of the herbicide atrazine in order to protect the aquatic ecosystems and human population.

A study carried out by Çavas [84] compared the genotoxic effects of the active ingredient atrazine and its commercial formulation Gesaprim, in the concentrations of 5, 10 and 15 μg/L, by the comet assay and micronucleus test, in erythrocytes of the fish gibel carp (*Carassius auratus*). The results showed that there was a significant increase in the frequencies of the micronuclei and DNA strand breaks in the erythrocytes treated with all the concentrations of the commercial formulation of atrazine, showing the genotoxic and mutagenic potential of Gesaprim for this species of fish. While the commercial formulation presented a high genotoxic potential, the assays showed that the active ingredient atrazine is not genotoxic, suggesting that the adjuvants present in Gesaprim must be the responsible for the genotoxic effects observed in this species of fish. Despite the comparative analysis of the genotoxicity between the active ingredient and the commercial product has showed to be a very effective tool for the discovery of genotoxic environmental risks, it is not easy to determine the exact identity of the products used as adjuvants and of the agents of surface action of pesticides due to the existence of the patent protection system.

Atrazine has also been tested to evaluate the ability to induce cytogenetic damages in rodents. Meisner et al. [85] submitted rats to 20 ppm of atrazine (by water ingestion) and did not observe, after exposure to the herbicide, an increase in the number of chromosome aberrations. In a similar study, Roloff et al. [86] reported that there was no significant increase of chromosome aberrations in cells of rat bone marrow, when they were fed with 20 ppm of atrazine.

Wu et al. [87] assessed the embryotoxic and teratogenic effects of atrazine, at the doses of 25, 100 and 200 mg/Kg/day, in Sprague-Dawley rats. Prenatal exposure to the highest dose of the herbicide tested caused hypospadias in 10.23% of male newborn rats, and the lowest dose induced diverse embryotoxic damages in some individuals. According to Modic et al. [88], high doses of atrazine (50 or 200 mg/kg/day), administered daily in male Wistar rats at 60 days of age, promoted alterations in the levels of several hormones in the serum of these individuals, observed by slight increases in the levels of androstenedione testosterone, estradiol, estrone, progesterone and corticosterone, quantified by radioimmunoassay.

To obtain more concise data on the genotoxicity of triazine herbicides, Tennant et al. [89] used the comet assay methodology, which showed to be highly sensitive for the detection of low rates of damages in the DNA. According to these authors, the comet assay showed that atrazine induced a small increase in the damages in the DNA in leukocytes of rats. Moreover, by the comet assay, Clements et al. [90] reported that atrazine induced a significant increase in the frequencies of damages in the DNA of erythrocytes of bullfrog tadpoles, noting the genotoxic potential of this herbicide for this species of amphibian, from the concentration of 4.8 mg/L.

Studies about the cytotoxicity, genotoxicity and mutagenicity of the atrazine herbicide (oral gavage - dose 400 mg/kg/day), carried out by Campos-Pereira et al. [91], have shown the induction of lipid peroxidation and liver damage, death of hepatocytes, and micronucleus formation in exposed Wistar rats. Tests performed by Ventura et al. [8] showed that the same triazine pesticide was able to induce significant DNA fragmentation when using the comet assay, and nuclear alterations and micronuclei using the micronucleus test in *Oreochromis niloticus* (Nile tilapia) erythrocytes exposed to different concentrations of atrazine (6.25, 12.5, 25 µg/L), thus corroborating the studies performed by Campos-Pereira et al. [91].

Ruiz and Marzin [92] assessed the genotoxic and mutagenic effects of the herbicide atrazine by two *in vitro* assays (*Salmonella* assay and SOS Chromotest), one to detect bacterial mutagenicity and the other to verify primary damages in the DNA. The assays were carried out both in the absence and in the presence of S9 fractions from rat liver homogenate (Sprague-Dawley). The authors found that the herbicide atrazine did not present genotoxic potential neither to the *in vitro* test with Salmonella/microsome nor by the SOS Chromotest, both in the absence and in the presence of the S9 fractions, when the strains were exposed to atrazine.

In vitro studies, performed with human lymphocytes, treated with 0.10 ppm of atrazine, detected a slight increase in the chromosome aberrations rates [85]. However, for concentrations below 0.001 ppm of this herbicide, chromosome aberrations were not detected [86] Lioi et al. [93] observed a small increase in the number of sister chromatid exchange but a great increase of chromosome aberrations in human lymphocytes exposed to atrazine. Meisner et al. [94] observed a significant increase in the frequency of chromosome breaks in human blood cells exposed to 1 ppm of the herbicide atrazine.

The genotoxicity of herbicides, such as atrazine, has also been evaluated by the comet assay by the use of human blood lymphocytes. According to Ribas et al. [69], blood cells treated with the herbicide atrazine, at concentrations of 50-200 µg/l, showed an extensive migration of DNA, mainly at concentrations of 100 and 200 µg/l.

In mammalian test systems, submitted to the action of the herbicide atrazine, most of the results seem to be negative, except for the results of Loprieno and Adler [95], who obtained a significant increase in the frequency of chromosome aberrations in bone marrow cells of rats, and the data obtained by Meisner et al. [94], who described an induction of chromosome aberrations in cultured human lymphocytes. While the results from bacteria and mammal test systems are almost all negative, atrazine exhibits clear mutagenic effects in different plant test systems, by inducing chromosome aberrations in *Hordeum vulgare* and *Vicia faba* [96, 97], in *Zea mays* [98], in *Sorghum vulgare* [99] and in *Allium cepa* [62] ; sister chromatid exchanges in maize [100] ; and point mutation in maize [98].

Studies performed by Zeljezic et al. [101] had already reported that atrazine does not present genotoxicity or capacity to induce apoptosis or necrosis in human lymphocytes, while the treatment of these cells with the commercial formulation, Gesaprim, significantly increased the rates of damages in DNA, observed by the comet assay. Srivastava and Mishra [102] observed results that are in agreement with the findings of Zeljezic et al. [101] and Çavaş [84], in which the exposure to different concentrations of Gesaprim inhibited the mitotic index and increased the frequencies of micronuclei and chromosome aberrations in somatic cells of *Allium cepa* and *Vicia faba*.

2.2. Atrazine and butachlor

Toxic effects of atrazine, alone or associated with the herbicide butachlor, for the freshwater species such as the green alga *Scenedesmus obliquus* and the cladoceran *Daphnia carinata*, were evaluated, showing values of 96 h-EC50 for *S. obliquus* (atrazine= 0.0147 mg/L and butachlor= 2.31 mg/L, and of 48h-LC50 for *D. carinata* (atrazine= 60.6 mg/L and butachlor= 3.40 mg/L) [20]. These results suggest that atrazine has a highly toxic potential for *S. obliquus* and slightly toxic for *D. carinata*, while butachlor exhibits a moderate toxic potential for both organisms. Now, the analysis of the mixture atrazine-butachlor allowed the authors to verify that the toxic effects were significantly antagonistic for *S. obliquus*, and that there was no significant synergism for *D. carinata* [20].

2.3. Atrazine, simazine ande cyanazine

Simazine and cyanazine, as well as atrazine, are widely used as triazine herbicides of pre- and post-emergence weed control, whose residues have been carried to the source of drinking water of several agricultural communities. These compounds also present a potential risk to humans, mainly due to their presence in food [103]. Studies on the effect of atrazine, simazine and cyanazine performed by Kligerman et al. [104], found that there was not a significant increase in the sister chromatid exchanges and chromosome aberrations in cultured human lymphocytes exposed to these herbicides, up to the solubility limit in aqueous solution using 0.5% of dimethyl sulfoxide. However, Adler [105] observed that doses of 1500 and 2000 mg/Kg of atrazine, administered by oral gavage in rats, induced dominant lethal mutations and chromatin breaks in the bone marrow of these organisms.

Kligerman et al. [103] observed that the association of the herbicides atrazine, simazine and cyanazine did not induce micronuclei in polychromatic erythrocytes of bone marrow of female rats (C57B1/6) exposed by intraperitoneal injection, even when very high doses of these herbicides were administered (125, 250 and 500 mg/Kg of atrazine; 500, 1000 and 2000 mg/Kg of simazine; 100, 200 and 400 mg/kg of cyanazine), showing an absence of genotoxic potential of these compounds for the organism tested.

On the other hand, Hrelia et al. [106] showed that males and females of Sprague-Dawley rats exposed by oral gavage to doses of 56, 112 and 224 mg/kg of cyanazine, did not present significant increases in chromosome aberrations.

Taets et al. [107] evaluated the clastogenic potential of environmental concentrations of the triazine herbicides simazine (0.001 to 0.004 µg/mL), cyanazine (0.003 to 0.012 µg/mL) and atrazine (0.003 to 0.018 µg/mL), in Chinese Hamster Ovary (CHO) cells, using flow cytometry assay. The authors proved the clastogenic action for the herbicides atrazine and cyanazine, proven by the high indices of damages in the cells exposed to atrazine and by the significant frequencies of damages observed in the cells exposed to cyanazine.

2.4. Terbutryn

The herbicide terbutryn is an s-triazine herbicide used pre- and post-emergence and widely used worldwide as an agent to control grass, sedge, and broadleaf weeds in vegetables, cereals and fruit trees. It is an herbicide persistent in the environment, which tends to dislocate by the flow of water and leachate [108].

An *in vitro* study performed by Moretti et al. [108] investigated the genotoxicity of the herbicide terbutryn, by analyzing the relationship between the cytogenetic damage, evaluated by the assays of SCE (sister chromatid exchanges) and MN (micronucleus), and the primary damage in the DNA, assessed by the comet assay, in leukocytes newly-isolated from peripheral human blood. The results showed that terbutryn did not produce significant increases of SCE or MN, both in the absence and in the presence of the metabolic activation system from rat liver (S9 fraction), although terbutryn has induced primary damages in the DNA in a more pronounced form in the absence of S9. The apparent lack of sensitivity of the assays of SCE and MN test for the genotoxicity of terbutryn, in comparison to the comet assay, can be attributed to the generation of specific types of damages, since the SCE and MN are determined in proliferative cells and are sensitive indicators of lesions that survive for, at least, one mitotic cycle, while the comet assay identifies repairable lesions in the DNA of on resting (G0) cells. According to these results, the authors suggest that terbutryn must be considered a genotoxic compound.

2.5. 2,4-D (2,4-dichlorophenoxyacetic acid)

The 2,4-D (2,4-dichlorophenoxyacetic acid) is an herbicide from the group of the polychlori-nated aromatic hydrocarbons that has been widely used throughout the world [109] since 1944, to control broadleaf weeds and woody plants [110]. Its action mimics the auxin of plants [111]. According to Martínez-Tabche et al. [112], this herbicide mimics the action of the hormone indole acetic acid, when used in small quantities but it is highly cytotoxic in high concentrations.

According to Ateeq et al. [113], the increase in the frequency of micronuclei and altered cells was significant, when erythrocytes of catfish (*Clarias batrachus*) were analyzed, after exposure to the herbicides 2,4-D and butachlor. There was a positive dose-response relationship in all exposures to the two herbicides and in all exposure periods tested.

Studies carried out by Suwalsky et al. [114] in nerve cells of *Caudiverbera caudiverbera* demonstrated the toxicity of the herbicide 2,4-D. The authors observed a reduction in the dose-dependent response to nerve stimulation in the simpact junction of the frog when they were exposed to this herbicide. This reduction is probably due to a mechanism of lipid perturbation and interference in the properties of the plasma membrane, such as protein conformation and/or interaction with protein receptors, which leads to an inhibition of the glandular chloride channel from the mucosal skin of this test organism.

According to Gómez et al [115], the main and most common entrance route of 2,4-D in fish is through gills. This herbicide can cause several adverse symptoms to these organisms, such as bleeding, increased damage to the kidneys and renal functions, as well as hepatic degeneration.

Martínez-Tabche et al. [112] evaluated the toxicity of different concentrations of the herbicides 2,4-D and paraquat (0, 5, 75 and 150 mg/L), using several assays (acute lethality test, lipid peroxidation assay by quantification of MDA – Malondialdehyde – and comet assay) in rainbow trout (*Oncorhynchus mykiss*). For the acute lethality tests, it was observed a more evident toxic action for the organisms exposed to the treatment of 24 h with the herbicide paraquat, which presented high indices of mortality, analyzed by the values of LC_{50} (LC_{50} of paraquat = 0.084 mg/L; LC_{50} of 2,4-D = 362.38 mg/L). The authors also showed that individuals exposed to the two higher concentrations of both herbicides had apnoea and white spots in their scales. All concentrations of 2,4-D and paraquat induced a significant increase in the DNA damages and the amount of MDA in the gills exposed.

González et al. [116] proved the genotoxicity of 2,4-D due to a significant increase of SCE in CHO cells treated with the concentrations of 2 to 4 ug/mL of this herbicide. Madrigal-Bujaidar et al. [117] also showed the genotoxic potential of 2,4-D, due to a clastogenc effect of this herbicide at the doses of 100 and 200 mg/Kg, detected by a significant increase of SCE in bone marrow cells and germ cells of rats. Soloneski et al. [118] studied the genotoxic effects of different concentrations (0, 10, 25, 50 and 100 mg/mL) of the herbicide 2,4-D (2,4-dichlorophenoxyacetic) and its commercial derivative 2,4-D DMA (Dimethylamine 2,4-D salt), by the SCE assay and analyses of cell cycle progression and mitotic index human lymphocytes maintained in culture, in the presence (human whole blood - WBC) and absence (plasma leukocyte cultures - PLC) of erythrocytes. These compounds did not induce significant frequencies of SCE and only the concentration of 100 mg/mL of 2,4-D caused alterations in the progression of the cell cycle in PLC, while the different concentrations of 2,4-D and 2,4-D DMA induced a significant increase in the frequency of SCE and a significant delay in the cell proliferation rates in WBC. Moreover, both 2,4-D and 2,4-D DMA presented a dose-response inhibition of the mitotic activity in PLC and WBC. Based on these results, the authors concluded that the herbicide and its commercial derivative presented genotoxic potential, which was higher in the presence of human erythrocytes.

Morgan et al. [119] showed, by embryotoxicity and teratogenicity assays carried out with *Xenopus* (FETAX - frog embryo teratogenic assay – *Xenopus*), that high concentrations of 2,4-D, induce potentially more embryotoxic effects than teratogenic in frog embryos, demonstrated by the values of EC50 and LC50 of 245 mg/L and 254 mg/L, respectively, and by the Teratogenic Index of 1.04. Moreover, the same authors compared the teratogenic action of the herbicide atrazine in relation to 2,4-D, showing that atrazine is potentially more teratogenic than 2,4-D, for frog embryos.

The estrogenic potential of 4 herbicides (triclopyr; 2,4-D; diquat dibromide and glyphosate), was evaluated by the *in vivo* de vitellogenin assay with rainbow trout. A significant estrogenic potential was shown for 2,4-D, since it induced a 93 fold increase in the levels of plasma vitellogenin of the fish treated with this herbicide during 7 days [120].

2.6. Glyphosate

Glyphosate is a non-selective organophosphorus, broad spectrum, post-emergence herbicide, widely used in agriculture, mainly to control grasses, sedges, and broadleaf weeds [121]. Its action occurs by the inhibition of the biosynthesis of aromatic amino acids [122]. Its main mode of action is by the inhibition of the enzyme 5-enolpyruvylshikimate-3-phosphate synthase (EPSPS), which is essential in plants for the synthesis of the referred amino acids. Since this enzyme is absent in animals, this herbicide should be relatively non toxic for these organisms [123]. There are many conflicting data on the toxicity of glyphosate and its commercial formulations.

According to Solomon and Thompson [124], environmental toxicology of glyphosate has been extensively reviewed by a series of international regulatory agencies. According to the authors, as glyphosate binds strongly with organic matter, it is considered immobile in soils and sediments. This binding also removes glyphosate from water, reducing efficiently, the exposure of aquatic organisms. As the acute exposures are most likely to occur, the measures of effect are the most adequate for the purpose of risk assessment. However, in general, the authors affirm that glyphosate presents a low potential of acute toxicity for wild animals, including mammals, birds, fish and aquatic invertebrates.

Williams et al. [125] carried out a critical review on the toxicity of the herbicide RoundUp™ and of its active ingredient glyphosate. The analysis of the toxicity data, carried out by pattern tests and evaluation criteria, indicated that there is no evidence that glyphosate causes direct damages in the DNA, both in assays performed *in vitro* and *in vivo*. The authors concluded that Roundup™ and its components do not represent a risk for the induction of inheritable/ somatic mutations in humans. Furthermore, the authors assert that, by the studies performed, glyphosate is not carcinogenic or teratogenic, nor does it cause significant adverse effects in the reproduction, development or in the endocrine system of humans and other mammals and, therefore, does not represent a risk for the health of human beings.

A study on the impact of the herbicide glyphosate and its commercial formulation Roundup™, in three microorganisms of food interest (*Geotrichum candidum*, *Lactococcus lactis subsp. cremoris* and *Lactobacillus delbrueckii subsp. bulgaricus*), showed that Roundup™ has an inhibitory effect on the microbial growth and a microbiocide effect at concentrations lower than the recommended for agricultural use. It was also observed in this study that glyphosate did not induce significant toxic effects for the three microorganisms studied. These differences between the toxic actions resulted from Roundup™ and glyphosate could be explained by a possible amplified effect of the commercial formulation due to the presence, according to Cox

[126] of adjuvants, such as polyethoxylated tallowamine (POEA), used for a better stability and penetration of the chemical compound [127].

Relyea [128] assessed the toxic potential of environmentally relevant concentrations of glyphosate on three species of tadpoles (wood frog [*Rana sylvatica* or *sylvaticus Lithobates*], leopard frog [*Rana pipiens pipiens* or L.], and American toad [*Bufo americanus* or *Anaxyrus americanus*]), by morphological analysis of individuals, before and after the application of the herbicide, showing that there is a significant induction of morphological alterations in the tadpoles of the three species. Specifically in the case of the wood frog and leopard frog, the exposure to the chemical compound has led to an evident alteration of the size of the tadpole tail, suggesting that the herbicide could be activating physiological mechanisms of development that are normally used as defence responses against predators. These results showed that glyphosate can have widespread and relevant effects on non target species, contradicting other studies, such as the one performed by Solomon and Thompson [124], who affirmed the inexistence or irrelevance of the toxicity of this compound on organisms and the environment.

Studies on the genotoxic potential of the active ingredient glyphosate, present in the commercial formulation Roundup, were performed on the roots of smooth hawksbeard (*Crepis capillaris* L.), in the concentrations of 0.05, 0.1, 0.5 and 1.0% of the active ingredient and for polychromatic erythrocytes (PCEs) of the bone marrow of C57BL rat, at doses inferior to half the LD_{50} (1080 mg/Kg). In these studies the chromosome aberrations assay and micronucleus test were used, which showed that this chemical compound did not induce significant responses for any of the biological systems tested [129].

Martini et al. [123] studied the effects of the commercial formulation of glyphosate in the proliferation, survival and differentiation of the 3T3-L1 fibroblasts (a mammal cell line), by the cell viability test with Trypan, MTT test, enzymatic activity assay of caspase-3 and staining assay with annexin-V and propidium iodide. The results showed that glyphosate inhibits the cell proliferation and induces apoptosis in a dose-dependent way, besides decreasing significantly the ability of the fibroblasts to differentiate to adipocytes. These data suggest the occurrence of important cell damages mediated by the action of this herbicide, indicating that glyphosate presents a potential risk factor for human health and the environment.

Dallegrave et al. [130] evaluated the teratogenicity of the herbicide glyphosate, marketed in Brazil as Roundup (36% of glyphosate and 18% of the surfactant polyoxyethyleneamine), to females of Wistar rats. The females treated orally with three different doses of glyphosate (500, 750, 1000 mg/Kg) from the 6th to the 15th day of gestation. After performing caesarean sections on day 21 of gestation, the number of corpora lutea, implantations, live and dead foetuses and reabsorptions, as well as the external malformations and skeletal malformation were recorded and analyzed. It was observed a mortality rate of 50% of the females treated with the highest concentration of glyphosate; the authors verified that there was a dose-response relationship directly proportional to the increase in the number of skeletal alterations found. These results led the authors to conclude that the commercial formulation of glyphosate (Roundup) is toxic for females of Wistar rats and is able to induce a delay in the fetal skeletal development of this species. It is important to consider that the toxicity and teratogenicity observed can result from both the action of glyphosate as well asthe surfactant present in the commercial formulation.

The oral administration of high doses of glyphosate (3500 mg/Kg) in Charles River COBS CD rats, between the 6th to the 19th day of pregnancy, and in rabbits, between the 6th to the 27th day of pregnancy, showed significant indices of maternal mortality for both species, as well as increase in the number of foetuses with reduced ossification of sternebrae [131], proving the toxicity and teratogenicity of this concentration of the herbicide for the organisms tested.

2.7. 2,4-D and glyphosate

Relyea [132] performed a study to observe the impact of two herbicides (glyphosate and 2,4-D) in the biodiversity of aquatic communities containing algae and more 25 species of animals. In this study the author observed that 2,4-D did not cause great impacts in the community and this is in agreement with previous studies that showed that this substance presents high LC-50 for several species. However, glyphosate had great impact in the community, causing a decrease of 22% of the species richness, while 2,4-D did not cause effects on this diversity. The authors also observed that neither of the two herbicides caused reduction in the periphyton biomass.

2.8. Diquat

Reglone is a bypiridylium herbicide, whose active ingredient is diquat (1,1'-ethylene -2,2'-ipyridyl dibromide), and of foliar application, used to eliminate weeds of different crops [133]. Reglone, in the concentrations tested (0.005, 0.01, 0.05 and 0.1% of the active ingredient for *Crepis capillaris* L.; 34.17 and 8.5 mg/Kg for mouse bone marrow polychromatic erythrocytes - PCEs), did not induce chromosome aberrations in any test system but promoted an increase in the frequency of micronuclei in both plant cells and PCEs [129], and thus is considered a potential mutagenic herbicide for these test organisms.

2.9. Pendimethalin

The herbicide Stomp 330, belongs to the dinitroanilines class, whose active ingredient is pendimethalin [N-(1-ethylpropyl)-2,6-dinitro-3,4-xylidine], it is applied as a systematic selective herbicide of the soil [133]. The responses of the two test systems for Stomp were very different: the concentrations tested (0.005, 0.1, 0.2 and 0.4% of the active ingredient for *Crepis capillaris* L.; 122.2, 244.5 and 489 mg/Kg for rats - PCEs) did not cause significant increases in the frequencies of chromosome aberrations in plant cells, but increased its incidence in rat cells, moreover, it induced an increase in the frequency of micronuclei in both test systems. This could be explained by the proven aneugenic effect of this herbicide, since all the concentrations tested produced C-mitoses in the assays with PCEs [129].

2.10. Paraquat

Paraquat (1,1'-dimethyl-4-4'-bipyridium dichloride) is a non-selective herbicide with fast action, widely used worldwide, mainly in the pre-harvest of cotton and potato and also to control a broad spectrum of weeds [134, 135, 136]. According to Tortorelli et al. [134], paraquat is able to modify the activity of several enzymes of fish, affecting the cardiac contraction and opercular ventilation, effects that can alter the initial development of these organisms. According to Tomita et al. [137], paraquat causes oxidative stress in different species of fish by generating elevated levels of superoxide ion.

A study conducted by D'Souza et al. [138] evaluated the toxicity of the herbicide paraquat for germ cells of male Sprague-Dawley rats by dermal exposure to this chemical. The authors verified that paraquat, even at low doses, significantly reduced the amount of spermatozoa, increased the frequency of spermatozoa bearing abnormalities and the mortality rate of these germ cells, as well as affected the mobility of the spermatozoa of the individuals studied, showing that the herbicide is a cytotoxic and genotoxic agent for the germ cells of this organism.

Hanada [136], analyzing the karyotype of species of *Rana ornativentris*, after exposure for 6 hours to the herbicide paraquat at the concentrations of 10^{-8} to 10^{-6} M, showed that this compound is able to induce genotoxic effects in this organism. The author observed that paraquat promoted, in a dose-dependent manner, a significant increase in the quantity of chromosome breaks in leukocytes of this test organism, suggesting that this species of anuran is highly sensitive to the genotoxic action of the herbicide.

According to Bus et al. [139], the genotoxic action of paraquat may be associated with the transference of a single electron of reduced oxygen to paraquat, forming superoxide ions. The singlet oxygen can be formed from the superoxide ion and subsequently react with lipids to form hydroperoxides and fatty acids. According to Tanaka and Amano [140], lipid peroxidation is responsible for the origin of several chromosome aberrations. Bauer Dial and Dial [141] still affirm that the oxidative stress induced by paraquat may be related to the teratogenic action of this compound to embryos and tadpoles of anurans.

Speit et al. [142] evaluated the genotoxic potential of the herbicide paraquat in Chinese hamster V79 cells, by chromosome aberrations and comet assays. Using a modified protocol of the comet assay with the modified protein FPG (formamidopyrimidine-DNA glycosylase), a repair enzyme that specifically nicks the DNA at sites of 8-oxo-guanines and formamidopyrimidines, it was not possible to detect oxidative damages in the bases of DNA after treatment with paraquat. Now, when the cells were treated directly on the slides, after lysis (i.e., after the cell membrane barrier has been eliminated), a significant increase in the migration of DNA was observed, only after treatment with high concentrations of the herbicide. Thus, the authors verified that the herbicide induced chromosome aberrations but was not able to induce relevant DNA lesions to promote mutations in the gene HPRT in cultured V79 cells.

Ribas et al. [135] assessed the cytotoxic, genotoxic and mutagenic potentials of different concentrations of the herbicide paraquat (0, 1, 5, 25, 50, 100, 250, 500, 1000, 2000 and 4000 µg/mL), by the assays of SCE, chromosome aberrations and micronuclei, in lymphocytes maintained in culture. The results showed that paraquat is an agent that induces cytotoxicity for lymphocytes, since it promoted the reduction in the nuclear division rate in all the concentrations tested and a significant decrease in the cell proliferation rates, when the cells were exposed to the highest concentration of the herbicide. In relation to the genotoxicity, the herbicide induced a significant increase in the frequencies of SCE of the lymphocytes treated, whose damage was not modified by co-treatment with the metabolic activation (S9 fraction of rat liver), but the data on the chromosome aberrations and micronuclei assays were not significant, which led the authors to conclude that paraquat is an inductor of primary damages in the DNA, although they have not shown that it has a clastogenic action.

A study performed by Hoffman and Eastin [143] evaluated the embryotoxic and teratogenic effects of two insecticides (lindane and toxaphene) and two herbicides (paraquat and 2,4,5-T), by external treatment of eggs of mallard duck (*Anas platyrhynchos*), using concentrations of field application. The authors showed that paraquat was the most significantly embryotoxic compound for this organism, independent of the type of vehicle in which the herbicide was associated, besides proving that paraquat impaired the growth of the organisms and was slightly teratogenic. The LC_{50} for this species was 1.5 Kg of the active ingredient/hectare in aqueous emulsion and 1 lb/acre in oil vehicles. When the organisms treated with paraquat were compared to the ones exposed to the herbicide 2,4,5-T, they presented little damages and it was observed few individuals bearing severe defects.

3. Harmful effects of herbicides on human health

The harmful effects of herbicides on human health are determined by several factors, such as the chemical class of those compounds, dose, time, and exposure route. Herbicides can be toxic to humans at high and lower doses [144]. The prolonged exposure can lead to a number of health effects, including the induction of diseases such as cancer and neurodegenerative [145, 146], reproductive and developmental changes [147] and respiratory effects [148].

Doll and Peto [149] estimated that 35% of all cases of cancer in the U.S. population originate from diet, and the herbicides present in foods are responsible. Estrogenicity assays made by Hernández et al. [150] show that organochlorine pesticides may act as endocrine disruption through more than one mechanism, including agonist or antagonist effects of different receptors. Chloro-s-triazize herbicides, pre-emergent pesticides used worldwide, have been generally considered as chemical compounds of low toxic potential for humans; however, there are many controversies on this issue. According to several international agencies, including the Environmental Protection Agency (EPA), Development for Environmental Assessment Center of the United States and IARC Monographs (International Agency for Research on Cancer), the herbicide atrazine, for example, was classified as a chemical agent probably carcinogenic to humans, although the basis for this conclusion is only evidenced in other animals [151, 152]. Due to the fact that atrazine induce mammary tumours in female Sprague-Dawley rats, the Peer Review Committee of the EPA Office of Pesticide Program (OPP) also concluded that atrazine should be considered in the Possibly Carcinogenic to Humans Group [153]. However, EPA [154] has classified this herbicide as a compound probably non carcinogenic to humans.

Some experimental studies have shown that exposure of humans to high doses of atrazine can result in an increased loss of body weight. However, a great number of epidemiological studies carried out with workers occupationally exposed to triazine herbicides indicate that these compounds do not have carcinogenic potential for these individuals. By analyses of different studies, it was observed that, although the chloro-s-triazine herbicides interfere in the endocrine responses of different species of mammals, their potential impact on humans seem to be mainly related to reproduction and development and not with human carcinogenesis [155].

Gammon et al. [83] discussed the extensive list of epidemiological studies with the herbicide atrazine, which describes that the carcinogenic potential of this compound to humans is not conclusive, although some studies have indicated a relationship between a high risk of prostate cancer and exposure to the herbicide.

Mladinic et al. [156] evaluated the genotoxic and mutagenic effects of low concentrations of the herbicides glyphosate and terbuthylazine, considered safe and, therefore, considered possible to occur in occupational and residential exposures (ADI – Acceptable Daily Intake, REL – Residential Exposure Level, OEL – Occupational Exposure Level, and 1/100 and 1/16 LD_{50} – Lethal Dose 50% - oral, rat), in human lymphocytes, with and without the use of metabolic activation (S9 fraction), by the FSH cytome assay, using pan-centromeric DNA probes to assess the content of micronuclei and other chromatinic instabilities. The authors verified that the frequencies of micronuclei, nuclear buds and nucleoplasmic bridges of cells treated with glyphosate slightly increased after the concentration of OEL 3.5 µg/mL, but no concentration induced an increase of the centromeric signals (C+) or DAPI (DAPI+). Now, the treatment with the herbicide terbuthylazine without metabolic activation showed a dose-response increase in the frequency of micronuclei of the lymphocytes exposed, and the significant data were from the concentration of 0.0008 µg/mL (REL) tested. The concentrations ADI (0.00058 µg/mL), REL (0.0008 µg/mL) and OEL (0.008 µg/mL) of terbuthylazine induced a significant occurrence of micronuclei hybridized with the centromeric probe (C+), regardless the presence or absence of S9, and of nuclear buds containing centromeric signals, only in the presence of S9. By the results obtained, it was suggested that the lowest concentrations of glyphosate do not have relevant harmful effects for the DNA molecule, while terbuthylazine presents a predominant aneugenic potential for the genetic material of human lymphocytes.

Terbuthylazine belongs to the chloro-s-triazine herbicides class, which inhibits the photosynthesis of weeds, by reaching the photosystem II. It is a chemical used for a variety of crops, such as maize, sugarcane, olive and pineapple [157]. Since the banishment of atrazine in European countries in 2006, terbuthylazine was recommended as its substitute. Due to the fact that the herbicide terbuthylazine is suspect of causing diseases in humans, such as non-Hodgkin lymphoma and lung cancer, Mladinic et al. [158] evaluated the effects of prolonged exposure (14 days) to low concentrations of this compound (0.58 ng/ml and 8 ng/ml) in human lymphocytes, using the comet assay and the comet-FISH assay (with the c-Myc and TP 53 genes). Treatment with terbuthylazine induced the migration of fragments of DNA in a significant manner, only for the highest concentration treated. The results showed an impairment of the structural integrity of c-Myc and TP 53, due to the prolonged exposure of human lymphocytes to terbuthylazine. The fact that several copies of TP53 were affected by the herbicide can indicate its ability to negatively interfere in the control of the cell cycle. However, the authors concluded that, for a more detailed assessment of the risk of cancer associated with exposure to terbuthylazine, it should be evaluated the impact of this pesticide on other housekeeping genes and markers.

Mladinic et al. [122] evaluated the genotoxic potential, by the comet assay and FISH, and oxidative damages, by the TBARS lipid peroxidation, of different concentrations of glyphosate (three similar to those observed in residential and occupational exposures and two related to LC_{50}) in human lymphocytes. The comet assay showed that the concentration of 580 µg/mL promoted a significant increase in the tail length, while the concentration of 92.8 µg/mL caused an increase in the tail intensity, both in relation to the control test. With the addition of the S9 fraction, the tail length was significantly increased for all the concentrations tested. When the lymphocytes were exposed to the three highest concentrations without S9, there was an increase in the frequency of micronuclei, nuclear buds and nucleoplasmic bridges. The addition of a metabolic activation system only promoted a significant increase of the nuclear instabilities for the highest concentration tested. The values of TBARS significantly increased with the increase of the concentrations tested, regardless the presence or absence of the S9 fraction. Due to the fact that dose-dependent effects for all the assays used were not observed, the authors concluded that these concentrations of glyphosate are not relevant for human exposure, since they did not present a significant risk for human health.

According to Mladinic et al. [122], the increase in the number of crops genetically modified used in assays and diagnosis of resistance to glyphosate, may be related to the fact that these crops tolerate increasingly higher concentrations of the active ingredient necessary for an effective control of weeds, which results from the introduction of increasing amounts of glyphosate into the environment. Thus, some epidemiological studies have shown that human exposure to glyphosate present in the environment is correlated to the development of diseases such as the non-Hodgkin lymphoma [159, 160].

According to He et al. [161], paraquat, the second most widely used herbicide in the world, is able to selectively accumulate in human lungs by causing oxidative injury and fibrosis, leading several individuals to death. Chronic exposure to this herbicide is also associated with hepatic lesions, kidney failure and Parkinson's disease [162, 163].

Studies carried out by He et al. [161] evaluated the toxicity of paraquat on BEAS-2B normal cells (human bronchial epithelial cells), showing that it is dose-dependent and results in mitochondrial damages, oxidative stress, death of lung cells exposed, as well as production of cytokines, pro-fibrogenic growth factors and transformation of myofibroblasts. The authors also proved that administration of resveratrol, a polyphenolic phytoalexin naturally produced by several plants, to control bacteria and fungi, was able to inhibit the production of reactive oxygen species, inflammations and fibrotic reactions induced by paraquat, by the activation of the Nrf2 signaling (Nuclear Factor Erythroid-2), revealing a new molecular mechanism for the intervention against oxidative damages and pulmonary fibrosis resulted from the action of toxic chemical compounds.

The study on the influence of a complex mixture of herbicides (atrazine, 2,4-D, alachlor, ciazine and malathion) in workers occupationally exposed to them, was carried out using cytogenetic methods standardly established (chromosome aberrations and micronucleus assay) and the comet assay technique. This assay showed a significant increase in the DNA migration (P<0.001), suggesting that long-term exposure to the pesticides could cause damages in the genome of somatic cells and, therefore, would represent a potential risk to human health [164].

4. Conclusion

The authors present in this manuscript the bioassays and the test-systems most commonly used to evaluate the effects of herbicides and the test-organisms to best suit the assessments of herbicide effects. In these considerations, the authors attempted to present the most sensitive and efficient organisms capable of detecting environmental contamination resulting from the action of these chemical agents. Additionally, we present in this paper the need to carry out research aimed at more effective methods to prevent and/or reduce the deleterious effects of such compounds on the environment, the biota potentially exposed, and especially to human health.

In this study it was addressed several studies that used different methodologies, which evaluated the toxicity and action of herbicides on different non-target organisms, including human species. The table below summarizes the main researches addressed in the text.

Herbicide	Test-organism	Endpoint	Results	Tested concentrations	References
Atrazine	Erythrocytes of Nile tilapia (Oreochromis niloticus)	micronucleus test; comet assay	increase in the DNA fragmentation; induction of micronuclei and nuclear abnormalities in all tested concentrations	6.25, 12.5, 25 µg/L	[8] Ventura et al., 2008
Atrazine	Wild leopard frogs (Rana pipiens)	toxicity assay	induction of abnormalities in the gonads; developmental delay and hermaphroditism (≥ 0.1 ppb)	0.01, 0.1, 0.4, 0.8, 1, 10, 25, 200 ppb	[82] Hayes et al., 2002
Atrazine	Sorghum vulgare	chromosome aberration assay	induction of multinucleated, aneuploid and polyploid cells; abnormalities in the mother cells of pollen grains; meiotic instability	2.7 Kg a.i./ha	[80] Liang et al., 1967
Atrazine	Human lymphocytes	chromosome aberration assay	increase in the chromosome aberrations frequency at 0.10 ppm	0.01, 1, 0.10 mg/ml	[85] Meisner et al., 1992
Atrazine	Human lymphocytes	chromosome aberration assay; SCE	increase in the frequency of chromosome aberrations; increase in the frequency of sister chromatid exchange in all tested concentrations	5, 8.5, 17, 51 µM	[93] Lioi et al., 1998
Atrazine	Human blood cells	chromosome aberration assay	Significant increase of chromosome breaks	1 ppm	[94] Meisner et al., 1993
Atrazine	Rat	chromosome aberration assay	there was no significant increase in the frequency of chromosome aberrations at 20 ppm	20 ppm	[85] Meisner et al., 1992
Atrazine	Bone marrow cells of rats	chromosome aberration assay	there was no significant increase in the frequency of chromosome aberrations	20 ppm	[86] Roloff et al., 1992
Atrazine	Human lymphocytes	chromosome aberration assay	Induction of chromosome aberrations	0.0001 µg/mL	[94] Meisner et al., 1993

Herbicide	Test-organism	Endpoint	Results	Tested concentrations	References
Atrazine	Rat leukocytes	comet assay	increase in the damages in the DNA for 500 mg/Kg	125, 250, 500 mg/Kg	[89] Tennant et al., 2001
Atrazine	Erythrocytes of bullfrog tadpoles	comet assay	significant increase in the DNA damages, from the concentration of 4.8 mg/L	4.8, 19.75, 77, 308 mg/L	[90] Clements et al., 1997
Atrazine	Human lymphocytes	comet assay	significant increase in the DNA damages, mainly at the concentrations of 100 and 200 µg/L	50, 100, 200 µg/L	[69] Ribas et al., 1995
Atrazine	Hepatocytes of Wistar rats	lipid peroxidation assay; micronucleus test	increase in the rates of lipid peroxidation, hepatic damages, death of hepatocytes and induction of micronuclei.	400 ppm	[91] Campos-Pereira et al., 2012
Atrazine	Erythrocytes and gill cels of the fish *Channa punctatus*	micronucleus test; comet assay	induction of damages in the DNA and micronuclei, in the tested concentrations, in all the exposure periods (from 1 to 35 days), with more significant effects in the highest concentrations and exposure periods; higher sensitivity for gill cells	4.24, 5.30. 8.48 mg/L	[22] Nwani et al., 2011
Atrazine	Erythrocytes of the gibel carp fish (*Carassius auratus*)	micronucleus test; comet assay	significant induction of DNA strand breaks and micronuclei, in all tested concentrations of the commercial product (Gesaprim), but there was not a induction of these genotoxic and mutagenic effects for the active ingredient.	5, 10, 15 µg/L	[84] Çavas, 2011
Atrazine	Human lymphocytes	comet assay	significant increase of damage in the DNA exposed to the commercial product Gesaprim, but there was no induction of genotoxicity for the active ingredient atrazine, for all tested concentrations.	0.047, 0.47, 4.7 ug/L	[101] Zeljezic et al., 2006
Atrazine	Somatic cells of *Allium cepa* and *Vicia faba*	chromosome aberration assay; micronucleus test	significant inhibition of the mitotic index, significant increase in the frequencies of micronuclei and chromosome aberrations of both test organisms, when exposed to the commercial product Gesaprim, but there was no induction of any significant effects when cells were exposed to the active ingredient atrazine, for all tested concentrations.	*A. cepa:* 15, 30, 60 mg/L; *V. faba:* 17,5, 35, 70 mg/L	[102] Srivastava and Mishra, 2009
Atrazine	*Salmonella* and hepatic cells of Sprague-Dawley rats	*Salmonella* assay and SOS Chromotest	there was no significant induction of genotoxic damages nor mutagenic	1 – 1000 µg/ plate	[92] Ruiz and Marzin, 1997

Herbicide	Test-organism	Endpoint	Results	Tested concentrations	References
Atrazine	Sprague-Dawley rats	embryotoxic and teratogenic tests	induction of hypospadias in male newborns at 200 ppm and diverse embryotoxic damages at 25 ppm.	25, 100, 200 mg/kg/d	[87] Wu et al., 2007
Atrazine	Wistar rats	Radioimmunoassay	alterations in the levels of testosterone, androstenedione, estradiol, estrone, progesterone and corticosterone to 50 or 200 ppm for 60 days	50, 200 mg / kg / day	[88] Modic et al., 2004
Atrazine, Simazine and Cyanazine	Human lymphocytes	chromosome aberration assay and SCE	there was no significant increase of chromosome aberrations and sister chromatid exchanges	0.5, 5, 50 ppb	[104] Kligerman et al., 1993
Atrazine, Simazine and Cyanazine	Polychromatic erythrocytes of the bone marrow of female C57B1/6 rats	micronucleus test	there was no significant induction of micronuclei	0, 125, 250, 500 mg/kg	[103] Kligerman et al., 2000
Atrazine, Simazine and Cyanazine	Chinese Hamster Ovary – CHO – cells	flow cytometry assay	significant induction of chromosome damages by atrazine for the tested concentrations, proven clastogenic potential of cyanazine	0.003 µg/mL, 0.018 µg/mL(atrazine); 0.003 µg/mL, 0.012 µg/mL (cyanazine)	[107] Taets et al., 1998
Atrazine and Butachlor	Green alga *Scenedesmus obliquus* and cladoceran *Daphnia carinata*	acute toxicity assay	atrazine is highly toxic for *S. obliquus* and slightly toxic for *D. carinata* and butachlor is moderately toxic for both; the toxic effects of the mixture of the herbicides were significantly antagonistic for *S. obliquus* and there was no significative synergism for *D. carinata*	*S. obliquus*: 0, 0.5, 1, 2, 4, 8 mg/L (butachlor) and 0, 0.008, 0.016, 0.032, 0.064, 0.128 mg/L (atrazine) / *D. carinata*: 0, 1, 1.8, 3, 5, 8 mg/L (butachlor) and 0, 7.5, 15, 30, 60, 120 mg/L (atrazine)	[20] He et al., 2012
Butachlor	Alpine cricket frog (*Fejervarya limnocharis*)	chromosome aberration assay	affected the survival, development and metamorphosis time of tadpoles in different concentrations; DNA damage (0.4-0.8 mg/L)	ranging from 0.025 to 3.2 mg/l	[43] Liu et al., 2011
Terbutryn	Human leukocytes	micronucleus test; comet assay; SCE	there was no significant induction of micronuclei and SCE; significant induction of DNA damages for all tested concentrations	0, 5, 10, 50, 100, 150 µg/mL	[108] Moretti et al., 2002
2,4-D	*Caudiverbera caudiverbera* frog	toxicity assay	dose-dependent reduction in the response of the simpatic junction to	0.01, 0.1, 1 mM	[114] Suwalsky et al., 1999

Herbicide	Test-organism	Endpoint	Results	Tested concentrations	References
			nerve stimulation due to inhibition of the glandular chloride channel in mucosa skin		
2,4-D	Gills of different species of fishes	toxicity assay	bleeding, renal increase, impairment of the renal functions and hepatic degeneration	400 mg/L	[115] Gómez et al., 1998
2,4-D	Chinese Hamster Ovary – CHO – cells	SCE	significant increase in the sister chromatid exchange at 2 and 4 µg/ml	2, 4, 6, 10 µg/mL	[116] González et al., 2005
2,4-D	Bone marrow and germ cells of rats	SCE	significant increase in the sister chromatid exchange at 100 and 200 ppm, for both cell types	50,100, 200 mg/kg	[117] Madrigal-Bujaidar et al., 2001
2,4-D	Frog Xenopus	FETAX - frog embryo teratogenic assay	significant induction of embryotoxic and teratogenic effects	245 mg/L	[119] Morgan et al., 1996
2,4-D and Butachlor	Erythrocytes of the catfish (Clarias batrachus)	chromosome aberration assay; micronucleus test	significant increase in the frequency of micronuclei and altered cells in a dose-response manner for both herbicides	2,4-D: 25, 50, 75ppm; Butachlor: 1, 2, 2.5ppm	[113] Ateeq et al., 2002
2,4-D and Paraquat	Rainbow trout (Oncorhynchus mykiss)	acute lethality test, lipid peroxidation assay by quantification of MDA; comet assay	toxic action more evident for paraquat (high indices of mortality); apnea and white spots in the scales of individuals exposed to the 2 herbicides; increase in the rates of MDA and damages in the DNA after exposure to all concentrations of the tested herbicides	2,4-D: 316, 346, 389, 436, 489 mg/L; Paraquat: 0.055, 0.066, 0.083, 0.116, 0.133 mg/L	[112] Martínez-Tabche et al., 2004
2,4-D and 2,4-D DMA	Humanh lymphocytes and erythrocytes	SCE; analysis of the cell cycle progression and mitotic index	alterations in the cell cycle and induction of SCE for some concentrations only with more significant genotoxic effects for erythrocytes	10, 25, 50, 100 µg/mL	[118] Soloneski et al., 2007
2,4-D; Triclopyr; Diquat dibromide; glyphosate	Rainbow trout (Oncorhynchus mykiss)	Vitellogenin estrogenic assay	significant increase in the levels of vitellogenin of the plasma of fishes exposed to 2,4-D	0.11, 1.64, 2.07, 1.25 mg/L	[120] Xie et al., 2005
Glyphosate	Geotrichum candidum, Lactococcus lactis subsp. Cremoris; Lactobacillus delbrueckii subsp. bulgaricus	microbial growth assay	inhibition of microbial growth by the commercial product Roundup; microbiocide effect at concentrations lower than the recommended for agricultural use for the commercial product Roundup; non induction of significant toxic effects for the three microorganisms by the active ingredient glyphosate	0.1, 1, 10, 100, 1000, 10000 ppm	[127] Clair et al., 2012
Glyphosate	Tadpoles of wood frog (Rana sylvatica or Sylvaticus lithobates),	acute toxicity assay	significant induction of morphological alterations of tadpoles of the three species; for the wood frogs and	0, 1, 2, or 3 mg acid equivalents [a.e.] /L of	[128] Relyea, 2012

Herbicide	Test-organism	Endpoint	Results	Tested concentrations	References
	leopard frog (*Rana pipiens pipiens* or L.), and American toad (*Bufo americanus* or *Anaxyrus americanus*)		leopard frogs, exposure to glyphosate affected the size of the tail of tadpoles, for all tested concentrations	Roundup Original MAX	
Glyphosate	Roots from the smooth hawksbeard (*Crepis capillaris* L.), polychromatic erythrocytes of the bone marrow of C57BL rat	chromosome aberration assay; micronucleus assay	there was no induction of genotoxic and/or mutagenic effects for any of the species	*Crepis capillaris*: 0.05, 0.1, 0.5, 1 %; erythrocytes: doses inferior to half the LD$_{50}$ (1080 mg/Kg)	[129] Dimitrov et al., 2006
Glyphosate	Female Wistar rats	acute toxicity assay; teratogenicity assay	high mortality index of females treated with the highest concentration of the commercial product Roundup; increase in the dose-response of fetal skeletal alterations	500, 750, 1000 mg/kg	[130] Dallegrave et al., 2003
Glyphosate	Human lymphocytes	comet assay; FISH; lipid peroxidation assay – TBARS	significant increase in the DNA migration at 580 µg/mL; significant increase of the comet tail intensity at 92.8 µg/mL; greater lesion in the DNA in the presence of S9; increase in the frequency micronuclei, nuclear buds and nucleoplasmic bridges, without S9; significant increase of nuclear instabilities in the highest concentration tested with S9; significant dose-response increase of the levels of TBARS	0.5, 2.91, 3.5, 92.8, 580 µg/mL	[122] Mladinic et al., 2009
Glyphosate adn 2,4-D	Algae and 25 species of aquatic animals	acute toxicity assay	there was no reduction in the biomass of periphyton by the 2 herbicides; there was no great impacts to the aquatic community by 2,4-D; high impact to the aquatic community by glyphosate by the significative decrease in the species richness	0, 1, 2, or 3 mg acid equivalents [a.e.] /L of Roundup Original MAX	[132] Relyea, 2005
Glyphosate and Terbuthylazine	Human lymphocytes	cytome FISH	glyphosate caused an increase in the frequencies of micronuclei, nuclear buds and nucleoplasmic bridges of clells treated (3.5 µg/mL onward), but without induction of centromeric signals; terbuthylazine induced an increase in the frequency of micronuclei hybridized with	0.5, 2.91, 3.50, 92.8, 580 µg/mL (glyphosate); 0,00058, 0,0008, 0,008, 25, 156,5 µg/mL (terbuthylazine)	[156] Mladinic et al., 2009

Herbicide	Test-organism	Endpoint	Results	Tested concentrations	References
			centromeric probe and nuclear buds with centromeric signals in the presence of S9 (0.008 ug/mL onward)		
Terbuthylazine	Human lymphocytes	comet assay; comet assay-FISH	induction of the migration of fragments of DNA, significant only at the highest concentration; impairment of the structural integrity of c-Myc and TP 53 due to prolonged exposure to terbuthylazine	Terbuthylazine: 0.58 ng/ml, 8 ng/ml; carbofuran: 8 ng/ml, 21.6 ng/ml	[158] Mladinic et al., 2012
Paraquat	Several species of fishes	acute toxicity assay; enzyme activity assay	alteration in the activity of different enzymes; negative effects on cardiac contraction and opercular ventilation	0.1-2.0 mg/L	[134] Tortorelli et al., 1990
Paraquat	Several species of fishes	enzyme activity assay	induction of oxidative stress; increase in the levels of SOD	0.2-50 mM	[137] Tomita et al., 2007
Paraquat	Germ cells of Sprague-Dawley rats	cytotoxicity assay	reduction in the quantity of spermatozoa; increase in the mortality rates and abnormalities in spermatozoa for the higher concentrations	0, 6, 15, 30 mg/kg	[138] D'Souza et al., 2006
Paraquat	Leukocytes of *Rana ornativentris*	conventional cytogenetics assay	genotoxic effects, such as chromosome breaks	10^{-6} M	[136] Hanada, 2011
Paraquat	Human lymphocytes	chromosome aberration assay; micronucleus test; SCE	reduction in the cell division index; decrease in the cell proliferation rates; significant increase in the frequencies of SCE (50 µg/mL for 24h treatment; 4000 µg/mL for 2h treatment), significant increase in the MN frequencies (concentrations ≥ 25 µg/mL)	0, 1, 5, 25, 50, 250, 500, 1000, 2000, 4000 µg/mL	[135] Ribas et al., 1998
Paraquat	BEAS 2B normal cells (human bronchial epithelial cells)	cytotoxicity assay, oxidative stress assay	mitochondrial damage; oxidative stress; cell death; production of cytokines, pro-fibrogenic growth facts and transformation of myofibroblast	10 uM	He et al., 2012
Diuron	Pacific oyster (*Crassostrea gigas*)	toxicity assay	irreversible damages to the genetic material, negative impacts in the reproduction of aquatic organisms	300 ng/L, 3 µg/L	[44] Bouilly et al., 2007
Diquat	Roots of smooth hawksbeard (*Crepis capillaris* L.); polychromatic erythrocytes of the bone marrow of C57BL rat	chromosome aberration test; micronucleus test	there was no induction of chromosome aberrations for any test system; significant increase of the frequency of micronuclei for both test systems	*Crepis capillaris*: 0.005, 0.01, 0.05, 0.1%; erythrocytes: 8.5, 34.17 mg/Kg	[129] Dimitrov et al., 2006

Herbicide	Test-organism	Endpoint	Results	Tested concentrations	References
Pendimethalin	Roots of smooth hawksbeard (*Crepis capillaris* L.); polychromatic erythrocytes of the bone marrow of C57BL rat	chromosome aberration test; micronucleus test	there was no significant increase in the frequencies of chromosome aberrations in plant cells, but an increase of their incidence in cells of rats; significant increase in the frequency of micronuclei for both test systems.	*Crepis capillaris*: 0.005, 0.1, 0.2, 0.4%; erythrocytes: 122.2, 244.5, 489 mg/Kg	[129] Dimitrov et al., 2006
Simetryn, mefenacet and thiobencarb	*Silurana tropicalis*	toxicity assay	toxic effects for tadpoles more significant for thiobencarb	Thiobencarb: 6.85-2.92 mM	[42] Saka, 2010
Complex mixture of pesticides (atrazine, 2,4-D, alachlor, ciazine and malathion)	Workers exposed	chromosome aberration assay; micronucleus test; comet assay	significant increase in the migration of the DNA	Mixture of various concentrations of pesticides	[163] Garaj-Vrhovac and Zeljezic, 2002

Table 2. List o the main researches carried out with several bioindicators to evaluate the toxicity of herbicides.

Author details

Maria Aparecida Marin-Morales*, Bruna de Campos Ventura-Camargo and Márcia Miyuki Hoshina

*Address all correspondence to: mamm@rc.unesp.br

Department of Biology, Institute of Biosciences, São Paulo State University (UNESP), SP, Brazil

References

[1] Guzzella, L.; Pozzoni, F.; Giuliano, G. Herbicide contamination of surficial groundwater in Northern Italy. Environmental Pollution, v. 142, p. 344-353, 2006.

[2] Kortekamp, A. Herbicides and Environment. Kroatia, 2011, 760 p.

[3] Bolognesi, C.; Merlo, F.D. Pesticides: Human Health Effects. Encyclopedia of Environmental Health, p. 438-453, 2011.

[4] Nehls, S.; Segner, H. Detection of DNA damage in two cell lines from rainbow trout, RTG-W1, using the comet assay. Environmental Toxicology, v. 16, p. 321-329, 2001.

[5] Spacie, A.; Hamelink, J.L. Bioaccumulation, in: RAND, G.M.; PETROCELLI, S.R. (Eds.), Fundamentals of Aquatic Toxicology: Methods and Applications, Hemisphere, New York, 1985, pp. 495-525.

[6] Grillo, R.; Santos, N.Z.P.; Maruyama, C.R.; Rosa, A.H.; De Lima, R.; Fraceto, L.F. Poly(Rmvarcpsilon-caprolactone)nanocapsules as carrier systems for herbicides: physico-chemical characterization and genotoxicity evaluation. Journal of Hazardous Materials, 2012, doi:10.1016/j.jhazmat.2012.06.019

[7] Fernandes, T.C.C; Mazzeo, D.E.C.; Marin-Morales, M.A. Mechanism of micronuclei formation in polyploidizated cells of *Allium cepa* exposed to trifluralin herbicide. Pesticide Biochemistry and Physiology, v. 88, n. 3, p. 252-259, 2007.

[8] Ventura, B.C.; Angelis, D.F.; Marin-Morales, M.A. Mutagenic and genotoxic effects of the Atrazine herbicide in *Oreochromis niloticus* (Perciformes, Cichlidae) detected by the micronuclei test and the comet assay. Pesticide Biochemistry and Physiology, v. 90, p. 42-51, 2008.

[9] Silva, J.; Fonseca, M.B. Genética Toxicológica. 1 ed. Brasil: Alcance, 2003. 471p.

[10] Bertoletti, E. Companhia de Tecnologia de Saneamento Ambiental: Ensaios biológicos com organismos aquáticos e sua ação no controle da poluição de São Paulo. 1 ed. Brasil: [s.n.] , 1996. 29p.

[11] Ribeiro, L.R.; Salvadori, D.M.F.; Marques, E.K. Mutagênese Ambiental. 1 ed. Brasil: ULBRA, 2003. 355p.

[12] RAND, G.M.; PETROCELLI, S.R. Fundamentals Of Aquatic toxicology: methods and applications. Hemisphere, Washington, v. 42, n. 1, p. 1-28, 1985.

[13] Arnaiz, R.R. Las Toxinas Ambientales y sus Efectos Genéticos. 2 ed. México: [s.n.] , 1995. 267 p.

[14] Vogel, E.W. Assessment of chemically induced genotoxic events. In: Prospectives and Limitations, The Netherlanlands: Universitaire Pers Leiden, 1982. p. 24.

[15] Tavares, D.C. Estudos da possível ação genotóxica do alcalóide boldina em sistemas de células de mamífero "in vitro" e "in vivo". 1991. 205 f. Tese (Mestrado em Medicina) - Faculdade de Medicina de Ribeirão Preto, Universidade do Estado de São Paulo, Ribeirão Preto.

[16] Ueta, J.; Pereira, N.L.; Shuhama, I.K.; Cerdeira, A.L. Biodegradação de herbicidas e biorremediação: Microrganismos degradadores do herbicida atrazine. 1 ed. Brasil: [s.n.] , 1997. 545p.

[17] Kudsk, P., Streibig, J.C. Herbicides: a two-edged sword. Weed Res., v. 43, p. 90-102, 2003.

[18] Alves, A. Usos e Abusos. Ciência Hoje, São Paulo, v. 4, n. 22, p. 49-52, 1986.

[19] Vasilescu, M.N. Medvedovici, A.V. Herbicides. Encyclopedia of Analytical Science. 2nd ed. Elsevier, Oxford, p. 243-260, 2005. http://dx.doi.org/10.1016/B0-12-369397-7/00256-9

[20] He, H., Yu, J., Chen, G., Li, W., He, J., Li, H. Acute toxicity of butachlor and atrazine to freshwater green alga *Scenedesmus obliquus* and cladoceran *Daphnia carinata*. Ecotox. Environ. Saf., v. 80, p. 91-96, 2012a.

[21] Zhang, W., Jiang, F., Ou, J. Global pesticide consumption and pollution: with Chinas as a focus. Proceedings of the International Academy of Ecology and Environmental Sciences, v. 1(2), p.125-144, 2011.

[22] Nwani, C.D., Nagpure, N.S., Kumar, R., Kushwaha, B., Kumar, P., Lakra, W.S. Mutagenic and genotoxic assessment of atrazine-based herbicide to freshwater fish *Channa puntatus* (Bloch) using micronucleus test and single cell gel electrophoresis. Environmental Toxicology and Pharmacology, v. 31, p. 314-322, 2011.

[23] Chevreuil, M.; Garmouma, M.; Teil, M.J.; Chesterikoff, A. Occurrence of organochlorines (PCBs, pesticides) and herbicides (triazines, phenylureas) in the atmosphere and in the fallout from urban and rural stations of Paris area. Science of the Environment, [S.l.] , v. 182, p. 25-37, 1996.

[24] Kim, J.H.; Feagley, S.E. Adsorption and leaching of trifluralin, metolachlor, and metribuzin in a commerce soil. Journal of Environmental Science and Health-B: Pesticides and Food Contaminants, New York, v. 33, p. 529-546, 1998.

[25] Abdel-Rahmam, A.R.; Wauchope, R.D.; Truman, C.C.; Dowler, C.C. Runoff and leaching of atrazine and alachlor on a sandy soil as affected by application in sprinkler irrigation. Journal of Environmental Science and Health-B: Pesticides and Food Contaminants, v. 34, p. 381-396, 1999.

[26] Munger, R.; Isacson, P.; Hu, S.; Burns, T.; Hanson, J.; Lynch, C.F.; Cherryholmes, K.; Vandorpe, P.; Hausler, Jr. W. J. Intrauterine growth retardation in Iowa communities with herbicides-contaminated drinking water supplies. Environmental Health Perspectives, v. 105, p. 308-314, 1997.

[27] Gorell, J.M.; Jhonson, C.C.; Rybicki, B.A.; Peterson, E.L.; Ricchardson, R.J. The Risk of Parkinson's disease with exposure to pesticides, farming, well water, and rural living. Neurology, Heidelberg, v. 50, p. 1346-1350, 1998.

[28] Vander Werf, H.M.G. Assessing the impact of pesticides on the environment. Agriculture, Ecosystems and Environment, The Netherlands, v. 60, p. 81-96, 1996.

[29] Timbrell, J.A. Introduction to Toxicology. 2. ed. Estados Unidos: Taylor & Francis, 1999. 167p.

[30] Linck, A.J. Effects on the cytology and fine structure of plant cells. Herbicides, [S.l.] , v. 1, p. 83-121, 1979.

[31] Natarajan, A.T. Chromosome Aberrations: past, present and future. Mutation Research, Leiden, v. 504, p. 3-16, 2002.

[32] Jurado, A.S., Fernandes, M.A.S., Videira, R.A., Peixoto, F.P., Vicente, J.A.F. Herbicides: The Face and the Reverse of the Coin. An *in vitro* Approach to the Toxicity of Herbicides in Non-Target Organisms. In: KORTEKAMP, A. (Ed.) Herbicides and Environment. Kroatia, 2011, p. 3- 45 p.

[33] Moreland, D.E. Mechanisms of action of herbicides. Ann. Rev. Plant Physiol., v. 31, p. 597-638, 1980.

[34] Rao, V.S. Principles of Weed Science, New Hampshire, USA, 2nd Ed, 2000, 559 p.

[35] Parsons, B.; Witt, J.M. Pesticides in groundwater in the U.S.A. A report of a 1988 survey of US States. EM8406, Oregon State University Extension Service. Archives of Environmental Contamination and Toxicology, v. 18, p. 734-747, 1989.

[36] Hance, R.J. Some continuing uncertainties in knowledge of herbicide behavior in the soil. Annals of Applied Biology, v.110, p.195-202, 1987.

[37] Hussain, S., Siddique, T., Saleem, M., Arshad, M., Khalid, A. Impact of pesticides on soil microbial diversity, enzymes and biochemical reactions. Advances in Agronomy, v.102, p.159-200, 2009.

[38] Moura, M.A.M., Franco, D.A.S., Matallo, M.B., Impacto de herbicidas sobre os recursos hídricos. Revista Tecnologia & Inovação Agropecuária, v. 1(1), p. 142-151, 2008.

[39] Roman, E.E., Beckie, H., Vargas, L., Hall, L., Rizzardi, M.A., Wolf, T.M. Como funcionam os herbicidas da biologia à aplicação. Passo Fundo, Brasil, 2007, 158 p.

[40] Ying, G.-G., Williams, B. Laboratory study on the interaction between herbicides and sediments in water systems. Environmental Pollution, v.107, p. 399-405, 2000.

[41] Toccalino, P.L., Norman, J.E., Scott, J.C. Chemical mixtures in untreated water from public-supply wells in the U.S. – Occurrence, composition and potential toxicity. Science of Total Environment, v. 431, p. 262-270, 2012.

[42] Saka, M. Acute toxicity of rice paddy herbicides simetryn, mefenacet, and thiobencarb to *Silurana tropicalis* tadpole. Ecotox. Environ. Saf., v. 73, p. 1165-1169, 2010.

[43] Liu, W.Y., Wang, C.Y., Wang, T.S., Fellers, G.M., Lai, B.C., Kam, Y.C. Impacts of the herbicide butachlor on the larvae of a paddy field breeding frog (*Fejervarya limnocharis*) in subtropical Taiwan. Ecotoxicology, v. 20, p. 377-384, 2011.

[44] Bouilly, K., Bonnard, M., Cagnaire, B., Renault, T., Lapègue, S. Impact of Diuron on Aneuploidy and Hemocyte Parameters in Pacific Oyster *Crassostrea gigas*. Arch. Environ. Contam. Toxicol., v.52, p.58-63, 2007.

[45] Hladik, M.L., Bouwer, E.J., Roberts, A.L. Neutral degradates of chloroacetamide her-
      bicides: Occurrence in drinking water and removal during conventional water treat-
      ment. Water Research, v.42, p.4905-4914, 2008.

[46] Bannink, A.D. How Dutch drinking water production is affected by the use of herbi-
      cides on pavements. Water Sci, Technol., v.49 (3), p.173-181, 2004.

[47] Rostad, C.E. From the 1998 drought to the 1993 flood: transport of halogenated or-
      ganic compounds with the Mississipi River suspended sediment at Thebes, Illinois.
      Environ. Sci. Tecnol., v.31, p.1308-1312, 1997.

[48] Jacomini, A.E., Camargo, P.B., Avelar, W.E.P., Bonato, P.S. Assessment of Ametryn
      Contamination in River Water, River Sediment, and Mollusk Bivalves in São Paulo
      State, Brazil. Arch. Environ. Contam. Toxicol., v. 60, p. 452-461, 2011.

[49] Duke, N.C., Bell, A.M., Pederson, D.K., Roelfsema, C.M., Nash, S.B. Herbicides impli-
      cated as the cause of severe mangrove dieback in the Mackay region, NE Australia:
      consequences for marine plant habitats of the GBR World Heritage Area. Marine Pol-
      lution Bulletin, v. 51, p.308-324, 2005.

[50] Jones, R., The ecotoxicological effects of Photosystem II herbicides on corals. Marine
      Pollution Bulletin, v.51, p.495-506, 2005.

[51] Lewis, S.E., Brodie, J.E., Bainbridge, Z.T., Rohde, K.W., Davis, A.M., Masters, B.L.,
      Maughan, M., Devlin, D.J., Mueller, J.C., Schaffelke, B. Herbicides: A new threat to
      the Great Barrier Reef. Environmental Pollution, v. 157, p. 2470-2484, 2009.

[52] Polard, T., Jean, S., Gauthier, L., Laplanche, C., Merlina, G., Sánches-Pérez, J.M., Pine-
      lli, E. Mutagenic impact on fish of runoff events in agricultural areas in south west
      France. Aquatic Toxicology, v.17, p. 126-134, 2011.

[53] Moraes, D.S.L. Avaliação dos potenciais tóxicos, citotóxicos e genotóxicos de águas
      ambientais de Corumbá-MS em raízes de Allium cepa. 2000. 158 f. Tese (Mestrado em
      Genética e Melhoramento) – Universidade Estadual de Londrina, Londrina.

[54] Mccarthy, J.F.; Shugart, L.R. Biomarkers of environmental contamination. 1 ed. Esta-
      dos Unidos: Lewis, 1990. 382p.

[55] Bombail, V.; Dennis, A.W.; Gordon, E.; Batty, J. Application of the comet and micro-
      nucleus assays to butterfish (Pholis gunnellus) erythrocytes from the Firth of Forth,
      Scotland. Chemosphere, Oxford, v. 44, p. 383-392, 2001.

[56] Peña, L.F.M. Uso do teste de micronúcleo em eritrócitos circulantes de peixes para
      monitorização de um local do rio Tibagi e avaliação da genotoxidade de agrotóxicos
      em bioensaios. 1996. 199 f. Tese (Mestrado em Genética e Melhoramento) – Universi-
      dade Estadual de Londrina, Londrina.

[57] Veiga, A.B. O uso do teste de *Allium cepa* para detectar a toxicidade do inseticida Nuvacron. 1995, 58 f. Monografia (Conclusão do Curso de Ciências Biológicas) - Universidade Estadual de Londrina, Londrina.

[58] Almeida, W. F. Acúmulo de inseticidas no homem e sua significação epidemiológica. O biológico, São Paulo, v. 6, p. 171-183, 1974.

[59] Pavanelli, E.A.S. Efeito de biocidas sobre a polinização e germinação de sementes de orquídeas dos gêneros *Cattleya* Lsl. e *Laelia* Lsl. (Orchidaceae). 1995. 179 f. Tese (Doutorado em Botânica) – Instituto de Biociências, Universidade Estadual Paulista, Rio Claro.

[60] Dassenoy, B.; Meyer, J.A. Mutagenic effects of benomyl on *Fucarion oxysporum*. Mutation Research, Amsterdam ,v .21, p. 119-120, 1973.

[61] Sakamoto, E.T.; Takahashi, C.S. Efeitos dos fungicidas Dithane M-45, Benlate e Vitavax 75 PM sobre os índices mitóticos dos meristemas radiculares de *Allium cepa*. In: Anais da SBPC, São Paulo, v. 31, p. 573, 1979.

[62] Ventura, B. C. Avaliação dos efeitos citotóxicos, genotóxicos e mutagênicos do herbicida atrazine, utilizando *Allium cepa* e *Oreochromis niloticus* como sistemas-testes. 2004. 105f. Dissertação (Mestrado em Biologia Celular e Molecular) – Instituto de Biociências, Universidade Estadual Paulista, Rio Claro, 2004.

[63] Terracini, B. Valutazione della carcinogenecita deghi idrocarburi clorutati usati come pesticide. Tumori, Milano, v. 53, p. 601-618, 1977.

[64] HAgMAR, L.; Bonassi, S.; Stromberg, U.; Brogger, a.; Knudsen, L.E.; Norppa, H.; Reuterwall, C. Chromosomal aberrations in lymphocytes predict cancer: a report from the European Study Group on Cytogenetic Biomarkers and Health (ESCTH). Cancer Research, Baltimore, v. 58, p. 4117-4121, 1998.

[65] ZAKERI, Z.; LOCKSHIN, R.A. Cell death during development. Journal of Immunological Methods, [S.l.] , v. 265, p. 3-20, 2002.

[66] KRISTEN, U. Use of higher plants as screens for toxicity assessment. Toxicology in vitro, United Kingdom, v. 11, p. 181-191, 1997.

[67] Grisolia, C.K.; Starling, F.L.R.M. Micronuclei monitoring of fishes from Lake Paranoá, under influence of sewage treatment plant discharges. Mutation Research, Amsterdam, v. 491, p. 39-44, 2001.

[68] Monteith, D.K.; Vanstone, J. Comparison of the microgel electrophoresis assay and other assays for genotoxicity in the detection of the DNA damage. Mutation Research, Amsterdam, v. 345, n. 3-4, p. 97-103, 1995.

[69] Ribas, G.; Frenzili, G.; Barale, R.; Marcos, R. Herbicide-induced DNA damage in human lymphocytes evaluated by the single-gel electrophoresis (SCGE) assay. Mutation Research, Amsterdam, v. 344, p. 41-54, 1995.

[70] Popa, N.E.; Zakrzhevskaya, A.M.; Kozhokaru, R.V.; Enaki, D.K. Cytogenetic effect of some herbicides on maize seedlings. Weed Abstracts, Farnham Royal, v. 35, n. 1, p. 50, 1986.

[71] Eldridge, J.C; Wetzel, L.T.; Stevens, J.T.; Simpkins, J.W. The mammary tumor response in triazine-treated female rats: a threshold-mediated interaction with strain and species-specific reproductive senescence. Steroids, Califórnia, v. 4, p. 672-678, 1999.

[72] Worthing, C.R.; Walker, S.B. The pesticide manual, 7 ed. U.K.: The Lavenham Press, 1983. 589p.

[73] Goldman, L.R. Atrazine, simazine and cyanazine: Notice of initiation of special review in federal register, Estados Unidos, s.n.: 60412-60443. 1994.

[74] Ribas, G.; Surrallés, J.; Carbonell, E.; Creus, A.; Xamena, N.; Marcos, R. Lack of genotoxicity of the herbicide atrazine in cultured human lymphocytes. Mutation Research, Amsterdam, v. 416, p. 93-99, 1998a.

[75] Summer, D.D.; Cassidy, I.M.; Szolics, I.M.; Marco, G.J. Evaluation of the mutagenic potential of corn (Zea mays L.) grown in untreated and a atrazine (A Atrex) treated soil in the field. Drug Chemical and Toxicology, [S.l.] , v. 7, p. 243-257, 1984.

[76] Kappas, A. On the mutagenic and the recombinogenic activity of ceratin herbicides in Salmonella typhimurium and in Aspergillus nidulans. Mutation Research, Amsterdam, v. 204, p. 615-621, 1988.

[77] Butler, M.A.; Hoagland, R.E. Genotoxicity assessment of atrazine and some major metabolities in the Ames test. Bulletin of Environmental Contamination and Toxicology, Florida, v. 43, p. 797-804, 1989.

[78] Murnik, M.R.; Nash, C.L. Mutagenicity of triazine herbicides atrazine, cyanazine, and simazine in Drosophila melanogaster. Journal of Toxicology Environmental Health, [S.l.] , v. 3, p. 691-697, 1977.

[79] Torres, C.; Ribas, G.; Xamena, N.; Creus, A.; Marcos, R. Genotoxicity of four herbicides in Drosophila wing spot tests. Mutation Research, Amsterdam, v. 280, p. 291-295, 1992.

[80] Liang, G.H.L.; Feltner, K.C.; Liang, Y.T.S.; Morrill, J.L. Cytogenetic effects and responses of agronomic characters in grain sorghum (Sorghum vulgare Pers.) following atrazine application. Crop Science, New York, v. 7, n. 3, p. 245-248, 1967.

[81] Grant, W.F.; Owens, E.T. Chromosome aberration assays in Pisum for the study of environmental mutagens. Mutation Research, Amsterdam, v. 188, p. 93-118, 2001.

[82] Hayes, T., Haston, K., Tsui, M., Hoang, A., Haeffele, C., Vonk, A. Herbicides: Feminization of male frogs in the wild. Nature, v.419, p.895-896, 2002.

[83] Gammon, D.W.; Aldous, C.N.; Carr Jr, W.C.; Sanborn, J.R.; Pfeifer, K.F. A risk assessment of atrazine use in California: human health and ecological aspects. Pest Manag Sci, v. 61, p. 331-355, 2005.

[84] Çavas, T. In vivo genotoxicity evaluation of atrazine and atrazine–based herbicide on fish *Carassius auratus* using the micronucleus test and the comet assay. Food and Chemical Toxicology, v. 49, p. 1431-1435, 2011.

[85] Meisner, L.F.; Belluck, D.A.; Rolloff, B.D. Cytogenetic effects of alachlor and/or atrazine *in vivo* and *in vitro*. Environmental and Molecular Mutagenesis, New York, v. 19, p. 77-82, 1992.

[86] Roloff, B.D.; Belluck, D.A.; Meisner, L.F. Cytogenetic studies of herbicide interactions *in vitro* and *in vivo* using atrazine and linuron. Environmental Toxicology, New York, v. 22, p. 267-271, 1992.

[87] Wu, Y.G.; Li, S.K.; Xin, Z.C.; Wang, Y.S.; Shou, K.R.; Gao, H.; Li, Y.Q. The establishment of hypospadias rat model and embryoteratogenic test of Atrazine. Zhonghua Zheng Xing Wai Ke Za Zhi, v. 23(4), p. 340-343, 2007.

[88] Modic, W.; Ferrell, J.; Wood, C.; Laskey, J.; Cooper, R.; Laws, S. Atrazine alters steroidogenesis in male Wistar rats. Toxicologist, v. 78, p. 117, 2004.

[89] Tennant, A.H.; Peng, B; Kligerman, A.D. Genotoxicity studies of triazine herbicides: in vivo studies using the alkaline single gel (SCG) assay. Mutation Research, Amsterdam, v. 493, p. 1-10, 2001.

[90] Clements, C.; Ralph, S.; Petras, M. Genotoxicity of selected herbicides in tadpoles *Rana catesbeiana*, using the alkaline single-cell gel DNA electrophoresis (comet) assay. Environmental and Molecular Mutagenesis, New York, v. 29, p. 277-288, 1997.

[91] Campos-Pereira, F.D.; Oliveira, C.A; Pigoso, A.A.; Silva-Zacarin, E.C.M; Barbieri, R.; Spatti, E.F.; Marin-Morales, M.A.; Severi-Aguiar, G.D.C. Early cytotoxic and genotoxic effects of atrazine on Wistar rat liver: A morphological, immuno-histochemical, biochemical, and molecular study. Ecotoxicology and Environmental Safety, v. 78, p. 170-177, 2012.

[92] Ruiz, M.J., Marzin, D. Genotoxicity of six pesticides by *Salmonella* mutagenicity test and SOS chromotest. Mutation Research, v. 390, p. 245-255, 1997.

[93] Lioi, M.B.; Scarfi, M.R.; Santoro, A.; Barbieri, R.; Zeni, O.; Salvemini, F.; Berardino, D.D.; Ursini, M.V. Cytogenetic damage and induction of pro-oxidant state in human lymphocytes exposed in vitro to gliphosate, vinclozolin, atrazine, and DPX-E9636. Environmental and Molecular Mutagenesis, New York, v. 32, p. 39-46, 1998.

[94] Meisner, L.F.; Roloff, B.D.; Belluck, D.A. *In vitro* effects of N-nitrosoatrazine on chromosome breakage. Environmental Toxicology, New York, v. 24, p. 108-112, 1993.

[95] Loprieno, N.; Adler, I.D. Cooperative Programme of the EEC on short-term assays for mutagenicity. In: MONTESANO, R.; BARTSCH, H.; TOMATIS, L. (eds.), Molecu-

lar and Cellular aspects of carcinogen screening tests, France: Science Publisher, 1980. p. 331-341.

[96] Wuu, K.D.; Grant, W.F. Morphological and somatic chromosomal aberrations induced by pesticides in barley (*Hordeum vulgare*). Canadian Journal of Genetic and Cytology, [S.l.] , v. 8, p. 481-501, 1966.

[97] Wuu, K.D.; Grant, W.F. Chromosomal aberrations in somatic cells of *Vicia faba* by pesticides. Nucleus, [S.l.] , v. 10, p. 37-46, 1967.

[98] Plewa, M.J.; Wagner, E.D.; Gentile, G.J.; Gentile, J.M. An evaluation of the genotoxic properties of herbicides following plant and animal activation. Mutation Research, Amsterdam, v. 136, p. 233-245, 1984.

[99] Lee, K.C.; Rao, G.M.; Barnett, F.L.; Liang, G.H. Further evidence of meiotic instability induced by atrazine in grain sorghum. Cytologia, Tokyo, v. 34, p. 697-702, 1974.

[100] Chou, T.S.; Weber, D.F. The effect of the atrazine on sister-chromatid exchanges in maize. Genetics, Califórnia, v. 97, p. 521, 1981.

[101] Zeljezic, D.; Garaj-Vrhovac, V.; Perkovic, P. Evaluation of DNA damage induced by atrazine and atrazine-based herbicide in human lymphocytes in vitro using a comet and DNA diffusion assay. Toxicology in Vitro, v. 20, p. 923-935, 2006.

[102] Srivastava, K.; Mishra, K.K. Cytogenetic effects of commercially formulated atrazine on the somatic cells of *Allium cepa* and *Vicia faba*. Pesticides Biochemistry and Physiology, v. 93, p. 8-12, 2009.

[103] Kligerman, A.D.; Doerr, C.L.; Tennant, A.H.; Peng, B. Cytogenetic studies of three triazine herbicides II. In vivo micronucleus studies in mouse bone marrow. Mutation Research, v. 471, p. 107-112, 2000.

[104] Kligerman, A.D.; Chapin, R.E., Erexson, G.L.; Germolec, D.R.; Kwanyuen, P.; Yang, R.S. Analyses of cytogenetic damage in rodents following exposure to simulated groundwater contaminated with pesticides and a fertilizer. Mutation Research, Amsterdam, v. 300, p. 125-134, 1993.

[105] Adler, I.D. A review of the coordinated research effort on the comparison of the test systems for the detection of mutagenic effects, sponsored by the E.C.C. Mutation Research, Amsterdam, v. 74, p. 77-93, 1980.

[106] Hrelia, P.; Vigagni, F.; Maffei, F.; Morotti, M.; Colacci, A.; Perocco, P.; Grilli, S.; Cantelli-Forti, G. Genetic safety evaluation of pesticides in different short-term tests. Mutation Research, v. 321, p. 219-228, 1994.

[107] Taets, C.; Aref, S.; Rayburn, A.L. The Clastogenic Potential of Triazine Herbicide Combinations Found in Potable Water Supplies. Environmental Health Perspectives, v. 106 (4), 1998.

[108] Moretti, M.; Marcarelli, M.; Villarini, M.; Fatigoni, C.; Scassellati-Sforzolini, G.; Pasquini, R. *In vitro* testing for genotoxicity of the herbicide terbutryn: cytogenetic and primary DNA damage. Toxicology in Vitro, v. 16, p. 81-88, 2002.

[109] Clausen, M.; Leier, G.; White, I. Comparison of the cytotoxicity and DNA-damaging properties of 2,4-D and U 46 D fluid (dimethylammonium salt of 2,4-D). Archives of Toxicology, v. 64, p. 497-501, 1990.

[110] IARC. Some fumigants, the herbicides 2,4-D and 2,4,5-T, chlorinated dibenzodioxins and miscellaneous industrial chemicals. IARC Monogr Eval Carcinog Risk Chem Man 1977;15:111-48.

[111] Osterloh, J.; Lotti, M.; Pond, S.M. Toxicologic studies in a fatal overdose of 2,4-D, MCPP, and chlorpyrifos. J Anal Toxicol, v. 7, p. 125-129, 1983.

[112] Martínez-Tabcge, L.; Madrigal-Bujaidar, E.; Negrete, T. Genotoxicity and lipoperoxidation produced by paraquat and 2,4-Dichlorophenoxyacetic acid in the gills of rainbow trout (*Oncorhynchus mikiss*). Bull Environ Contam Toxicol, v. 73, p.146-152, 2004.

[113] Ateeq, B.; Abdul-Farah, M.; Ali, M.N.; Ahmad, W. Induction of micronuclei and erythrocyte alterations in the catfish *Clarias batrachus* by 2,4-dichlorophenoxyacetic acid and butachlor. Mutation Research, Amsterdam, v. 518, p. 135-144, 2002.

[114] Suwalsky, M.; Quevedo, L.; Norris, B.; Benites, M. Toxic Action of the Herbicide 2,4-D on the Neuroepithelial Synapse and on the Nonstimulated Skin of the Frog *Caudiverbera caudiverbera*. Bull. Environ. Contam. Toxicol., v. 62, p. 570-577, 1999.

[115] Gómez, L.;, Masot, J.; Martinez, S.; Durán, E.; Soler, F.; Romero, V. Acute 2,4-D poisoning in tench (*Tinca tinca* L.): lesions in the hematopoietic portion of the kidney. Arch Environ Contam Toxicol, v. 35, p. 479-483, 1998.

[116] González, M.; Soloneski, S.; Reigosa, M.A.; Larramendy, M.L. Effect of the herbicide 2,4-dichlorophenoxyacetic acid (2,4-D) and its derivative 2,4-D dichlorophenoxyacetic acid dimethylamine salt (2,4-D DMA). I. Genotoxic evaluation on Chinese hamster ovary (CHO) cells. Toxicology in Vitro, v. 19, p. 289-297, 2005.

[117] Madrigal-Bujaidar, E.; Hernandez-Ceruelos, A.; Chamorro, G. Induction of sister chromatid exchanges by 2,4-dichlorophenoxyacetic acid in somatic and germ cells of mice exposed *in vivo*. Food Chem Toxicol, v. 39, p. 941-946, 2001.

[118] Soloneski, S.; González, N.V.; Reigosa, M.A.; Larramendy, M.L. Herbicide 2,4-dichlorophenoxyacetic acid (2,4-D)-induced cytogenetic damage in human lymphocytes *in vitro* in presence of erythrocytes. Cell Biology International, v. 31, p. 1316-1322, 2007.

[119] Morgan, M.K.; Scheuerman, P.R.; Bishop, C.S.; Pyles, R.A. Teratogenic potential of atrazine and 2,4-D using FETAX. J Toxicol Environ Health, v. 48 (2), p.151-168, 1996.

[120] Xie, L; Thrippleton, K.; Irwin, M.A.; Siemering, G.S.; Mekebri, A.; Crane, D.; Berry, K.; Schlenk D. Evaluation of estrogenic activities of aquatic herbicides and surfac-

tants using an rainbow trout vitellogenin assay. Toxicology Science, v. 87(2), p. 391-398, 2005.

[121] Smith, E.A.; Oehme, F.W. The biological activity of glyphosate to plants and animals; a literature review. Vet. Hum. Toxicol., v. 34, p. 531-543, 1992.

[122] Mladinic, M.; Berend, S.; Vrdoljak, A.L.; Kopjar, N.; Radic, B.; Zeljezic, D. Evaluation of genome damage and its relation to oxidative stress induced by glyphosate in human lymphocytes *in vitro*. Environmental and Molecular Mutagenesis, v. 50, p. 800 807, 2009a.

[123] Martini, C.N.; Gabrielli, M.; Vila, M.D.C. A commercial formulation of glyphosate inhibits proliferation and differentiation to adipocytes and induces apoptosis in 3T3-L1 fibroblasts. Toxicology in Vitro, v. 26, p. 1007-1013, 2012.

[124] Solomon, K.; Thompson, D. Ecological Risk Assessment for Aquatic Organisms from Over-Water Uses of Glyphosate. Journal of Toxicology and Environmental Health, Part B: Critical Reviews, v. 6 (3), p. 289-324, 2003.

[125] Williams, G.M.; Kroes, R.; Munro, I.C. Safety Evaluation and Risk Assessment of the Herbicide Roundup and Its Active Ingredient, Glyphosate, for Humans. Regulatory Toxicology and Pharmacology, v. 31, p. 117-165, 2000.

[126] Cox, C. Herbicide factsheet – Glyphosate. J Pest Reform, v. 24, p. 10-15, 2004.

[127] Clair, E.; Linn, L.; Travert, C.; Amiel, C.; Séralini, G.E.; Panoff, J.M. Effects of Roundup(®) and glyphosate on three food microorganisms: *Geotrichum candidum, Lactococcus lactis subsp. cremoris and Lactobacillus delbrueckii subsp. bulgaricus*. Curr Microbiol., v. 64 (5), p. 486-491, 2012.

[128] Relyea, R.A. New effects of Roundup on amphibians: predators reduce herbicide mortality; herbicides induce antipredator morphology. Ecol. Appl., v. 22 (2), p. 634-647, 2012.

[129] Dimitrov, B.D.; Gadeva, P.G.; Benova, D.K.; Bineva, M.V. Comparative genotoxicity of the herbicides Roundup, Stomp and Reglone in plant and mammalian test systems. Mutagenesis, v. 21 (6), p. 375-382, 2006.

[130] Dallegrave, E.; Mantese, F.D.; Coelho, R.S.; Pereira, J.D.; Dalsenter, P.R.; Langeloh, A. The teratogenic potential of the herbicide glyphosate-Roundup in Wistar rats. Toxicology Letters, v. 142, p. 45-52, 2003.

[131] WHO (World Health Organization), 1994. Glyphosate. Environmental Health Criteria. 159, pp. 1-177.

[132] Relyea, R.A., The impact of insecticides and herbicides on the biodiversity and productivity of aquatic communities. Ecological Applications, v. 15 (2), p. 618-627, 2005.

[133] Fetvadjieva, N.; Straka, F.; Michailova, P.; Balinov, I.; Lubenov, I.; Balinova, A.; Pelov, V.; Karsova, V.; Tsvetkov, D. In: Fetvadjieva,N. (ed.), Handbook of Pesticides. 2nd revised edn. Zemizdat Inc., Sofia, pp. 330, 1994.

[134] Tortorelli, M.C.; Hernandez, D.A.; Rey Vazquez, G.; Salibian, A. Effects of paraquat on mortality and cardiorespiratory function of catfish fry *Plecostomus commersoni*. Arch Environ Contam Toxicol, v. 19, p. 523-529, 1990.

[135] Ribas, G.; Surrallés, J.; Carbonell, E.; Xamena, N.; Creus, A.; Marcos, R.. Genotoxic Evaluation of the Herbicide Paraquat in Cultured Human Lymphocytes. Teratogenesis, Carcinogenesis, and Mutagenesis, v. 17, p. 339-347, 1998b.

[136] Hanada, H. Dl-α-tocopherol enhances the herbicide 1,1'-dimetyl-4,4'–bipyridium dichloride (Paraquat, PQ) genotoxicity in cultured anuran leukocytes. Hereditas, v. 148, p. 118-124, 2011.

[137] Tomita, M.; Okuyama, T.; Ishikawa, T.; Idaka, K.; Nohno, T. The role of nitric oxide in paraquat-induced cytotoxicity in the human A549 lung carcinoma cell line. Free Rad Res, v. 34, p. 193-202, 2001.

[138] D'souza, U.J.; Narayana, K.; Zain, A.; Raju, S.; Nizam, H.M.; Noriah, O. Dermal exposure to the herbicide-paraquat results in genotoxic and cytotoxic damage to germ cells in the male rat. Folia Morphol (Warsz), v. 65 (1), p. 6-10, 2006.

[139] Bus, J.S.; Aust, S.D.; Gibson, J.E. Superoxide and singlet oxygen-catalyzed lipid peroxidation as a possible mechanism for paraquat (methyl viologen) toxicity. Biochem. Biophys. Res. Comm., v. 58, p. 749-755, 1974.

[140] Tanaka, R.; Amano, Y. Genotoxic effects of paraquat and diquat evaluated by sister-chromatid exchange,chromosomal aberration and cell-cycle rate. Toxicology in Vitro, v. 3, p. 53-57, 1989.

[141] Bauer Dial, C.A.; Dial, N.A. Lethal effects of consumption of fi eld levels of paraquat-contaminated plants on frog tadpoles. Bull. Environ. Contam. Toxicol., v. 55, p. 870-877, 1995.

[142] Speit, G.; Haupter, S.; Hartmann, A. Evaluation of the genotoxic properties of paraquat in V79 Chinese hamster cells. Mutation Research, v. 412 (2), p. 187-193, 1998.

[143] Hoffman, D.J.; Eastin Jr, W.C. Effects of lindane, paraquat, toxaphene, and 2,4,5-trichlorophenoxyacetic acid on mallard embryo development. Arch Environ Contam Toxicol., v. 11, p. 79-86, 1982.

[144] Zeliger, H.I., Human toxicology of chemical mixtures. In: Toxic Consequences Beyond the Impact of One-component Product and Environmental Exposures. 2nd ed. Elsevier, Oxford, 2011.

[145] Bassil, K.L.; VAKIL, C.; SANBORN, M.; COLE, D.C.; KAUR, J.S.; KERR, K.J. Cancer health 585 effects of pesticides: systematic review. Can. Fam. Physician, v. 53, p. 1704-1711, 2007.

[146] Parrón, T.; Requena, M.; Hernández, A.F.; Alarcón, R. Association between environmental exposure to pesticides and neurodegenerative diseases. Toxicol. Appl. Pharmacol., v. 256, p. 379-385, 2011.

[147] Hanke, W.; Jurewicz, J., The risk of adverse reproductive and developmental disorders due to occupational pesticide exposure: an overview of current epidemiological evidence. Int. J. Occup. Med. Environ. Health, v. 17, p. 223-243, 2004.

[148] Hernández, A.F.; Parrón, T.; Alarcón, R. Pesticides and asthma. Curr. Opin. Allergy Clin. Immunol., v. 11, p. 90-96, 2011.

[149] Doll, R.; Peto, R. The causes of cancer: quantitative estimates of avoidable risks of cancer in the United States today. Journal National Cancer Institute, v. 66, p. 1191-1308, 1981.

[150] Hernández, A.F.; Parrón, T.; Tsatsakis, A.M.; Requena, M.; Alarcón; López-Guarnido, O. Toxic effects of pesticide mixtures at a molecular level: Their relevance to human health. Toxicology, 2012 (in press).

[151] Jones, F.; Fawell, J.K. Lessons learnt from the river DEE Pollution Incident. Public Health in Proceedings of the Word conference on chemicals accidents, Harvard, v. 4, p. 223-226, 1987.

[152] Waters, M.D.; Stack, H.F.; Jackson, M.A. Genetic toxicology data in the evaluation of potential human environmental carcinogens. Mutation Research, Amsterdam, v. 437, p. 21-49, 1999.

[153] USEPA. Atrazine: Carcinogenicity characterization and hazard assessment, office of pesticide programs, health effects division. 1999. Http:/ www.epa.gov/scipoly/sap/ #jan

[154] EPA, 2003. US Environmental Protection Agency. October 31, 2003, revised atrazine interim reregistration eligibility decision (IRED). Office of Prevention, Pesticides and Toxic Substances, 2003.

[155] Jowa, L.; HOWD, R. Should atrazine and related chlorotriazines be considered carcinogenic for human health risk assessment? J Environ Sci Health C Environ Carcinog Ecotoxicol Rev., v. 29 (2), p. 91-144, 2011.

[156] Mladinic, M.; Perkovic, P.; Zeljezic, D. Characterization of chromatin instabilities induced by glyphosate, terbuthylazine and carbofuran using cytome FISH assay. Toxicology Letters, v. 189, p. 130-137, 2009b.

[157] Gebel, T.; Kevekordes, S.; Pav, K.; Edenharder, R.; Dunkelberg, H. *In vivo* genotoxicity of selected herbicides in the mouse bone-marrow micronucleus test. Arch. Toxicol., v. 71, 193-197, 1997.

[158] Mladinic, M.; Zeljezic, D.; Shaposhnikov, S.A.; Collins, A.R. The use of FISH-comet to
detect c-Myc and TP 53 damage in extended-term lymphocyte cultures treated with
terbuthylazine and carbofuran. Toxicology Letters, v. 211, p. 62-69, 2012.

[159] Hardell, L.; Eriksson, M.; Nordstrom, M. Exposure to pesticides as risk factor for
non-Hodgkin's lymphoma and hairy cell leukemia: Pooled analysis of two Swedish
case-control studies. Leuk Lymp, v. 43, p. 1043-1049, 2002.

[160] De Roos, A.J.; Zahm, S.H.; Cantor, K.P.; Weisenburger, D.D.; Holmes, F.F.; Burmeis-
ter, L.F.; Blair, A. Integrative assessment of multiple pesticides as risk factors for non-
Hodgkin's lymphoma among men. Occup Environ Med, v. 60, p.11, 2003.

[161] He, X.; Wang, L.; Szklarz, G.; Bi, Y.; Ma, Q. Resveratrol Inhibits Paraquat-Induced
Oxidative Stress and Fibrogenic Response by Activating the Nuclear Factor Eryth-
roid 2-Related Factor 2 Pathway. The Journal of Pharmacology and Experimental
Therapeutics, v. 342, p. 81-90, 2012b.

[162] Ossowska, K; Smiałowska, M.; Kuter, K.; Wieron' Ska, J.; Zieba, B.; Wardas, J.; Now-
ak, P.; DABROWSKA, J.; BORTEL, A.; BIEDKA, I. et al. Degeneration of dopaminer-
gic mesocortical neurons and activation of compensatory processes induced by a
long-term paraquat administration in rats: implications for Parkinson's disease. Neu-
roscience, v. 141, p. 2155-2165, 2006.

[163] Tanner, C.M.; Kamel, F.; Ross, G.W.; Hoppin, J.A.; Goldman, S.M.; Korell, M.; Mar-
ras, C.; Bhudhikanok, G.S.; Kasten, M.; Chade, A.R. et al. Rotenone, paraquat, and
Parkinson's disease. Environ Health Perspect, v. 119, p. 866-872, 2011.

[164] Garaj-Vrhovac, V.; Zeljezic, D. Assessment of genome damage in a population of
Croatian workers employed in pesticide production by chromosomal aberration
analysis, micronucleus assay and Comet assay. Journal of Applied Toxicology, Chichester, v. 22, n. 4, p. 249-255, 2002.

Chapter 3

Pesticide Tank Mixes: An Environmental Point of View

Valdemar Luiz Tornisielo, Rafael Grossi Botelho,
Paulo Alexandre de Toledo Alves,
Eloana Janice Bonfleur and
Sergio Henrique Monteiro

Additional information is available at the end of the chapter

1. Introduction

During the last decades, human activity has affected the aquatic and terrestrial ecosystems' sustainability. None of these activities has damaged the environment as severely as agricultural practices.

Current agricultural practices have negatively affected aquatic and terrestrial ecosystems by destroying habitats, deforesting to increase cropping areas and applying pesticides.

Pesticides are a heterogeneous category of chemical products destined to pest, disease and weed control including several types, such as insecticides, fungicides, herbicides, nematicides and others.

Nowadays, such chemical product applications have been considered the most efficient plant protection procedures and have significantly contributed to the improvement of crop productivity.

Nevertheless, the claimed objective of supplying the population with enough food does not justify damaging the environment, just because small quantities of pesticides are known to efficiently control pests, diseases and weeds. However, most of them are rapidly spread out affecting all living beings (flora and fauna, including humans).

The use of chemical molecules in agriculture increased after the Second World War with the advent of DDT (dichloro-diphenyl-trichloroethane). DDT was discovered in 1939 by Paul Müller (Swiss entomologist) and its worldwide use was rapidly expanded due to its large action range, low cost and efficiency in the control of tropical disease vectors, such as typhoid fever and malaria [1].

After the release of DDT, a large range of molecule groups destined to crop protection were developed and commercialized. In 1962, the book "Silent Spring" was the first act of environment manifest against DDT, describing the bird population decrease (from the top of the food chain) attributed to its indiscriminate use.

After the 1960's, the use of chemical products in agriculture rapidly increased and it was associated with the appearance of environmental and human health problems.

The frequent and incorrect use of pesticides have caused soil, atmosphere, food and water resource (superficial/underwater) contaminations, negatively affecting aquatic and terrestrial organisms as well as frequently causing toxicity to the human population.

Therefore, studies are urgently needed to make environmental monitoring procedures viable in order to detect potential contamination risks and give support to public actions for environmental safety and agriculture sustainability.

Currently, product mixtures (associations between one or more molecules) are applied in agriculture instead of individual molecules; therefore, previous studies that focused on only one molecule should now consider molecule mixtures.

The existence of such a large variety of pests, diseases and weeds affecting yields have led farmers to use product mixtures, aiming at efficiently managing crop protection. Such mixtures, also called product associations, enter the environment in a different way compared to the individual product application. Thus, more studies are required about these mixture-environment interactions and possible interactions between molecules and consequent interferences in the environment.

Although mixtures have been intensively studied concerning their agronomic efficacy, little information is found about their implications on environmental safety.

In this chapter, the tank mixture subject is approached from an environmental point of view, explaining the chemical product mixture interactions and the possible contaminant effects. Studies on the product-environment interactions are presented to provide the main available information as support to future studies and decisions in environmental sustainability and safety.

2. Agronomic characteristics of tank mixtures

Tank mixtures are associations among two or more chemical products (pesticides) or among chemical products and fertilizers in a unique tank for application in crops. This practice is common in Australia, Canada, U.S.A and United Kingdom, where there are recommendations on application procedures, incompatibilities, and safety [2].

Concerning agricultural practices, the tank mixture of two or more chemical products might be a good application strategy, saving fuel and labor-hours, causing less soil compaction, and possibly providing a larger pest control range and efficacy, when compared to the single product application. For these reasons, this technique is preferred by farmers [3].

Nevertheless, the herbicide mixture might induce, for instance, interactions before or after reaching the target-plant, by altering the product action in synergistic, antagonistic or additive ways. One common practice is the simultaneous application of herbicides with and without residual effect in order to increase the weed species control range and/or the control period. Another practice is the addition of adjuvants to improve herbicide performance to control weeds. The simultaneous application of pesticides (concerning the species-target to be controlled) might induce undesirable (antagonistic, synergistic or additive) reactions, depending on the herbicide type and plant species [4]. When the mixture induces an antagonistic reaction, it means that a lower weed control action than expected is observed. When the mixture induces a synergistic reaction, it means that a higher weed control than expected is observed. And, finally, when the mixture induces an additive reaction, it means that no change in weed control is observed.

Several studies have elucidated the questions about synergistic and antagonistic effects of active ingredient mixtures on weed control, for instance, the studies with glyphosate reported by Vidal et al. [4], Shaw and Arnold [3], Selleck and Baird [5].

The application of pesticides plus adjuvants has also been a usual practice. The adjuvant enhances the active ingredient action [6]. In other words, the adjuvant substance induces the herbicide molecule uptake by leaf tissues, by accelerating the product penetration through plant cuticles. The most common types are the biosurfactants, mineral or vegetal oils, synthetic or natural polymers, humectants, organic salts, buffer solutions, and others [7].

The tank mixture practice or different individual pesticide applications at short intervals might result in multiple pesticide residues on foods, as observed by Gebara et al. [8], when monitoring food samples in São Paulo metropolis, Brazil, during the period between 1994 and 2001. The authors found multiple pesticide residues in 5.8% of vegetable samples analyzed and 11.4% of fruit samples.

Gebara et al. [9] alerted for the violation risk of the Theoretical Maximum Dietary Intake (TMDI), which is calculated by the relationship between the Limit of Maximum Residues (LMR, mg kg^{-1}) established for a pesticide in a food and the daily consumption (DC, kg day^{-1}), based on the individual diet. The presence of multiple pesticide residues in foods due to the use of tank mixtures, might lead to the extrapolation of toxicological parameters for the acceptable daily intake (ADI), mainly for children and nursing women.

3. Pesticide tank mixtures environmental effects

3.1. Soil

Weed control with pesticide tank mixtures has been widely studied concerning mixture effectiveness, component antagonism and/or synergism. However, there is little information on environmental issues.

Knowledge on soil-herbicide interactions when herbicide mixtures are applied is extremely relevant. However, few studies on herbicide associations and their soil interactions can be found, because most studies are restricted to the individual molecule behavior.

When a pesticide is released in the environment, it will probably enter the soil by direct application, or indirectly, by crop residue incorporation into the soil and molecule transport by spraying derivation. In the soil, several processes might occur, that is, molecule retention (adsorption, absorption), transformation (decomposition, degradation) and transport (spraying derivation, volatilization, lixiviation, superficial runoff). Such processes will determine the molecule destiny, persistence and agronomic efficiency. The main factors influencing those processes are the climatic conditions, the pesticide physical-chemical properties and the soil physical-chemical attributes. According to Oliveira [10], the complex molecule retention process by soil sorption/desorption directly or indirectly influences other factor activities.

Knowledge on pesticide physical-chemical properties is fundamental to predict soil interactions, potential contamination and transport risks when in the soil solution or associated to sediments. Studies on pesticide mixtures have been restricted to their phytotoxicity effects and few were dedicated to the interactions between two or more associated molecules.

Alves [11] demonstrated that ametryn mineralization half-life is longer when associated to glyphosate than when applied alone; but there was a synergistic effect in the soil, because ametryn half-life was 15 days for the ametryn + glyphosate mixture and 20 days for isolated ametryn in the soil. In the same study, the author observed increased glyphosate mineralization half-life from 55 to 119 days, when comparing single glyphosate and glyphosate + ametryn treatments, respectively; the glyphosate soil half-life could not be determined due to its strong soil sorption during extractions.

Yet in studies of soil microbial activity, Alves [11] observed that glyphosate (at a higher rate) enhanced microbial activity; meanwhile isolated ametryn (at a lower rate) negatively affected microbial activity, but a less negative effect of ametryn + glyphosate mixture (at a lower rate) was observed compared with single ametryn at the same rate. The ametryn + glyphosate mixture (at a higher rate) increased the microbial activity, evidencing a stronger mixture synergistic effect.

Alves [11] also studied the herbicide sorption/desorption in a red Ultisol. High glyphosate and low ametryn sorption were observed when herbicides were applied alone. Higher soil sorption was observed for both herbicides in mixture than for the single molecules. Low glyphosate desorption occurred at all rates in both application procedures (alone or in mixture), but ametryn desorption decreased when applied in mixture.

White et al. [12] studied the effects of chlorothalonil, tebuconazole, flutriafol and cyproconazole fungicides on the metolachlor herbicide dissipation kinetics. Significantly lower metolachlor dissipation was observed with chlorothalonil, when compared with soil treatments without chlorothalonil or with other fungicides. The authors observed significant reduction in metolachlor metabolites probably attributed to the fungicide effect on glutathione S-transferase enzyme activity. Overall, chlorothalonil fungicide induced a two-fold increase in metolachlor persistence.

Ke-Bin et al. [13] observed that atrazine and bentazon herbicides showed longer lag-phase and lower degradation rate when applied in tank mixture in a maize crop. Therefore, the association of atrazine-bentazon had longer soil persistence which means that higher environmental potential contamination risks might be expected.

The effect of glyphosate on atrazine degradation was studied by Krutz et al. [14] in a silt clayey soil (pH 8.3 and 10.6 g kg^{-1} of organic-C) from the Texas region in USA. Atrazine degradation was inversely related to glyphosate rate and microbial activity during an eight-day period, evidencing that glyphosate enhanced microbial activity and inhibited atrazine degradation. The authors discussed that atrazine degradation, when in association, is mainly a microbial mechanism, and the degradation reduction might be explained by a lower enzymatic activity and/or by microbial population suppression by glyphosate.

Similar results were reported by Haney et al. [15] for the same soil type, demonstrating the atrazine and glyphosate effects on soil microbial activity evaluated through the soil carbon (C) and nitrogen (N) mineralization. Soil plots treated with the herbicide mixture showed higher microbial activity than plots treated with single atrazine. The evaluated soil C and N flows allowed understanding of the microbial preference for glyphosate because this herbicide's complete mineralization occurred in 14 days, followed by fast atrazine degradation.

Zablotowicz et al. [16] studied the effects of glufosinate (herbicide), ammonium sulfate (fertilizer) and both products in mixture on atrazine mineralization. The authors observed decreased atrazine mineralization when the product mixture was applied. The authors explained that an alteration in ^{14}C-atrazine molecule partition into its metabolites and residues would occur caused by ammonium sulfate that would restrict the triazine ring cleavage. Such results evidenced that the application of glufosinate combined to a mineral N source might increase soil atrazine persistence, increasing its residual effect.

Lancaster et al. [17] observed that glyphosate increased soil C mineralization and fluometuron microbial degradation. The authors suggested that the increasing C mineralization might be related to the increasing fluometuron degradation or to a priming glyphosate effect.

Concerning the glyphosate and diflufenican association, Tejada [18] observed longer degradation periods for both herbicides in mixture than for the individual molecules. Furthermore, the glyphosate-diflufenican association increased both herbicide toxicities to the soil biological activity (measured by the microbial C biomass and enzyme activities - dehydrogenase, urease, β-glycosidase, phosphatase and arylsulfatase) and the individual herbicide persistence.

Pereira et al. [19] evaluated the application of isolated glyphosate and associated to endosulfan on the soil microbial activity in soybeans and observed reduced microbial activity and biomass, and also, reduced metabolic quotient.

In genetically modified glyphosate-tolerant maize cultivars, it is possible to mix glyphosate and atrazine. In the USA, there are a number of commercially available associations, among them, glufosinate or glyphosate mixed with atrazine [20]. Bonfleur et al. [21] observed that glyphosate mineralization was not affected by atrazine presence in a tropical soil. However, increased atrazine mineralization (measured by the ^{14}CO$_2$ release) was observed with increasing glyphosate rates. The authors observed a 100-day variation in the atrazine half-life when associated with a two-fold glyphosate rate. Therefore, the glyphosate-atrazine tank mixture allowed atrazine persistence reduction in the soil. The authors said that a possible explanation is the glyphosate contribution to the microorganisms as source of N, and this N supply might decrease the initial atrazine immobilization when this is the only substrate, and then, increasing its mineralization.

Fogg and Boxall [22] observed inhibitory effects of an isoproturon-chlorothalonil mixture on the isoproturon degradation in soils. Isoproturon half-life (DT50) values varied from 18.5 to 71.5 days when combined with chlorothalonil. This might be explained by the TPN-OH chlorothalonil metabolite inhibition and the reduction in the soil microorganism population involved in isoproturon degradation.

The soil degradation of pendimethalin (herbicide) was significantly reduced when mixed with mancozeb (fungicide) or mancozeb+thiamethoxam (insecticide) [23]. Pendimethalin herbicide half-life increased from 26.9 to 62.2 days when in single and combined (mancozeb + thiamethoxam) applications, respectively, in a sandy soil. On the other hand, the same authors observed that pendimethalin degradation is not affected by the presence of isolated metribuzin or thiamethoxam.

Several studies have pointed out the adjuvant influence on pesticide destiny in the environment, specifically their persistence and bioavailability. Cabrera [24], in laboratory studies, affirmed that metazachlor herbicide added to oil and surfactant showed reduced degradation rates and increased residues in the soil. Similar results to other pesticides were reported by Kucharski and Sadowski [25] and Rodríguez-Cruz et al. [26]. In a field experiment, Kucharski et al. [27] observed a 43% increase in lenacil herbicide residues in the superficial soil layer, with the addition of adjuvants (oil and surfactant).

High mobility pesticides used together with adjuvants present decreased movement along the soil profile. Reddy and Singh [28] evaluated bromacil and diuron herbicides lixiviation in soil columns. In treatments with adjuvant addition, the authors observed significant lower bromacil vertical movement and no effect on diuron movement. These two herbicides present distinct physical-chemical characteristics that explain their differential movement abilities in the soil. Thus, bromacil is an acidic molecule with high water solubility (815 mg L^{-1}); meanwhile diuron is a non-ionic herbicide of low water solubility (42 mg L^{-1}). From the environmental point of view, the adjuvant effect was positive in the case of bromacil, but the agronomic efficacy was restricted.

The results found in the literature have highlighted the interactions existing among several molecules, especially in the soil, but such interactions might be different under other environment compartments. For this reason, studies on environmental pesticide behavior and destination must include all aspects, bringing together laboratory and field approaches.

3.2. Water: An ecotoxicological approach for pesticide mixtures

According to Botelho et al. [29], water resource contamination has currently been considered one of the greatest environmental problems on Earth.

Pesticides applied to field crops are released in the environment mainly through lixiviation (when molecules move into the soil and reach the underground waters), superficial runoff (when molecules move together with soil and water runoff), and spraying derivation (when molecules are carried by wind during pesticide spraying).

The situation is complex once crop diversity allied to the high number and diversity of pesticide products usually applied to field crops, and the short distances between fields and aquatic areas have exposed the water resources not only to individual products but also to all their associations [30].

Several products, mainly herbicides and insecticides, are common superficial water contaminants, due to their large application in agriculture and residential areas. Therefore, there is an increasing concern about superficial and underground water contamination, due to the lack of information on pesticide impacts mainly in aquatic systems.

In Brazil, several studies have been carried out to determine the presence of pesticides in aquatic ecosystems. Armas et al. [31] evaluated the presence of herbicides in the superficial water and sediments of Corumbataí River (State of São Paulo, Brazil). The authors found several herbicides - ametryn, atrazine, simazine, hexazinone, glyphosate and clomazone – and triazines were specifically found in higher levels, above the limits allowed for potable water by Brazilian legislation. Dores et al. [32] found herbicide residues from the triazine group and their metabolites, as well as metribuzin, metolachlor and trifluralin residues. Among the Brazilian literature, the research works of Caldas et al. [33], Lanchote et al. [34], Filizola et al. [35], Laabs et al. [36], Dores et al. [37], Jacomini et al. [38] are pointed out.

Other interesting results can be found in the literature: Benvenuto et al. [39] determined the presence of eleven pesticides in superficial waters of Italy and Spain and observed concentration values varying between 0.002 and 0.087 μg L^{-1}. Yu et al. [40] determined the presence of nine (among eleven pesticides evaluated) herbicides of the triazine group in all water samples analyzed. Similar determinations were made by Ma et al. [41], Palma et al. [42], Balinova and Mondesky [43] and Segura et al. [44].

Understanding of how pesticides affect aquatic environments has been a challenge to researchers, and the science of ecotoxicology has helped to answer many questions on this subject.

The "ecotoxicology" term was first suggested by the French toxicologist René Truhaut, during the *Committee of the International Council of Scientific Unions* (ICSU) meeting, in June 1969, in Stockholm (Sweden) [45]. According to this author, Ecotoxicology is the science that studies the effects of natural or synthetic substances on living beings, populations and communities, animal or vegetal, terrestrial or aquatic, constituting the biosphere, including the substance interaction with the environment where they live in an integrated context [46].

Usually, ecotoxicological experiments follow standardized protocols developed by international organizations, for example, the Environmental Protection Agency (EPA); the Organization for Cooperation and Economical Development (OCDE); and the Brazilian Agency of Technical Norms (ABNT).

The toxicity tests allow evaluating the environmental contamination by different pollutant sources, such as agricultural, industrial and domestic residues, sediments, medicines and chemical products overall, as well as the results of their synergistic and antagonistic effects [47-48]. The ecotoxicological tests can also detect the toxic agent or mixture capacity of causing deleterious effects on living organisms, allowing determination of the harmful concentration ranges, and how and where the effects are expressed [49].

Several parameters have been used to determine the xenobiotic effects in different organisms. Among these variables, the lethality [50-51], immobility [52], gill alterations [53-56], and reproduction [57-59] are pointed out.

The ecotoxicological experiments consist of exposing living organisms to several concentrations of a specific product and evaluating the results that might be expressed according to the test type. For instance, the acute test consists of short-term exposure of organisms to several product concentrations, and then, the species life cycle is evaluated; the toxicity indicative parameters more frequently used are: lethality (expressed by the average lethal concentration - LC_{50}), and immobility (expressed by the observable toxic concentration effect - EC_{50}). It is important to highlight that both parameters take into consideration the effects for 50% of the organisms tested under the specific experiment conditions [60-61]. In the case of a chronic test, the organism is submitted to long-term product exposure and the observable effects are usually focused on organism reproduction, behavior, morphology, and size, among others.

Water quality tests have been important tools aiming to minimize the pollution effects on aquatic ecosystems and to implement remediation and monitoring programs, and for that, the ecotoxicological tests have been used.

In the case of pesticide mixtures, the ecotoxicological tests to determine toxicity effects are difficult to interpret, because toxicity symptoms might depend on interactions occurring among different chemical molecules in solution and their accumulative quantities in organisms [61].

When analyzing mixture toxicity effects, some approaches and definitions must be established. In the aquatic ecotoxicology, two different models have been used to describe the relationships between single compound effects and their mixtures: concentration addition model (CA) and independent action model (IA) [62]. In the CA model, each mixture component toxicity effect is induced through a same mechanism, meanwhile in the IA model, the combined components show different actions, inducing a unique toxicological response, but via distinct reactions within the organisms [63]. Nevertheless, both models are used as references to predict the expected mixture toxicity effect, based on the known toxicity of the individual compounds [62].

For a long time, there has been concern about mixture impacts on aquatic ecosystems, not only from pesticides but also from other compound groups, and several discussions and reviews have been reported. In 1984, Hermens and collaborators investigated organic mixture effects on mortality and reproduction of *Daphnia magna* microcrustacean, after exposure to 14 products with different modes of action. The authors observed more severe toxicity effects on mixture-treated organisms than with individual products, although the chronic test results with the mixture showed less severe symptoms [64]. Strmac and Braunbeck [65] observed several structure and biochemical alterations in rainbow trout hepatocytes submitted to a 20-component mixture, including pesticides. Delorenzo and Serrano [66] evaluated the effects of atrazine (herbicide), chlorpyrifos (insecticide) and chlorothalonil (fungicide) on the *Dunaliella tertiolecta* algae growth; the results of atrazine - chlorpyrifos mixture showed an additive toxicity pattern, meanwhile atrazine - chlorothalonil mixture showed a synergistic effect. Yet, the authors observed a two-fold higher toxicity effect of atrazine – Chlorothalonil mixture than the individual products. Choung et al. [67] observed that relatively high atrazine rates increased the terbufos (insecticide) toxicity to *Ceriodaphnia dubia* microcrustacean.

4. Final remarks

Pesticide tank mixtures are currently and frequently used not only in developed countries with specific regulatory legislation for the practice, but also in all agricultural countries where information on harmful effects do not directly reach farmers.

From the agronomic point of view, an effective pest control with pesticide mixtures will depend on the molecule compatibility and also on specific control tests. When the farmer uses two chemically incompatible substances in tank mixture, high losses in crop yield and equipment problems might occur, for example, sprayer nozzle obstruction due to chemical reaction between molecules and subsequent compound precipitation.

Although the pesticide tank mixture may appear to be an efficient pest control practice with synergistic results, the aspects concerning environmental safety must be considered. Little specific information on associated pesticide residues is available in the literature concerning withholding periods and overall environmental behavior.

When a single pesticide is applied, the expected environmental results should be similar to previous results reported for the pesticide registration and before its commercial release. The environment (mainly aquatic and soil medium) is a large contaminant reservoir, where the chemical compounds used in agriculture can be found together. In spite of that, it is important to reinforce that a single pesticide interacts quite differently with the medium, compared to the mixture interaction, as already discussed in this chapter.

In light of the large global demand for food and the increasing crop productivity in the same cropping area, it is imperative to consider the environmental safety questions concerning tank chemical mixture applications in agriculture.

This is a relatively new science area that demands urgent studies on environmental safety, ecotoxicology and toxicology, in order to make highly prevalent the declaration of the United Nation Organization about the planet environment: *"The man has the fundamental right to liberty, equality and enjoyment of adequate life conditions, under an environment of such quality that allows him living a dignifying life and well-being, and he is carrier of the solemn duty of protecting and improving the environment for the present and future generations"* [68].

Acknowledgements

The authors are grateful to the Research Foundation of the State of São Paulo (FAPESP) and to the National Council for Scientific and Technological Development (CNPQ).

Author details

Valdemar Luiz Tornisielo, Rafael Grossi Botelho, Paulo Alexandre de Toledo Alves, Eloana Janice Bonfleur and Sergio Henrique Monteiro

Laboratory of Ecotoxicology, Center for Nuclear Energy in Agriculture, University of São Paulo, Piracicaba, SP, Brazil

References

[1] Amato, D. C, Torres JPM, Malm O. DDT (dicloro difenil tricloroetano): toxicidade e contaminação ambiental- uma revisão. Química Nova (2002). a) , 995-1002.

[2] CanadáMinistry of Agriculture. Safety precautions- mixing and loading pesticides. Vancouver, (2011). http://www.al.gov.bc.ca/pesticides/d_5.htmaccessed 15 February 2012).

[3] Shaw, D. R, & Arnold, J. C. Weed control from herbicide combinations with glyphosate. Weed Technology (2002). , 16(1), 1-6.

[4] Vidal, R. A, Machry, M, Hernandes, G. C, & Fleck, N. G. Antagonismo na associação de glyphosate e triazinas. Planta Daninha (2003). , 21(2), 301-306.

[5] Selleck, G. W, & Baird, D. D. Antagonism with glyphosate and residual herbicide combinations. Weed Science (1981). , 29(2), 185-190.

[6] International Union of Pure and Applied ChemistryIUPAC: Pesticide Formulations. http://agrochemicals.iupac.org/index.php?option=com_sobi2&sobi2Task=sobi2Details&catid=3&sobi2Id=38&Itemid=19accessed 15 September (2012).

[7] Cronfeld, P, Lader, K, & Baur, P. Classification of Adjuvants and Adjuvant Blends by Effects on Cuticular Penetration. In: Viets AK, Tann RS, Mueninghoff JC. (Eds.) Pesticide Formulations and Application Systems: Twentieth Volume, ASTM STP 1400, American Society for Testing and Materials. West Conshohocken PA; (2001). , 81-94.

[8] Gebara, A. B. Ciscato CHP, Ferreira M da S, Monteiro SH. Pesticide Residues in Vegetables and Fruits Monitored in São Paulo City, Brazil, 1994-2001. Bulletin of Environmental Contamination and Toxicology (2005). , 75(1), 163-169.

[9]   Gebara, A. B. Ciscato CHP, Monteiro SH, Souza GS. Pesticide Residues in some Com-
      modities: Dietary Risk for Children. Bulletin of Environmental Contamination and
      Toxicology (2011). , 86(5), 506-510.

[10] Oliveira, M. F. Comportamento de herbicidas no ambiente. In: Oliveira Jr., Constan-
      tin RSJ. (Eds.) Plantas daninhas e seu manejo. Guaíba: Agropecuária; (2001). ,
      315-362.

[11] Alves PATComportamento dos herbicidas ametrina e glifosato aplicados em associa-
      ção em solo de cultivo de cana-de açúcar PhD Thesis. University of São Paulo;
      (2012).

[12] White, P. M, Potter, T. L, & Culbreath, A. K. Fungicide dissipation and impact on me-
      tolachlor aerobic soil degradation and soil microbial dynamics. Science of the Total
      Environment (2010). , 408(6), 1393-1402.

[13] Ke-bin, L. I, Cheng, J, Wang, X, Zhou, Y, & Liu, W. Degradation of herbicides atra-
      zine and bentazone applied alone and in combination in soils. Pedosphere (2008). ,
      18(2), 265-272.

[14] Krutz, L. J, Senseman, S. A, & Haney, R. L. Effect of Roundup Ultra on atrazine deg-
      radation in soil. Biology and Fertility of Soils (2003). , 38(2), 115-118.

[15] Haney, R. L, Senseman, S. A, Krutz, L. J, & Hons, F. M. Soil carbon and nitrogen min-
      eralization as affected by atrazine and glyphosate. Biology and Fertility of Soils
      (2002). , 35(1), 35-40.

[16] Zablotowicz, R. M, Krutz, L. J, Weaver, M. A, Accinelli, C, & Reddy, K. N. Glufosi-
      nate and ammonium sulfate inhibit atrazine degradation in adapted soils. Biology
      and Fertility of Soils (2008). , 45(1), 19-26.

[17] Lancaster, S. H, Haney, R. L, Senseman, S. A, Kenerley, C. M, & Hons, F. M. Microbi-
      al degradation of Fluometuron is influenced by Roundup Weather MAX. Journal of
      Agricultural and Food Chemistry (2008). , 56(18), 8588-8593.

[18] Tejada, M. Evolution of soil biological properties after addition of glyphosate, diflufe-
      nican and glyphosate + diflufenican herbicides. Chemosphere (2009). , 76(3), 365-373.

[19] Pereira, J. L, Picanço, M. C, Silva, A. A, & Santos, E. A. Tomé HVV, Olarte JB. Effects
      of glyphosate and endosulfan on soil microorganisms in soybean crop. Planta Dani-
      nha (2008). , 26(4), 825-830.

[20] Owen MDKCurrent use of transgenic herbicide-resistant soybean and corn in the
      USA. Crop Protection (2000).

[21] Bonfleur, E. J, Lavorenti, A, & Tornisielo, V. L. Mineralization and degradation of
      glyphosate and atrazine applied in combination in a Brazilian oxisol. Journal of Envi-
      ronmental Science and Health: Part B-Pesticides Food Contaminants and Agricultur-
      al Wastes (2011). , 46(1), 69-75.

[22] Fogg, P. Boxall ABA. Degradation of Pesticides in Biobeds: The Effect of Concentra-
      tion and Pesticide Mixtures. Journal of Agricultural and Food Chemistry (2003). ,
      51(18), 5344-5349.

[23] Swarcewicz, M. K, & Gregorczyk, A. J. The effects of pesticide mixtures on degrada-
      tion of pendimethalin in soils. Environmental Monitoring and Assessment (2012). ,
      184(5), 3077-3084.

[24] Cabrera, D, Lopez-pineiro, A, Albarran, A, & Pena, D. Direct and residual effects on
      diuron behaviour and persistence following two-phase olive mill waste addition to
      soil. Geoderma (2010).

[25] Kucharski, M, & Sadowski, J. Influence of adjuvants on behavior of phenmedipham
      in plant and soil environment. Polish Journal of Agronomy (2009). , 1(1), 32-36.

[26] Rodríguez-cruz, M. S, Sánchez-martín, M. J, Andrades, M. S, & Sánchez-camazano,
      M. Retention of pesticides in soil columns modified in situ and ex situ with a cationic
      surfactant. Science of the Total Environment (2007).

[27] [27]Kucharski, M, Sadowski, J, Wujek, B, & Trajdos, J. Influence of adjuvants addition
      on lenacil residues in plant and soil. Polish Journal of Agronomy (2011). , 5(5), 39-42.

[28] Reddy, K. N, & Singh, M. Effect of Acrylic Polymer Adjuvants on Leaching of Bro-
      macil, Diuron, Norflurazon, and Simazine in Soil Columns. Bulletin of Environmen-
      tal Contamination and Toxicology (1993). , 50(3), 449-457.

[29] Botelho, R. G, Cury, J. P, Tornisielo, V. L, & Santos, J. B. Herbicides and the Aquatic
      Environment. In: Mohammed N A El-G H. (Ed.) Herbicides- Properties, Synthesis
      and Control of Weeds. Rijeka: InTech; (2012). , 149-164.

[30] Gilliom, R. J, Barbash, J. E, Crawford, C. G, Hamilton, P. A, Martin, J. D, Nakagaki,
      N, Nowell, L. H, Scott, J. C, Stackelberg, P. E, Thelin, G. P, & Wolock, D. M. (2006).
      The quality of our nation's waters. Pesticides in the nation's streams and ground wa-
      ter, US Geological Survey, Reston, VA, 1992-2001.

[31] Armas, E. D. Monteiro RTR, Antunes PM, Santos MAPF, Camargo PB, Abakerli RB.
      Diagnóstico espaço-temporal da ocorrência de herbicidas nas águas superficiais e
      sedimentos do rio corumbataı' e principais afluentes. Química Nova (2007). , 30(5),
      1119-1127.

[32] Dores EFGCCarbo L, Ribeiro ML, De-Lamonica-Freire EM. Pesticide Levels in
      Ground and Surface Waters of Primavera do Leste Region, Mato Grosso, Brazil. Jour-
      nal of Chromatographic Science (2008). , 46(7), 585-590.

[33] Caldas, E. D, & Coelho, R. Souza LCKR. Organochlorine pesticides in water, sedi-
      ment and fish of Paranoá Lake of Brasilia, Brazil. Bulletin of Environmental Contami-
      nation and Toxicology (1999). , 62(2), 199-206.

[34] Lanchote, V. L, Bonato, P. S, & Cerdeira, A. L. Santos NAG, Carvalho D, Gomes MA. HPLC screening and GC-MS confirmation of triazine herbicides residues in drinking water from sugar cane area in Brazil. Water, Air, Soil Pollution (2000).

[35] Filizola, F. F, & Ferracini, V. L. Sans LMA, Gomes MAF, Ferreira CJA. Monitoramento e avaliação do risco de contaminação por pesticidas em água superficial e subterrânea na região de Guairá. Pesquisa Agropecuária Brasileira (2002). , 37(5), 659-667.

[36] Laabs, V, Amelung, W, Pinto, A. A, Wantzen, M, Silva, C. J, & Zech, W. Pesticides in Surface Water, Sediment, and Rainfall of the Northeastern Pantanal Basin, Brazil. Journal of Environmental Quality (2002). , 31(5), 1636-1648.

[37] Dores EFGCNavickiene S, Cunha MLF, Carbo L, Ribeiro ML, De-Lamonica-Freire EM. Multiresidue determination of herbicides in environmental waters from "Primavera do Leste" region (Middle West of Brazil) by SPE-GC-NPD. Journal of Brazilian Chemical Society (2006). , 17(5), 866-873.

[38] Jacomini, A. E, Camargo, P. B, & Bonato, P. S. Determination of ametryn in river water, river sediment and bivalve mussels by liquid chromatography-tandem mass spectrometry. Journal Brazilian Chemical Society (2009). , 20(1), 107-116.

[39] Benvenuto, F, Marín, J. M, Sancho, J. V, Canobbio, S, Mezzanotte, V, & Hernández, F. Simultaneous determination of triazines and their main transformation products in surface and urban wastewater by ultra-high-pressure liquid chromatography-tandem mass spectrometry. Analytical and Bioanalytical Chemistry (2010). , 397(7), 2791-2805.

[40] Yu, Z. G, Qin, Z, Ji, H. R, Du, X, Chen, Y. H, Pan, P, Wang, H, & Liu, Y. Y. Application of SPE Using Multi-Walled Carbon Nanotubes as Adsorbent and Rapid Resolution LC-MS-MS for the Simultaneous Determination of 11 Triazine Herbicides Residues in River Water. Chromatographia (2010).

[41] Ma, W. T, Fu, K. K, Cai, Z, & Jiang, G. B. Gas chromatography/mass spectrometry applied for the analysis of triazine herbicides in environmental waters. Chemosphere (2003). , 52(9), 1627-1632.

[42] Palma, G, Sánchez, A, Olave, Y, Encina, F, Palma, R, & Barra, R. Pesticide levels in surface waters in an agricultural-forestry basin in Southern Chile. Chemosphere (2004). , 57(8), 763-70.

[43] Balinova, A. M, & Mondesky, M. Pesticide contamination of ground and surface water in Bulgarian Danube plain. Journal of Environmental Science and Health, Part B. Pesticides, Food Contaminants, and Agricultural Wastes (1999). , 34(1), 33-46.

[44] Segura, P. A, Mcleod, S. L, Loemoine, P, Sauvé, S, & Gagnon, C. Quantification of carbamazepine and atrazine and screening of suspect organic contaminants in surface and drinking waters. Chemosphere (2011 8). , 2011(84), 8-1085.

[45] Truhaut, R. Ecotoxicology: objectives, principles and perspectives. Ecotoxicology and Environmental Safety (1977). , 1(2), 151-173.

[46] Plaa, G. L. Present status: toxic substances in the environment. Canadian Journal of physiology and Pharmacology (1982). , 60(7), 1010-1016.

[47] Lombardi, J. V. Fundamentos de toxicologia aquatica. In: Ranzani-Paiva MJT, Takemota RM, Lizama MAP. (Eds.) Sanidade de organismos aquáticos. São Paulo: Varela; (2004). , 263-272.

[48] Marschner, A. Biologische Bodensanierung und ihre Erfolgskontrolle durch Biomonitoring. In: Oehlmann J., Markert B. (Eds.) Okotoxikologie-Okosystemare Ansatze und Methoden. Ecomed: Landsberg; (1999). , 568-576.

[49] Magalhães, D. P. Ferrão Filho AS. A ecotoxicologia como ferramenta no biomonitoramento de ecossistemas aquáticos. Oecologia Autralis (2008). , 12(3), 355-381.

[50] Botelho, R. G, Santos, J. B, Oliveira, T. A, & Braga, R. R. Byrro ECM. Toxicidade aguda de herbicidas a tilápia (Oreochromis niloticus). Planta Daninha (2009). , 27(3), 621-626.

[51] Novelli, A, Vieira, B. H, Cordeiro, D, Cappelini, L. T, Vieira, E. M, & Espíndola, E. L. *Lethal effects of abamectin on the aquatic organisms Daphnia similis, Chironomus xanthus and Danio rerio. Chemosphere* (2012 8). , 2012(86), 1-36.

[52] Botelho, R. G, Inafuku, M. M, & Maranho, L. A. Machado Neto L, Olinda RA, Dias CT, Tornisielo VL. Toxicidade aguda e crônica do extrato de nim (Azadirachta indica) para Ceriodaphnia dubia. Pesticidas: revista de ecotoxicologia e meio ambiente (2010). , 20(1), 29-34.

[53] Poleksic, V, & Mitrovic-tutundzic, V. Fish gills as a monitor of sublethal and hronic effects of pollution. In: Muller R., Lloyd R. (Eds.) Sublethal and Chronic Effects of Pollutants on Freshwater Fish. Oxford; (1994). , 339-352.

[54] Camargo MMPFernandes MN, Martinez CBR. How aluminium exposure promotes osmoregulatory disturbances in the neotropical freshwater fish Prochilus lineatus. Aquatic Toxicology (2009). , 94(1), 40-46.

[55] Botelho, R. G, Santos, J. B, Fernandes, K. M, & Neves, C. A. Effects of atrazine and picloram on grass carp: acute toxicity and histological assessment. Toxicological and Environmental Chemistry (2011). , 94(1), 121-127.

[56] Paulino, M. G. Souza NES, Fernandes MN. Subchronic exposure to atrazine induces biochemical and histopathological changes in the gills of a Neotropical freshwater fish, Prochilodus lineatus. Ecotoxicology and Environmental Safety (2012). , 80(1), 6-13.

[57] Brausch, J. M, & Salice, C. J. Effects of an environmentally realistic pesticide mixture on Daphnia magna exposed for two generations. Archives of Environmental Contamination Toxicology (2011). , 2011(61), 2-272.

[58] Botelho, R. G. Machado Neto L, Olinda RA, Dias CTS, Tornisielo VL. Water Quality Assessment in Piracicamirim Creek Upstream and Downstream a Sugar and Ethanol Industry Through Toxicity Tests With Cladocerans. Brazilian Archives of Biology and Technology (2012). , 55(4), 631-636.

[59] Manar, R, Vasseur, P, & Bessi, H. Chronic toxicity of chlordane to *Daphnia magna* and *Ceriodaphnia dubia*: a comparative study. *Environmental Toxicology* (2012 2). , 2012(27), 2-90.

[60] Cooney, J, D, Freshwater tests. In: Rand GM. (ed.) Fundamentals of Aquatic Toxicology: Effects, Environmental Fate And Risk Assessment. New York: CRC; (1995). , 71-102.

[61] Walker, C. H, Hopkin, S. P, Sibly, R. M, & Peakall, D. B. Principles of ecotoxicology. Taylor and Francis; (2001).

[62] Mccarty, L. S, & Borgert, C. J. Review of the toxicity of chemical mixtures containing at least one organochlorine. Regulatory Toxicology and Pharmacology (2006). , 45(2), 104-118.

[63] Bliss, C. I. The toxicity of poisons applied jointly. Annals of Applied Biology (1939). , 26(3), 585-615.

[64] Hermens, J, Canton, H, Steyger, N, & Wegman, R. Joint effects of a mixture of 14 chemicals on mortality and inhibition of reproduction of Daphnia magna. Aquatic Toxicology (1984). , 5(4), 315-322.

[65] Strmac, M, & Braunbeck, T. Cytological and Biochemical Effects of a Mixture of 20 Pollutants on Isolated Rainbow Trout (*Oncorhynchus mykiss*) Hepatocytes. Ecotoxicology and Environmental Safety (2002). , 53(2), 293-304.

[66] Delorenzo, M. E, & Serrano, L. Individual and Mixture Toxicity of Three Pesticides; Atrazine, Chlorpyrifos, and Chlorothalonil to the Marine Phytoplankton Species *Dunaliella tertiolecta*. Journal of Environmental Science and Health Part B- Pesticides, Food Contaminants, and Agricultural Wastes (2003). , 38(5), 529-538.

[67] Choung, C. B, Hyne, R. V, Stevens, M. M, & Hose, G. C. Toxicity of the Insecticide Terbufos, its Oxidation Metabolites, and the Herbicide Atrazine in Binary Mixtures to *Ceriodaphnia dubia*. Archives of Environmental Contamination and Toxicology (2011). , 60(3), 417-425.

[68] Organização Das Nações Unidas- ONUDeclaração sobre o meio ambiente. Estocolmo, (1972).

Chapter 4

Characterization, Modes of Action and Effects of Trifluralin: A Review

Thaís C. C. Fernandes, Marcos A. Pizano and
Maria A. Marin-Morales

Additional information is available at the end of the chapter

1. Introduction

The use of chemicals to control human diseases, plagues and weeds in agriculture started in the late 19th century, but only after the Second World War did this practice follow rather scientific criteria [1]. According to targets against which they are designated, the chemicals used in agriculture are called insecticides, fungicides, herbicides, nematicides, among others [2].

All pesticides have the common priority of stopping a metabolic process essential to undesirable organisms, for which they are toxic. These chemicals act directly upon the organisms, eliminating or controlling them, such as interfering in their reproductive process [3].

Among agricultural pesticides, herbicides comprise the most employed group in agriculture. The main function of these chemicals is to control weeds, weed competition reduces productivity, without significantly impacting crop yield. Weeds tend to compete with crops by extracting essential elements from the soil, water, intercepting light and CO_2, interfering in the culture development and affecting agricultural production practices including harvest [4]. Herbicides are also used for eliminating plants from both road, railways, and riversides [3].

The mechanism of action of some herbicides on organisms is not completely understood [5]. Lack of detailed information about the action of herbicides on the biological environment may cause damage to human health [1], [6] and [7].

Herbicides may be classified according to different criteria related to their properties, characteristics, use, efficiency, permanence in the environment and mechanism of action. As for their chemical features, herbicides may be classified as carbamates, amides, diphenyl ethers, amino phosphates, and dinitroanilines, among others [8].

Classification of herbicides based on their mechanism of action has changed over time, both according to the discovery of new herbicides and the elucidation of site of action of the herbicide on plants. The internationally accepted classification is the one proposed by the Herbicide Resistance Action Committee (HRAC). In it, the herbicides are classified in alphabetical order in accordance with their sites of action and chemical classes (Table 1). Herbicides having unknown site of action are grouped under Z until identification. The (numeric) Weed Science Society of America (WSSA) classification system is also listed in Table 1 [5].

HRAC	SITES OF ACTION	CHEMICAL GROUP	WSSA
A	Inhibition of Acetyl-CoA Carboxylase (ACCase)	Aryloxyphenoxypropionates (FOPs)	1
		Ciclohexanodiones (DIMs)	1
		Phenylpyrazolones (DENs)	1
B	Inhibition of Acetolactate Synthase (ALS) (or acetohydrxy acid synthase AHAS)	Sulfonylureas	2
		Imidazolinones	2
		Triazolopyrimidines	2
		Pirimidinil(tio)benzoates	2
		Sulfonylaminocarbonyl-triazolinones	2
C1	Inhibition of Phtosynthesis in photosystem II	Triazines	5
		Triazinones	5
		Triazolinones	5
		Uracils	5
		Pyridazinone	5
		Phenyl Carbamates	5
C2	Inhibition of Phtosynthesis in photosystem II	Ureas	7
		Amides	7
C3	Inhibition of Phtosynthesis in photosystem II	Nitriles	6
		Benzotiadiazinones	6
		Phenyl-pyridazines	6
D	Inhibition of Phtosynthesis in photosystem I	Bipiridiliuns	22
E	Inhibition of Protoporphyrinogen Oxidase (PPO)	Diphenyl ethers	14
		Phenylpyrazoles	14
		N-phenylftalimidas	14
		Thiadiazoles	14
		Oxadiazoles	14
		Triazolinones	14
		Oxazolidinediones	14
		Pyirimidinediones	14
		Others	14
F1	Inhibition of carotenoid biosynthesis in naphytoenedesaturase (PDS)	Pyridazinones	12
		Pyridine Carboxamides	12
		Others	12
F2	Inhibition of carotenoid biosynthesis in 4-hydroxyphenyl-pyruvate-dioxygenase (4HPPD)	Triacetones	27
		Isoxazoles	27
		Pyrazoles	27

HRAC	SITES OF ACTION	CHEMICAL GROUP	WSSA
		Others	27
F3	Inhibition of carotenoid biosynthesis (unknown target)	Triazoles	11
		Isoxazolidinones	13
		Diphenyl ethers	11
G	Inhibition of EPSP synthase	Glycines	9
H	Inhibition of glutamine synthase	Phosphinic acid	10
I	Inhibition of DHP (dihydropteroate synthase)	Carbamates	18
K1	Inhibition of microtubule assembly	Dinitroanilines	3
		Phosphoramidates	53
		Pyridines	3
		Benzamides	3
		Benzoic acid	3
K2	Inhibition of mitosis	Carbamates	23
K3	Inhibition of cell cycle	Chloroacetamides	15
		Acetamides	15
		Tetrazolinones	15
		Others	15
L	Inhibition of cell wall (cellulose) synthesis	Nitriles	20
		Benzamides	21
		Triazolocarboxamides	27
		Quinolinocarboxylic acid	26/27
M	Decouplers (cell membrane disruptors)	Dinitrophenols	24
N	Inhibition of lipid synthesis (different from ACCase inhibitors)	Tiocarbamates	8
		Phosphoroditioates	8
		Benzofurans	16
		Chlorocarbonic acid	26
P	Auxin mimics	Phenoxicarboxylic acid	4
		Benzoic acid	4
		Pyridinecarboxylic acid	4
		Quinolinocarboxylic acid	4
		Others	4
Q	Auxin transport inhibitors	Ftalamates	19
		Semicarbazones	19
R	...	...	
S	...	...	
.	...	...	
Z	Unknown	Arylamino Propionic acid	25
		Pirazoliuns	26
		Organoarsenicals	17
		Others	

WSSA. Weed Science Society of America; **HRAC.** Herbicide Resistance Action Committee.

Table 1. Herbicide Classification in accordance with their mechanism of action.

2. Trifluralin identification and characterisitcs

Trifluralin belongs to the dinitroaniline group which has the aniline structure as a basis, containing NO_2 molecules at 2 and 6 or 3 and 5 positions of the benzene ring. This group has more than ten different herbicides, among which are trifluralin, dinitramine, oryzalin and pendimethalin [8].

Trifluralin has been used in agriculture since 1963 [9]. This herbicide is registered separately or in mixtures, and used in the following crops: *Glycine max, citrus, Coffea arábica* under formation, *Gossypium hirsutum, Arachis hypogaea, Phaseolus vulgaris, Allium sativum, Ricinus communis, Manihot esculenta, Helianthus annuus, Solanum melongena, Daucus carota, Abelmoschus esculentus, Brassica oleracea, Brassica oleracea capitata, Brassica oleracea botrytis, Capsicum annuum, Lycopersicon esculentum,* and ornamental plants [10].

Trifluralin is available either in emulsifiable concentrate or in crystalline solid both formulations of the yellow-orange color. It is not quite soluble in water (0.3 to 0.6 mg/L solubility at 25°C) [9], it is mildly volatile (1.1. 10^{-4} mmHg pressure vapor at 25°C), its density is 1.36 g/cm³ at 22°C, it is considered alkaline and long-lasting in the environment (120-240 days) [8]. Trifluralin has a high affinity to soil [11], is relatively immobile and has a half-life of 3 to 18 weeks, depending on the soil and the geographical location [12].

Trifluralin chemical composition is α,α,α–trifluoro–2–6–dinitro–N–N– dipropyl–p–toluidine [13]. The chemical structure formula is shown in Figure 1.

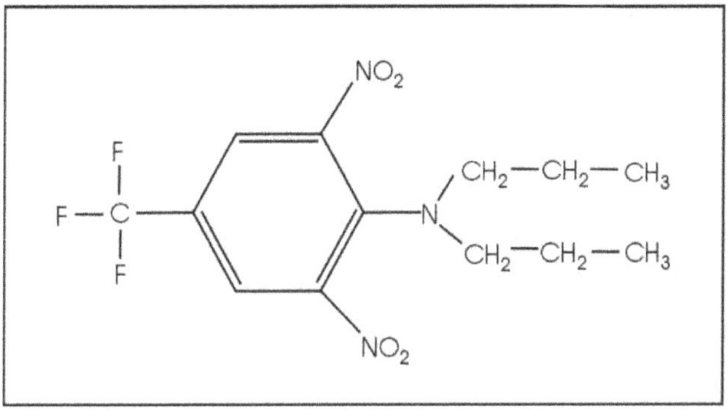

Figure 1. Trifluralin chemical structure formule.

Trifluralin commercial products contain nitrosodipropylamine, a carcinogenic contaminant (NDPA) [14]. This compound reacts with 0^6-guanine DNA and may cause mutation [15]. On account of concerns about this characteristic, the *Environmental Protection Agency* (EPA) demanded that industries make sure products containing trifluralin active principle had nitrosodipropylamine 0.5 ppm concentrations at the most [14].

USEPA (1999) [16] classifies trifluralin as group C: possibly carcinogenic to humans, based on
evidences with animals, not with humans.

3. Trifluralin behavior in the environment

3.1. Behavior in soil

Trifluralin is strongly adsorbed by organic matter colloids and not much by clay ones. In
organic matter rich soils, adsorption prevents absorption of the product by plant roots.
Therefore, the use of this herbicide under such conditions is not advisable [10]. Leaching, as
well as soil lateral movement is quite reduced compared to some pesticides [17]. Its main
characteristic is soil persistence resulting from low mobility, which can cause damage to crops
following its application [12].

Such herbicides as trifluralin, applied in pre-emergence, act better when soil humidity is
between high and elevated. Therefore, the herbicide may at least be partially solubilized and
distributed in the first layers of the soil surface, which will protect it from losses [8].

This herbicide degradation in soil occurs through chemical, microbial pathways and photolysis. Chemical degradation promotes dealkylation of the amino group, reduction from the
nitro to the amino group, partial oxidation from the trifluoromethyl to the carboxyl group and,
subsequently, degradation into smaller fragments (Figure 2).

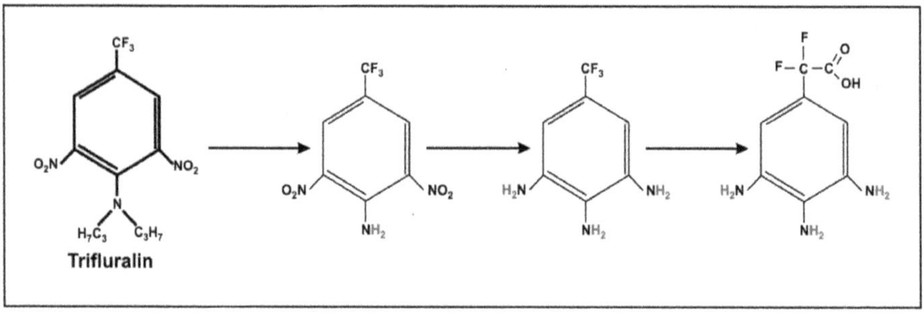

Figure 2. Possible sequence of events that occur during trifluralin chemical degradation.

Microbial degradation may occur under aerobic and anaerobic conditions (Figure 3). However,
it is observed that degradation occurs mainly under anaerobic conditions, as the ones observed
in poorly drained soils, when there is subsequent rainfall. Under anaerobic conditions, within
the same time period, 98% of trifluralin degrades, whereas under aerobic conditions only 25%
of the product decomposes. Among the fungi capable of decomposing trifluralin are *Sclero-
tiumrolfsii, Aspergillusniger, Fusarium*sp and *Tricoderma*sp [10]. According to Carter and Camper
[18], trifluralin may also be degraded by *Pseudomonas* sp.

Trifluralin is also sensitive to degradation by ultraviolet rays, and its volatility is one of the main factors of product loss in the soil as well [19] and [20]. Trifluralin photodecomposition generally involves three processes: propylamine oxidative dealkylation, cyclization and nitro group reduction (Figure 4) [21].

The first product of trifluralin photolysis, according to Dimou et al. [21] and illustrated in Figure 3, seems to be a mono-dealkylate deriving from the main compound, originating compound 1. Dealkylation is attributed to the free radical oxidation. Another intermediate of photodegradation appears to be formed by cyclization reactions. The compounds 4 and 5 are apparently formed by reaction among trifluralinpropylamine α carbon and the NO₂ group of compound 1, ant they are identified as 2– ethyl -7nitro-1-propyl-5 (trifluoromethyl)-1H-benzimidazole and 2-ethyl-4 nitro-6- (trifluoromethyl)-1H-enzimidazole, respectively. The benzimidazoledealkylate (compound 4) is the most stable photoproduct, which can last in the environment longer, making its detection possible. This product may be formed by the reaction of compound 5 dealkylation.

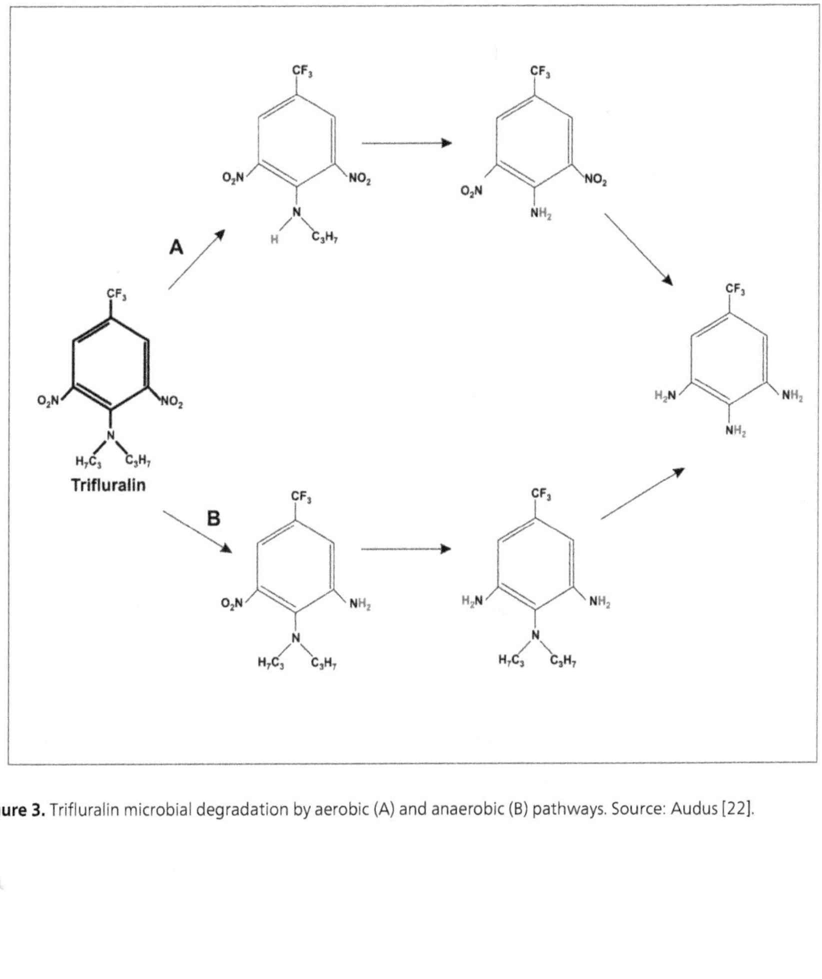

Figure 3. Trifluralin microbial degradation by aerobic (A) and anaerobic (B) pathways. Source: Audus [22].

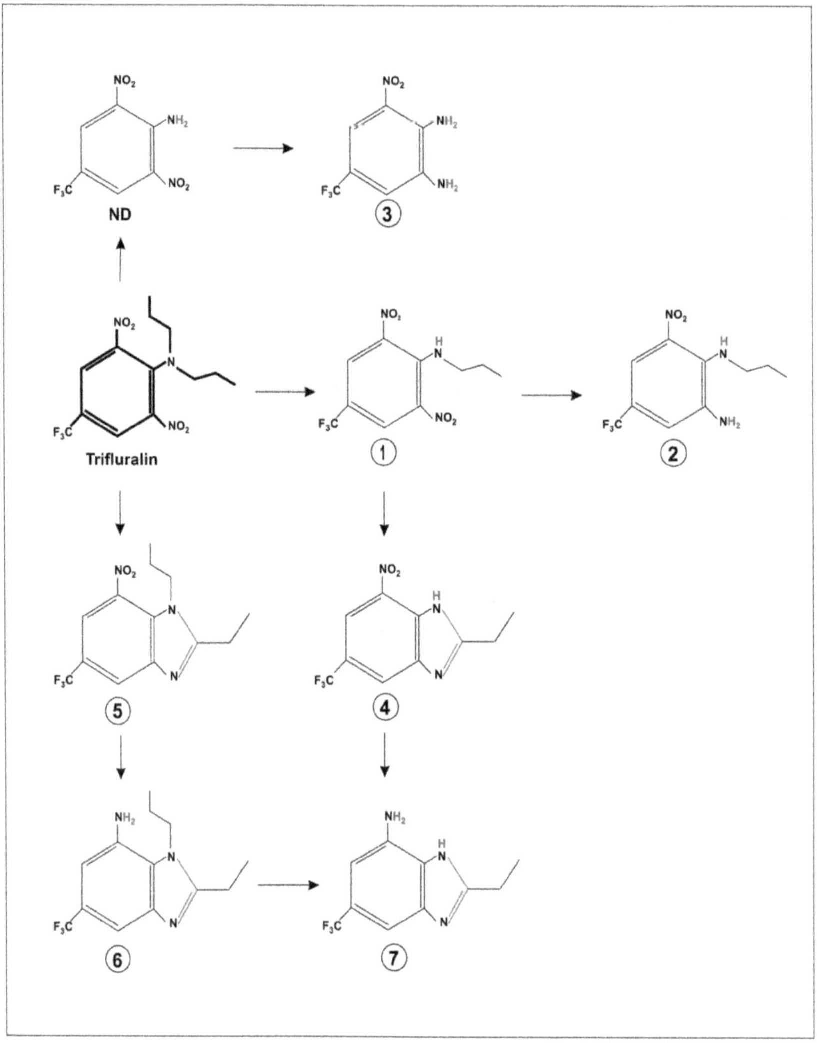

Figure 4. Trifluralin photodegradation. *ND = Substance not detected in the source. Modified scheme by Dimou et al. [21].

Compounds 4 and 5 can be reduced in water by not so clear mechanisms [23], straight from the aryl hydroxylamine formation [24] to form compound 7 and 6, respectively. According to the same author these products have also been formed during trifluralin chemical degradation. Compound 2 and 3 are formed from NO_2 to NH_2 group reduction of compound 1 and 2,6-dinitro-4- (trifluoromethyl) benzenamine (compound ND), respectively. These compounds have also been identified during trifluralin chemical degradation [24], showing that this pathway also happens in other processes, besides photodegradation [21].

Trifluralin average persistence in soil for the recommended doses under field conditions is of 1.8 ppm residue after 180 days following application [25]. However, according to the same author, this persistence may vary in accordance with the kind of soil and climatic conditions.

3.2. Herbicide behavior in water

Water contamination with trifluralin may occur by sediment leaching while equipment is being cleaned, or due to accidental spills. Nevertheless, only 0.5% of the quantity applied to the soil in field conditions is leached and may consequently contaminate water sources. This percentage means a rather low water contamination, representing smaller concentrations than 1.0 μg L^{-1}. As a consequence, trifluralin is not commonly detected in surface water [9] and [26].

While Zimmerman et al. [26], Dayama and Coupe [27], Thurman et al. [28] and were carrying out analyses in the Mississippi River, they detected extremely low levels of trifluralin (lower than 0.1 g/L). Once this herbicide is widely used, the authors ascertain that low concentrations of it detected in surface water may be attributed to its low mobility in soil and low solubility in water (lower than 1 mg/L). USEPA [29] and the European Community legislation [30] established limits of $2\mu g/$ L and $0.1\mu g/$ Ltrifluralin in drinking water, respectively. According to Dimou et al. [21], trifluralin degradation in water is influenced by the presence of nitrate ions, which accelerate photolysis reaction. Products derived from this reaction have either low or no toxicity, when compared to the whole product.

3.3. Herbicide behavior in the air

Grover et al. [31] ascertain that trifluralin is quickly dissipated in the atmosphere. Depending on the season of the year, about 25% of the product applied is volatilized, but only 2-3 $\mu g/m^3$ at the most of trifluralin is found in the air, soon after its application, to less than 100ng/m^3 a few hours later [32]. According to the United States Environment Protection Agency (1993) [33], an average 0.27 ng/m^3 concentration of herbicide, varying from 0 to 3.4 ng/m^3, was found in the Canadian atmosphere between 1988 and 1989.

Mongar and Miller [34] state that low concentrations of this herbicide found in the atmosphere are due to both trifluralin quick reaction with the hydroxyl radical (OH) and the photolysis reaction, which promotes the product degradation.Nonetheless, Waite et al. [32] verified that of the five most used herbicides on the Canadian prairies, trifluralin was the most frequently found in the air (79% of samples).

3.4. Herbicide behavior in plants

Trifluralin is a pre-emergence herbicide which must be incorporated into the soil and applied soon after sowing, when the plant seeds are beginning the germination process [36]. The herbicide absorption occurs mainly by the hypocotyl, then by the seedling radicles, at the beginning of germination [10].

Trifluralin's main mechanism of action is the inhibition of cell mitosis. This herbicide typically acts on the meristems and tissues of underground organs, such as roots, epicotyls, hypocotyls, plumules, rhizomes, bulbs and seeds [8].

The inhibition of radicle development by trifluralin action, both on main root growth and the emission of secondary roots, is quite evident in some dicotyledons. Thickening of the hypocotyls also commonly occurs [8], as well as swollen root tips [36]. According to Almeida [25], trifluralin induces several biochemical changes in higher plants, including alterations of carbohydrate, lipid, nitrogen concentrations and, especially, nucleic acid alterations. Therefore, the product affects cell division in meristematic tissues, thus inhibiting seed germination and the formation of new radicle and hypocotyl cells.

Bayer et al. [37] report that trifluralin promotes a decrease in the zone of meristematic tissues and the interruption of mitosis in the roots of wheat, cotton and onions. The onion cells treated with trifluralin showed to be small, dense and multinucleated, abnormal, weak and aberrant [38].Studies conducted by Fernandes [39] using *Allium cepa* showed that the toxicity of trifluralin residual concentrations might induce changes in that plant. The author observed that the herbicide promoted plant growth inhibition, higher turgidity, weakness and thickness of the roots, in relation to the control treatment.

Plants grown in soils treated with trifluralin exhibited residues on the roots only. No residue was found on the leaves, fruit and seeds [25]. These results indicate that trifluralin is not transported by sap into other plat tissues.

4. Trifluralin mechanisms of action

Plant growth and development depend on mitosis in their meristematic regions. Cell division is a process that requires different cell organelles, structures and the products of many genes to be working correctly. Dinitroanilines, the family to which trifluralin, phosphoride amides and N-phenyl carbonates belong, are microtubule-depolymerizing chemical compounds [5], [40], [41], [42] and[43]. According to Senseman [36], the herbicide-trifluralin complex inhibits microtubule polymerization, leading to physical misconfiguration and loss of function. As a consequence, the mitotic spindle does not form, causing misalignment and chromosome separation during mitosis. In addition to that, the so-called spindle apparatus is not formed.

Microtubules are subcellular structure filaments, basically made up of heterodimeric tubulin protein (Figure 5A) [44]. They have important cellular functions, which are directly related to mitosis and indirectly related to organism development. These structures are involved in several cellular processes such as chromosome migration, cellular structure maintenance, cellulose microfibril orientation and organization, cell wall formation, intracellular movement, as well as cellular differentiation [42] and [45]. Most sets of cell microtubules are labile and their functions depend on this lability.The mitotic spindle is one of the most extraordinary examples, whose formation is brought about after disorganization of cytoplasmic microtubule at the beginning of mitosis. For this reason, the mitotic spindle is targeted by various specific anti-mitotic drugs, which interfere in the exchange of tubulin subunits between the microtu-bules and the pool of free tubulins [46].

In-vitro analyses of *Chlamydomonas reinhardii* showed that trifluralin specifically binds tubulins, demonstrating that it is the first subcellular target of dinitroaniline action [47]. Trifluralinsub-micromolar concentrations totally blocked cytokinesis and inhibit nuclear division in *Toxo-plasma gondii* by interfering in intracellular spindle and in other cytoskeletal components [48].

According to Anthony and Hussey [47], the herbicide-tubulin complex is related to the suppression of microtubule growth. With minus-end specific microtubule depolymerization, the tubules progressively start to get shorter, eventually leading to total loss of microtubule (Figure 5B). The author still states that cortical microtubules are among the most resistant to trifluralin action and microtubule spindles and fragments are among the most sensitive to the herbicide action.

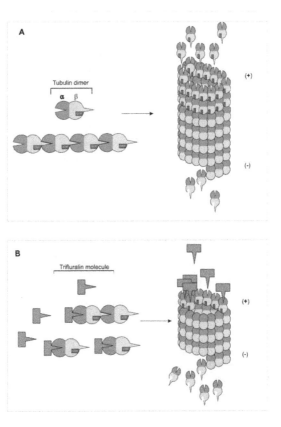

Figure 5. A. tubulin dimers forming the microtubule; **B.** herbicide-tubulin complex preventing microtubule polymerization.

Anthony et al. [49] ascertained that, as a rule, the tubulin sequence is the most preserved among the different organisms; and this preservation is related to the basic functions of microtubules. Mahresh and Larry [50], however, believe that, depending on the organism, dinitroaniline herbicides have different affinities to tubulins, since they do not interact with vertebrate tubulins, although they interact with plant and *Chlamydomonas* tubulins. This situation is reinforced with data from Anthony and Hussey [47], Baird et al. [51], Breviário and Nick [52] and Yemets and Blume [53], who ascertain that dinitroaniline herbicides are compounds with higher specificity for binding plant tubulins than to those of vertebrates.

Studies on plant resistance to dinitroanilines showed that some plant species own a natural mutation which bring about a change in base pairs, and consequently in their genetic code. One of these alterations of base causes a change in the amino acids of the tubulin protein. Threonine, a normal amino acid at position 239, is changed into isoleucine, stopping group NO_2 of the dinitroaniline herbicides from binding the tubulin molecule, thus preventing its mechanism of action (Figure 6) [47].

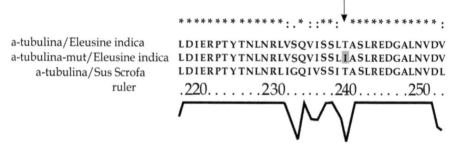

Figure 6. Alignment of amino acid sequence of α-tubulins, evidencing the position of substitution in the mutating tubulin from *Eleusine indica* (Thr 239 into Ile- represented in black and indicated with an arrow). Modified from Blume et al. [54].

From these pieces of information, it would be intuitive to hypothesize the idea that the smallest affinity of trifluralin to vertebrates should be owing to the fact they do not have the amino acid at position 239, seemingly the herbicide target site. Nevertheless, it can be seen in Figure 7 that the threonine amino acid at position 239 of the α-tubulin protein is present in plants, parasites and vertebrates, including man.

However, Hashim et al. [58] found mutations in the α-tubulin gene expression which changed the amino acid synthesis at a different position than that found by Anthony and Hussey [47]. According to Hashim et al. [58], *Alopecurus aequalis* plants that underwent mutations, which altered the amino acid synthesis at positions 202, 136 and 125 of the α-tubulin, also brought about resistance to trifluralin.

Sree et al. [59], Hansen et al. [60] and Vidakovié-Cifrek et al. [61], ascertain that trifluralin can inhibit microtubule polymerization by binding tubulin. However, it can also cause changes in

the ion calcium concentration in cytoplasm and influence polymerization and depolymerization regulation of microtubules. According to Hertel et al. [62], changes in the quantity of free
Ca^{2+} in cytoplasm, due to trifluralin action, can alter calcium-dependant biochemical and
physiological processes, in addition to causing problems to microtubules, either in animals or
in plants. Vidakovié-Cifrek et al. [61] report that trifluralin may increase the concentration of
Ca^{2+} ions in cytoplasm, influencing onion root mitosis.

Due to trifluralin chemical structure, this herbicide tends to receive two electrons, which
significantly increases its toxicity, since the group NH_2 hydrogen of trifluralin tends to bind
the polar group of cellular membranes and cause disorganization to its structure, eventually
bringing function disorders [63]. This disorganization in the membrane structure seems to
interfere mainly in the permeability of plasma and mitochondrial membranes. Trifluralin
changes the permeability of membranes because it promotes a collapse in their electric
potential, making Ca^{+2} efflux of the mitochondrial inner membranes and Ca^{+2} go from the outer
to the inner surface of the cell membrane via uniporters, thus increasing the concentrations of
such ions in the inner cytoplasmic membrane.

Since low levels of calcium are needed for polymerization, Hepler [64] ascertains that mitotic
spindles may undergo disorders due to the high levels of this ion. Low concentrations of free
calcium in the cytoplasm (0.1-0.2 μM) are essential to prevent phosphorus precipitation,
compete with Mg^{2+} for binding sites and act as a secondary messenger [65].

According to Alberts et al. [46], Ca^{+2} is important for regulating mitochondrial enzyme activity,
and it is imported from the cytosol through an H+ electrostatic gradient. It is also believed that
this process is important to remove Ca^{+2} from the cytosol when cytosolic Ca^{+2} levels get
dangerously high.

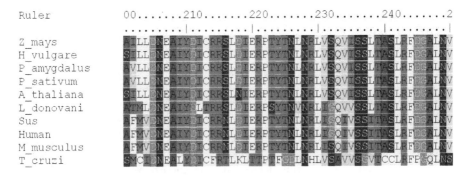

Figure 7. Comparisons among sequences of α-tubulin amino acids of species *Zea mays* (vegetable), *Hordeumvulgare*
(vegetable), *Arabidopsis thaliana* (vegetable) *Prunus amygdalus* (vegetable), *Pisum sativum* (vegetable),Leishmania
donovani (parasite), *Trypanosoma cruzi* (vegetable),*Mus musculus* (vertebrate), *Sus scrofa* (vertebrate) and *Homo sapi-
ens*. The sequences were obtained from the data base at NCBI (National Center of Biotechnology Information) in accordance with the codes P14641, Y08490, P29511, P33629, U12589, U09612, M97956, P05213, P02550 and P04687,
respectively [55]. The sequences were aligned by means of the ClustalW program [56], using default parameters. The
alignment was then analyzed using the MPALign program [57].

Another important factor to be considered is the derivate generation through pesticide biodegradation [66] and [67]. One of the byproducts of trifluralin biodegradation is an aniline: 2,6dinitroaniline (Figure 8) [68].

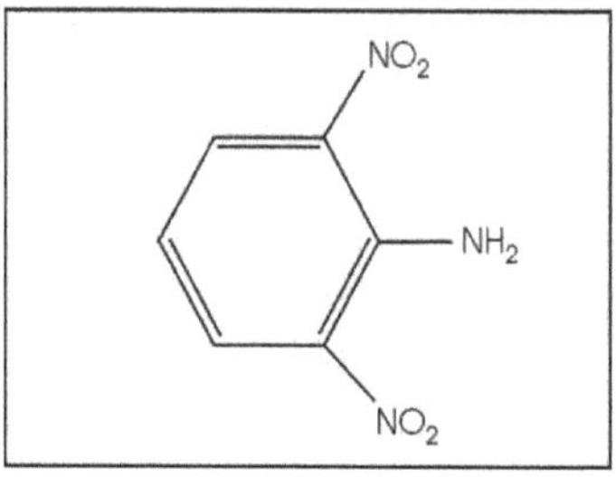

Figure 8. Chemical structure of 2,6 dinitroaniline.

Anilines are compounds that cause a variety of toxic effects depending on the structural changes they undergo. Several studies demonstrate that anilines and halogens can induce metahemoglobin formation and also be toxic to the kidneys and the liver, either treated *in vitro* or *in vivo* [69] and [70]. Aminophenols, the primary products of aniline metabolism, are compounds related to neurotoxicity induction [69].

5. Trifluralin toxic effect

Although many researchers and international governmental agencies have investigated and published trifluralin toxic effects on different fields, whether they are related to either acute or chronic toxicity, cytotoxicity, genotoxicity, mutagenicity and carcinogenicity, the results shown are confusing and often contradictory.

According to the W.H.O (World Health Organization) [70], trifluralin causes hemoglobin oxidation (by forming metahemoglobin), red blood cell destruction, besides being toxic to the kidneys and the liver, and stimulating depression in the central nervous system. It may cause vomiting, diarrhea, weakness, profuse sweating, loss of sight, memory and concentration, and dermatitis as well. This herbicide is considered to be neurotoxic and gastrointestinal irritant. It can lead to death because of ventricular fibrillation [71], although several authors [10], [72], [73], [74], [75] and [76], ascertain that trifluralin is a low toxicity substance.

Trifluralin lethal concentrations and doses for vertebrates and invertebrates are shown in Table 2.

Treatment	Species	Group	Popular Name	Toxicity
CI50 (48h)	Lepomis macrochirus	Fish	Bluegill	19 µg L⁻¹
CL50 (48h)	Mola mola	Fish	Ocean sunfish	19µg L⁻¹
CL50 (48h)	Cyprinus carpio	Fish	Common carp	1.0mg L⁻¹
CL50 (96h)	Oncorhynchus mykiss	Fish	Rainbow trout	0,21mgL⁻¹
CL50 (96h)	Oncorhynchus mykiss, Lepomis macrochirus, Mola mola	(Young)fish	Rainbow trout, Bluegill, Ocean sunfish	10-90µg L⁻¹
CL50 (48h)	Daphnia magna	Micro-crustacean	-	0,56 mgL⁻¹
CL50 (96h)	Procambarus clarkia	Crustacean	Lobster	1?mgL⁻¹
DL50 (oral)	Apis mellifera	Insect	Honey bee	0,011mg bee-1
DL50 (oral)	Mus musculus	Mammal	Laboratory mice	>500 mg kg⁻¹
DL50 (oral)	Ratus norvegicus	Mammal	Laboratory mice	> 10.000 mg kg⁻¹
DL50 (oral)	-	Mammal	Dog	> 200 mg kg⁻¹
DL50 (oral)	-	Mammal	Rabbit	> 200 mg kg⁻¹
DL50 (oral)	-	Bird	Hen	> 200 mg kg⁻¹

Data extracted from Gangolli [77].

Table 2. TrifluralinCL50 and DL50 for different organisms

Meister [78] conducted tests with animals and verified that trifluralin does not have any toxic effect on them when they are exposed to the product either through ingestion, inhalation or when in contact with the skin. Nauseas and severe gastrointestinal discomfort may occur after trifluralin ingestion. When placed in the rabbit eyes, it produced a mild irritation, which was reverted within seven days. In humans, it may induce skin allergies and, when inhaled, it may irritate the throat and the lungs.

Table 3 shows some information regarding trifluralin chronic, sub-acute and sub- chronic toxicity to different organisms.

Treatment	Species	Group	Popular Name	Toxicity	Symptoms
LOEC	Amphiprion percula	Fish	clownfish	5µg L⁻¹	-
NOEL	Amphiprion percula	Fish	clownfish	2µ L⁻¹	-
CE50 (10 days)	Chlorococcum sp	Protozoa	-	2,5 mg L⁻¹	-
Sub-acute(dermis -14 days)	Oryctolagus caniculus	Mammal	Rabbit	2mL Kg⁻¹	diarrhea and mild erythema
Sub-chronic(ingestion - 3 months)	Ratus norvegicus	Mammal	Mouse	25, 50 e 100 mg kg⁻¹ dia⁻¹	no effects produced on either survival or appearance *

*Liver weight of the animals submitted to the 50 and 100mg Kg⁻¹ diet somehow showed to be higher, when compared to the control animals. Data extracted from Gangolli [77].

Table 3. Data on trifluralin sub-acute, chronic and sub-chronic toxicity.

According to the Occupational Health Service [79], prolonged skin contact with trifluralin may cause allergic dermatitis. The WSSA [80] states that administering trifluralin to dogs while washing them for two years does not cause toxic effects. However, in trifluralin chronic assays conducted with 60 animals (F344 mice), which received 0.813, 3250 and 6500 ppm dietary does for two years, damage to their liver and kidneys were observed [81].

Worthing [71] states that trifluralin is highly toxic and neurotoxic. The author ascertains that the herbicide is capable of accumulating in the adipose tissue and inhibiting the immunologic function of the thymus. Trifluralin is regarded as possibly teratogenic and fetal toxicity.It has the property of altering the endocrine and reproductive system, and it reduces the quantity of semen, besides increasing the number of abnormal sperm.

In studies conducted by Ovidi et al. [82], they tested trifluralin concentration of 1.53 mg/ml and observed that the herbicide exerts a specific effect on the reproductive system in plants, by direct action on the formation of the pollinic tubes, since it causes complete microtubule depolymerization. The authors even suggest that pollinic microtubule cytoskeleton may be used as bioindicators for studies on toxicity induced by aneugenic agents such as trifluralin.

As a general rule, the effects of pesticides may be diversified, such as the direct reaction with nuclear DNA; incorporation of DNA during cellular replication; interference in mitosis or meiosis, resulting from incorrect cell division [83].

Genotoxic effects may lead to DNA breaks, causing loss of genetic material and mutations which lead to cell death or result in carcinogenesis. Genotoxicity is assessed by different tests, carry out with several organisms and provide safe, precise information regarding their potential to damage the DNA.There are a number of reports evaluating trifluralin for geno-toxicity, immunotoxicity, and reproductive toxicity, although the results are not entirely consistent, trifluralin does not appear to be strongly genotoxic [84].

Chromosome aberration tests have shown evidences of trifluralin mutagenicity for different plant species [85], [86], [87], [88], [89] and [90]. Könen and Çavas [91], Peña [92] and Canevari [93] ascertained that the herbicide is capable of inducing significant microtubule rates in Oreochromis niloticus. Kaya et al. [94] also ascertained that the herbicide may be considered genotoxic to Drosophila melanogaster, since it exhibited positive outcomes for the Somatic Mutation and Recombination Test (SMART). Tests conducted in the bone marrow of mice exposed to trifluralin showed that it is potentially genotoxic [95] and it is also capable of influencing serum concentration of reproductive and metabolic hormones, especially thyroxin [96]. Nonetheless, tests performed on bacteria [14], on Drosophyla melanogaster conducted by Bryant and Murnik [97] and Foureman [98], on cells taken from the bone marrow of mice conducted by Nehéz et al. [99], Pilinkaya [100], Gebel et al., [95], and on cell culture conducted by IARC [101] and Ribas et al. [35 and 102] demonstrated contradictory results. According to Chan and Fong [103], Bhattacharya et al. [104] and Esteves et al. [105], due to its characteristics, mechanisms of action and, especially its reduced effects on human cells, trifluralin can be regarded as a promising substance for fighting Leishmaniasis. There is also research that confirms the use of trifluralin as a powerful antiparasitic to treat Trypanosoma [106] and [107], Toxoplasma [48] and Plasmodium [108].

Studies carried out by Peña [92] and Canevari [93] indicate that low trifluralin concentrations may induce mutagenic effects. These authors observed significant presence of micronuclei in erythrocytes of fish submitted to acute treatments with this herbicide. When the micronuclei diameters were measured by Canevari [93], data indicated that they could be derived from losses of whole chromosomes, thus proving the aneugenic effect of the herbicide due to the pesticide interference in the mitotic spindle.

*Allium cepa*meristematic cells treated with trifluralin also presented problems during mitosis, such as polyploidies, C-metaphases, multipolar anaphases, anaphase-telophase chromatin bridges, chromosome delay and loss of genetic material [89]. (Figure 9).

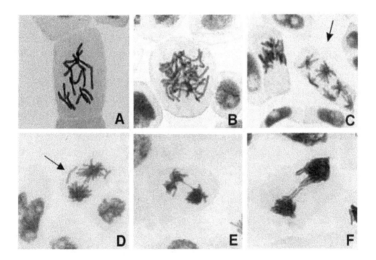

Figure 9. Meristematic cells of *Allium cepa* treated with trifluralin. **A.** C-metaphase; **B.**polyploid cell; **C.** multipolar cell; **D**. loss of genetic material; **E.**chromosome bridge; **F.**telophase with chromosome delay.

According to Fernandes et al. [88], in the bioassays with root meristems of *Allium cepa* treated with trifluralin, a large amount of interphase cells with more than one nucleus and cells with micronuclei and a mini cell were observed (Figure 10).

Lignowski and Scott [85] observed C-metaphases, micronuclei, amoeboid nuclei and poly-ploidies in root meristems of wheat and onion submitted to trifluralin action. Due to the occurrence of irregular metaphases, they concluded that the mitotic spindle might have been broken owing to the herbicide action on it.

Bioassays performed with trifluralin, using *Pisum sativum* as test material revealed a positive action of the herbicide with the increase in chromosome alterations, C-mitosis and anti-mitosis effects [87].

Fernandes et al. [89] ascertained that, among the root meristems of *Allium cepa* under division, trifluralin promotes a significant increase in the irregular metaphase rate. These data corroborate the statement of Lignowski and Scott [85], Lee et al. [109], Dow et al. [110], Werbovetz et al. [111] and Ovidi et al. [82], who characterized trifluralin as a powerful microtubule inhibitor, which is therefore capable of accumulating a large amount of meristematic cells in metaphase.

Genotoxicity tests using the comet assay in human lymphocyte cultures showed that trifluralin produced a significant increase in the length of the comet's tail. This increase is due to DNA breaks, since there is an induction of nucleotide excision repair, resulting from damage caused by the herbicide action [103]. As for the frequency of comet-bearing cells, the author observed that, after 48 hours of exposure to the herbicide, few tailed nucleoides were found. These results proved to be statistically significant, though.

According to Ribas et al. [35], trifluralin has a genotoxic effect on human cell cultures because it causes a decrease in cell proliferation. The same author ascertains that this herbicide has not revealed carcinogenic effects, since it caused little induction exchange between sister chromatids. The micronucleus test conducted by Ribas et al. [35], used for detecting aneugenic activity, has also produced a negative response, which contradicts studies carried out by several other authors [88], [89], [91], [92], [97], [112], among others) who ascertain that trifluralin brings about chromosome aberrations and nuclear alterations resulting from problems in the mitotic spindle

According to Kang et al. [113], trifluralin is not associated with bladder, kidney, liver, leukemia, colorectal or hematopoietic-lymphatic cancers. The authors only suggest a possible connection between trifluralin exposure and the risk of colon cancer in human beings, but the inconsistency per exposure level and a small number of colon cancers indicate that this could be an incidental finding.

Data from the National Cancer Institute (NCI) [114] report that mice subjected to trifluralin chronic exposure, at low concentrations, had an increase in hepatocellular carcinoma and higher incidence of alveolar bronchial adenomas. An increase in bladder cancer was also verified in mice exposed to low trifluralin concentrations. It was observed that, when male mice were submitted to high doses of trifluralin, they presented higher incidence of follicular cell and thyroid gland tumors [115]. Trifluralin has been reported to cause a significant increase

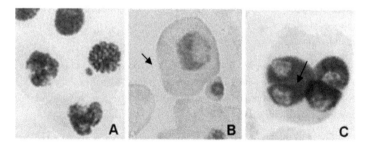

Figure 10. Meristematic cells of *Allium* cepa treated with trifluralin. **A.** cell with micronucleus; **B.** cell with micronucleus and an adjacent mini cell; **C.** polynucleated cell.

in thyroid follicular cell tumors in male Fischer 344 rats only at the highest dietary dose of 6500ppm in a 2-year chronic study [115].

6. Final considerations

The increase in agricultural productivity has occurred thanks to several factors, among which are improvements in genetics, agricultural machinery and the use of substances that allow control of weeds in agriculture.

The use of pesticides has generated discussions and controversy among the scientific community and its users, registering advantageous and disadvantageous recommendations in different ways. Among contrary recommendations to the use of pesticides, we can point out lack of detailed studies on the action of such chemicals on the exposed organisms, making it impossible to associate their action with the emergence of eventual problems. In the soil, trifluralin is moderately persistent, which might jeopardize organisms that are eventually exposed to it. Trifluralin is a substance that has a microtubule-depolymerizing activity, which prevents cell division, a fact that might compromise organism development.

Existing reports characterize trifluralin as a highly acute toxic substance to fish, but there are not enough descriptions of its chronic toxicity and cytotoxic effect. Studies mainly related to its genotoxic, mutagenic and carcinogenic potential are mostly inconclusive or even contradictory. There is little information about the toxicity of products derived from trifluralin degradation and its effects on the organisms.

Author details

Thaís C. C. Fernandes, Marcos A. Pizano and Maria A. Marin-Morales*

*Address all correspondence to: mamm@rc.unesp.br

Universidade Estadual Paulista, IB-Campus de Rio Claro, Rio Claro/SP, Brasil

References

[1] Londres, F. Agrotóxicos no Brasil: um guiaparaaçãoemdefesa da vida. – Rio de Janeiro: AS-PTA – Assessoria e Serviços a ProjetosemAgriculturaAlternativa, 2011. 190 p.

[2] Kotaka, E. T., Zambrone, F.A.D. Contribuições para a construção de diretrizes de avaliação do risco toxicológico de agrotóxicos. Campinas, SP: ILSI Brasil, 2001.

[3] Baird, C. QuímicaAmbiental. Porto Alegre: Bookman, 2002.

[4] Lorenzi, H. Manual de identificação e controle de plantas daninhas. Nova Odessa: EditoraPlantarum. 1990.

[5] Oliveira Jr., R.S. Mecanismo de açãoherbicidas. In: Biologia e Manejo de PlantasDaninhas. Oliveira Jr., R.S., Constantin, J., Inoue, M.H (Eds) Omnipax, 2011.

[6] Munger, R., Isacson, P., Hu, S., Burns, T., Hanson, J., Lynch, C.F., Cherryholmes, K., Vandorpe, P., Hausler, Jr. W. J. Intrauterine growth redardation in Iowa communities with herbicides-contaminated drinking watersupplies. Environ. Health Perspect., Research Triangle Park v. 105, p. 308-314, 1997.

[7] Gorell, J.M., Jhonson, C.C., Rybicki, B.A., Peterson, E.L., Ricchardson, R.J. The risk of Parkinson's disease with exposure to pesticides, farmin, well water, and rural living. Neurology, Madras v.50, p.1346-1350, 1998.

[8] Deuber, R. Botânica das plantasdaninhas. In: DEUBER, R. Ciência das plantasdaninhas. Jaboticabal: FUNEP, 1992.

[9] Grover, R., Wolt, J.D., Cessna, A. J., Schiefer, H.B. Environmental fate of trifluralin. Rev. Environ. Contam. Toxicol. v. 153, p. 1-64, 1997

[10] Rodrigues, B. N., Almeida, F. S.Guia de herbicidas, 5ª ed., Grafmarke: Londrina, 2005.

[11] Sanders Pf, Seiber Jn. A chamber for measuring volatilization of pesticides for model soil and water disposal system. Chemosphere, Oxford v.12, p. 999-1012, 1983.

[12] Calderon, M.J., Hermosín, M.C., Cornejo, J. Y Moreno, F. Movilidad de trifluralina en laboreo tradicional y de conservación. Estudios de la Zona No Saturada del Suelo. Eds. R. Muñoz-Carpena, A. Ritter, C. Tascón: 1999. Tenerife, p.83-88.

[13] Bellinaso M De. L., Henrique L.A., Gaylarde C.C., Greer C.W. Genes similar tonaphthalenedioxygenase genes in trifluralin-degrading bacteria. Pest Manag. Sci., Sussexv. 5, p. 474-478, 2004.

[14] U.S. Environmental Protection Agency. 1987. Trifluralin health advisory. Office of Drinking Water, Washington, DC.

[15] Cooper, M.T., Porter, T.D. Mutagenicity of nitrosamines in methyltransferase-deficient strains of Salmonella typhimuriumcoexpressing human cytochrome P450 2E1 and reductase. Mut. Res., AmsterdamV. 6, P.45-52, 2000.

[16] U.S. Environmental Protection Agency. 1999. Chemicals Evaluated for Carcinogenic Potential Science Information Management Branch Health Effects Division Office of Pesticide Programs.

[17] Laabs, V., Amelung, W., Pinto, A., Altstaedt, A., Zech, W. Leaching and degradation of corn and soybean pesticides in an Oxisol of the Brazilian Cerrados Chemosphere. v. 41, p. 1441-1449, 2000.

[18] Carter, N. D., Camper, N. D. Soil enrichment studies with trifluralin. Weed. Sci, Champaign. v. 23, p. 71-74, 1975.

[19] Selim H.M., Zhu H. Retention and mobility of deltamethrin in soils. Transport. Soil Sci. v. 167, p. 580-589, 2002.

[20] Cooke, C. M., Shaw G., Collins, C. Determination of solid-liquid partition coefficients (K_d) for the herbicides isoproturon and trifluralin in five UK agricultural soils.Environ. Pollut., Barking v. 132, p. 541-552, 2004.

[21] Dimou, A. D., Sakkas, V. A., Albanis, T. A. Trifluralin photolysis in natural waters and under the presence of isolated organic matter and nitrate ions: kinetics and photoproduct analysis. J. of Photochem. Photobiol., A, Chem, Lausanne v. 163, p. 473-480, 2004.

[22] Audus , L.J. Herbicides. London:Academic Press,1980, p. 608.

[23] Crosby, D. G. Fate of organic pesticides in the aquatic environment. Adv. Chem. Ser., Washingtonv. 111, p. 173, 1972.

[24] Klupinski, T. P., Chin, Y. P. Abiotic Degradation of Trifluralin by Fe(II): Kinetics and Transformation Pathways. Environ. Sci. Technol., Easton v. 37, p. 1311-1318, 2003.

[25] Almeida, F.S. Guia de herbicidas; recomendações para o uso adequado em plantio direto e convencional. Londrina, PR. 1985.

[26] Zimmerman, L.R., Thurman, E.M., Bastian, K.C. Detection of persistent organic pollutants in the Mississippi Delta using semipermeable membrane devices. Sci. Total Environ., Amsterdam v. 248, p. 1, 2000.

[27] Dayama, A., Coupe, R.H. Jr. Pesticides in the Yazoo River and BoguePhalia, February through September 1996. In: Daniel JB, editor. Proceedings of the 27th Mississippi Water Resources Conference, Jackson, MS, March 25–26, 1997. Mississippi Water Resources Institute, Starkville MS. p.127–132 , 1997.

[28] Thurman, E.M., Zimmerman, L.R., Scribner EA, Coupe RH Jr. Occurrence of Cotton Pesticides in Surface Water of the Mississippi Embayment. US Geological Survey Fact Sheet. v.4, p. 22-98, 1998.

[29] U.S. Environmental Protection Agency. 2001 (nov). Environmental Law Institute Research Report na Opportunities for Advancing Environmental Justice: Na Analusis of US-EPA, Washington, DC.

[30] E. C. (European Communities). Directive Relating to the Quality of Water Intended for Human Comsumption 1982, 80/778/EEC, oficce for official. Publications of the European Communities, L-2985 Luxemborg.

[31] Grover, R., Cessna, A.J., Waite, D.T. Volatilization losses na transport in air of tria-
      zine herbicides. In: Le Baron, H.M., Gianessi, L.P., Mcfarland, J., Burnside, O.C., edi-
      tors. The triazine herbicides. Amsterdam, the Netherlands: Elscrvier Science B.V.,
      2000.

[32] Waite, A.D.T., Bailey, A. P. Sproull, B. J.F., Quiring, A. D.V., Chau, B.D..F J., Bailey,
      C. J. Cessna, C. Atmospheric concentrations and dry and wet deposits of some herbi-
      cides currently used on the Canadian Prairies. Chemosphere, Oxford. v. 58, p.693–
      703, 2005.

[33] U.S. Environmental Protection Agency. 1993. Health advisories for drinking waters
      contaminants, Lewis Publishers, Boca laton, FL, USA .

[34] Mongar, K., Miller, G.C. Vapor phase photolysis of trifluralin in an outdoor chamber:
      Chemosphere, Oxford v. 17, p. 2183–2188, 1988.

[35] Ribas, G. J. S., Carbonell, E. N. X., Creus, A., Marcos, R. Genotoxic evaluation of the
      herbicide trifluralin on human lymphocytes exposed *in vitro*. Mutat. Res., Amster-
      dam v. 371, p. 15-21, 1996.

[36] Senseman, S.A. Herbicide Handbook, Ninth Edition. Weed Sci. Soc. Am. Champaign,
      IL: 458 pp. 2007.

[37] Bayer D.E., Foy C.L., Mallory T.E., Cutter E.G. Morphological & histological effects of
      trifluralin on root development. Am. J. Bot. v.54, p. 945-952, 1967.

[38] Hacskaylo J., Amato V.A. Effect of trifluralin on roots of corn & cotton. Weed Sci. v.
      16, p. 513-515, 1968.

[39] Fernandes, T.C.C. Uso do teste de Allium cepanadetecção da toxicidade e genotoxici-
      dade do herbicidatrifluralina. Monografia (Bacharel), UniversidadeEstadualPaulista,
      Rio Claro/SP, 2002.

[40] Morejohn, L.C., Bureau, T.E., Molé-Bajer, J., Bajer, A., Fosket, D. E. Oryzalin, a dini-
      troaniline herbicide, binds to plant tubulin and inhibits microtubule polymerization
      in vitro. Plant.,Berlim v. 172, p.41-147, 1987.

[41] Verhoeven, H.A., Ramulu, K.S., Dijkhuis, P.A. comparison of the effects of various
      spindle toxins on metaphase arrest and formation of micronuclei in cell-suspension
      cultures of *Nicotianaplumbaginifolia*. Plant., Berlin v. 182, p. 408-411, 1990.

[42] Morejohn, L.C. The molecular pharmacology of plant tubuline and microtubules:
      The cytoskeletal basis of plant growth and form In: ed. Lloyd C.W.: 1991. Academic
      Press, London, p.29-43.

[43] Ramulu, K.S., Verhoeven, H.A., Dijkhuis, P. Mitotic blocking, micronucleation, and
      chromosome doubling by oryzalin, amiprophos-methyl and colchicine in potato.
      Protoplasma, New York v. 160, p. 65-71, 1995.

[44] Quader, H. Cytoskeleton: Microtubules. Prog. Bot., Berlin v. 59, p. 375-395, 1997.

[45] Jordan, M.A., Wilson, L. The use and action of drugs in analyzing mitosis. Methods in Cell Biol., New York v. 61, p. 267-295, 1999.

[46] Alberts, B., Johnson, A., Lewis, J., Raff, M., Roberts, K., Walter, P. Molecular Biology of the Cell. Garland Science, New York , 4thed, 2002.

[47] Anthony, R. G., Hussey, P. J. Dinitroaniline herbicide resistance and the microtubule cytokeleton. Trends Plant Sci., Oxford v. 4, n.3, p. 112-116, 1999.

[48] Stokkermans, T.J.W., Artzman, J.D.S., Keenen, K., Morrissette, N.S., Tilney, L.G., Roos, D.S. inhibition of *Toxoplasma gondii* replication by dinitroaniline herbicides. Exp. Parasitol., San Diego v. 84, p. 355-370, 1996.

[49] Anthony, R. G., Waldin, T. R., Ray J. A., Bright, S. W. J., Hussey, P. J. Herbicide resistance caused by spontaneus mutation of the cytoskeletal protein tubulin. Nature, New York v. 393, p. 260-263, 1998.

[50] Mahresh, K. U., Larry D. N. Mode of dinitroaniline herbicide action. Plant. Physiol., Minneapolis v. 66, p. 1048-1052, 1980.

[51] Baird, W. V., Blume, YaB., WICK, S. Microtubular and cytoskeletal mutants. In: ed. Nick, P. Springer: 2000, p. 159-91.

[52] Breviário, D., Nick, P. Plant tubulins: a melting pot for basic questions and promising applicarions. Transgenic Res., London v. 9, p. 383-93, 2000.

[53] Yemets, A.I., Blume, Y.A.B. Resistence to herbicides with antimicrotubular activity: from natural mutants to transgenic plants. Russ J. Plant. Physiol., New York v. 46, p. 899-907, 1999.

[54] Blume, Ya.B., Nyporko, A. Yu., Yemets, A. I., Baird, W.V. Structural modeling of the interaction of plant α-tubulin with dinitroaniline and phosphoroamidate herbicides. Cell Biol. Int., London v. 27, p. 171-174, 2003.

[55] Yamamoto, E., Zeng, L., Baird, W.V. a-Tubulin missense mutations correlate with antimicrotubule drug resistence in *Eleusineindica*. Plant Cell, Berlin v. 10, p. 297-308, 1998.

[56] Higgins, D., Thompson, J., Higgins D. G., Gibs, T. J. ClustalW: improving the sensitivy of progressive multiple sequeraligment through sequence weighting, position-specific gap penaltie weight matrix choice. Nucleic Acids Res., Oxford, v. 22, p. 4673-4680. 1994.

[57] Arnold, F. C., Debonzi, D. H.MPAlign: Graphical and multiplatform tool for molecular alignments. Proceedings of II International Conference on Bioinformatics and Computational Biology, Angra dos Reis,2004.

[58] Hashim, S., Jan, A., Sunohara, Y., Hachinohe, M., Ohdan, H., Matsumoto, H. Mutation of alpha-tubulin genes in trifluralin-resistant water foxtail (Alopecurusaequalis). Pest. Manag. Sci. v. 68, p. 422–429, 2012.

[59] Sree, K.R., Verhoeven, H.A., Dijkhuis, P. Mitotic dynamics of micronuclei induced by amiprophos-metyl and prospects for chromosome-mediated gene transfer in plants. Tag, Berlin v. 75, p. 575-584, 1988.

[60] Hansen, A. L., Gertz, A., Joersbo, B.,Andesrsen, S.B. Antimicrotubule herbicide for in vitro chromosome doubling in *Beta vulgaris* L. ovule culture. Euphytica, Wageningen v. 101, p. 231-237, 1998.

[61] Vidakovié-Cifrek, M., Pavlica, I., Regula, D.P. Cytogenetic damage in shollot (*Allium cepa*) root meristems induced by oil industry "high-density brines". Arch. Environ. Contam. Toxicol. v. 43, p. 284-291, 2002.

[62] Hertel C., Quader H, Robinson D. G., Roos I., Carafoli E., Marme D. Herbicides and fungicidesstimulate Ca2+ effluxfromratlivermitochondria. FebsLett. Amsterdam v. 127, n.1 p. 37-39, 1981.

[63] Argese, E., Bettiol, C., Fasolo, M., Zambon, A., Agnoli, F. Substituted aniline interation with submitochondrial particles and quantitative struture-activity relationships. Biochim. Biophy. Acta, Amsterdam v. 1558, p. 151-160, 2002.

[64] Hepler, P.K. Calcium and mitosis. IntVerCytol. v. 2, p. 1273-1282, 1992.

[65] Marschner, H. Mineral nutrition of higher plants. London: Academic Press, Harcourt Brace, 1988.

[66] Fishbein, L. The Handbook of Environmental Chemistry. Part C- Anthrop. Comp. Berlin, v. 3, p 1-40, 1984.

[67] Hong, S.K., Anestis, D.K., Henderson, T.T., Rankin, G.O.Haloaniline induced in vitro nephrotoxicity: Effects of 4-haloanilines and 3,5-dihaloanilines. Toxicol. Lett. v. 114, 125-133. 2000.

[68] Wang S., Arnold W.A. Abiotic reduction of dinitroaniline herbicides. Water Res. Supl. 37, v. 17, p. 4191-201. 2003.

[69] Valentovick, M.A., Ball, J.G., Hong, S.K., Rogers, B.A., Meadows, M.K., Harmon, R.C., Rankin, G.O. In vitro toxicity of 2-and 4-chloroaniline: comparisons ith 4-ami-no-3-chlorophenol, 2-amino-5-chlorophenol e aminophenols. Toxicol. In Vitro, Ox-ford v. 10, p. 713-720, 1996.

[70] W.H.O. World Health Organization: Public health impact of pesticides in agriculture. Geneva, 1992.

[71] Worthing C.R, ed. *The pesticide manual*, 9th ed. Farnham, British Crop Protection Council, 1991.

[72] Worth, H.M. The toxicological evaluation of benefin and trifluralin. I: Pesticides Sim-posia: Inter-American Conference on toxicology and Occupational Medicine, Dei-chmann, W.B., Penalver, R.A., Radomski, J.L., Eds. Halos and Associates, Miami, 1970.

[73] Bem-Dyke, R., Sanderson, D.M., Noakes, D.N. Acute toxicity data for pesticides-1970. Pest Control., London v. 9, p. 119-127, 1970.

[74] Landonin, V. F., Hassan, A., Winteringham, F. P. W. Dinitroaniline pesticides. Chemosphere, Oxford v. 9, p. 67-69, 1980.

[75] Gaines, T. B., And Linder, R. E. Acute toxicity of pesticides in adult and weanling rats. Fundam. Appl. Toxicol., Akron v. 7, p. 299-308, 1986.

[76] Royal Society Of Chemists. Trifluralin. In The Agrochemical Handbook, 2nd ed., Update 5, p. A412. Graham, Cambridge. 1990.

[77] Gangolli, S. The dictionary of substances and their effects. Cambridge: Royal Society of Chemistry, v. 7. 1999, 998p.

[78] Meister, R.T. Farm Chemical Handbook '92. Willoughby: Meister Publishing Company, 1992.

[79] Occupational Health Services.MSDs for Trifluralin. OHS Inc., Secaucus, NJ. 1991.

[80] Wssa Herbicide Handbook Committee. Herbicide Handbook of the Weed Science Society of America, 6th Ed. WSSA, Champaign,1989.

[81] U.S. Environmental Protection Agency. 1989 (jan). Health Advisory Summary: Trifluralin. USEPA, Washington, DC.

[82] Ovidi, E., Gambellini, G., Taddei, A.R., Cai, G., Casino,C.D., Ceci, M., Rondíni, S., Tiezzi, A. Herbicides and themicrotubularapparatus of Nicotianatabacumpollentube: immunofluorescence and immunogoldlabellingstudies. Toxicol. in Vitro, Oxford v. 15, p.143-151, 2001.

[83] Timbrell, J.A. Introduction to Toxicology. London: Taylor & Francis, 1999.

[84] Garriott, M.L., Adams, E.R., Probst, G.S., Emmerson, J.L., Oberly, T.J., Kindig, D.E.F., Neal, S.B., Bewsey, B.J., Rexroat, M.A. Genotoxicity studies on the preemergence herbicide Trifluralin. Mutat. Res. v. 260, p. 187–193, 1991.

[85] Lignowski, E.M., Scott, E.G. Effect of trifluralin on mitosis. Weed Sci. v. 20, p. 267-270, 1972.

[86] Wu, T.P. Some cytological effects of treflan and mitomycin C on root tips of Viciafaba. Taiwania, Taipiei v. 17, p. 248-254, 1972.

[87] Grigorento, N.K., Fasilchenko, V.F., Merezhinski, Y.G., Morgun, V.V., Logvinenko, V.F., Sharmankin, S.V. Cytogenetic activity of a herbicide treflan, and its metabolites as applied to maize.Tsiol. Genet.v. 20, p. 294-298, 1986.

[88] Grant, W.F., Owens, E.T. Chromosome aberration assays in Pisum for the study of environmental mutagens. Mutat.Res., Amsterdam. v. 188, p. 93-118, 2001.

[89] Fernandes, T.C.C., Mazzeo, D.E.C., Marin Morales, M.A. Mechanism of micronuclei formation in polyploidizated cells of *Allium cepa* exposed to trifluralin herbicide. Pesticide Biochemistry and Physiology. v. 88, p. 252-259, 2007.

[90] Fernandes, T.C.C., Mazzeo, D.E.C., Marin Morales, M.A.Origin of nuclear and chromosomal alterations derived from the action of an aneugenic agent—Trifluralin herbicide Ecotoxicology and Environmental Safety. v. 72, p. 1680–1686, 2009.

[91] Könen, S., Çavas, T. Genotoxicity testing of the herbicide Trifluralin and its commercial formulation treflan using the piscine micronucleus test. Environmental and Molecular Mutagenesis, v.49, p.434-438, 2008.

[92] Peña, L.F.M. Uso do teste de micronúcleoemeritrócitoscirculantes de peixesparamonitorização de um local do rioTibagi e avaliação da genotoxidade de agrotóxicosembioensaios. Londrina. 1996. [Tese de mestradoemGenética e Melhoramento – UniversidadeEstadual de Londrina].

[93] Canevari, R.A. Avaliação dos efeitosgenotóxicos e diâmetro dos micronúcleosobtidosem*Prochiloduslineatus* (Pisces, *Prochilodontidae*) submetidos a tratamentosagudos com o inseticidaazodrin e o herbicidatrifluralina. Londrina. 1996. [Monografia (Bacharelado) embiologiageralUniversidadeEstadual de Londrina].

[94] Kaya B., Marcos R., Yanikoglu A., Creus A. Evaluation of the genotoxicity of four herbicides in the wing spot test of *Drosophila melanogaster* using two different strains. Mutat. Res., Amsterdam v. 557, p. 53-62, 2004.

[95] Gebel, T., Kevekordes, S., Pav, K., Edenharder, R., Dunkelberg, H. *In vivo*genotoxicity of selected herbicides in the mouse bone-marrow micronucleus test. Arch. Toxicol. v. 71, p. 193–197, 1997.

[96] Rawlings, N.C., Cook, S.J., Waldbillig, D. Effects of the pesticides carbofuran, chlorpyrifos, dimethoate, lindane, triallate, trifluralin, 2,4-D, and pentachlorophenol on the metabolic endocrine and reproductive endocrine system in ewes. J. Toxicol. Environ. Health A. v. 54, p. 21–36, 1998.

[97] Bryant, M.L., Murnik, M.R. Mutagenicity of the herbicide trifluralin inDrosophila melanogaster. Mutat. Res., Amsterdam v. 53 , p. 235, 1977.

[98] Foureman, P.A. Identification of aneuploidy inducing chemicals in *Drosophila*.Environ. Mutagen., New York v. 3, p. 319, 1981.

[99] Nehéz, M.; Páldy, A. Selypes, A. And Berencsi, G. Experiments on the mutagenic effects of two pesticides, DNOC and trifluralin. Mutat Res., Amsterdam v. 74, p. 202-203, 1980.

[100] Pilinskaya, M.S.A Evaluation of the cytogenetic effect of the herbicide treflan and of a number of its metabolites on mammalian somatic cells. Tsitol. Genetic. v. 21, p. 131-135, 1987.

[101] Iarc (1991) Iarc Monographs on the evolution of carcinogenic risks to humans. Occupational Exposures insecticide Application, and Some Pesticides. v. 53. Lyon, France.

[102] Ribas, G., Frenzilli, G., Barale, R., Marcos, R. Herbicide-induced damage in human lymphocytes evalueated by the single-cell gel eletrophoresis (SCGE) assay. Mutat. Res., Amsterdam v. 344, p. 41-54, 1995.

[103] Chan, M.M; Fong, D. Inhibition of Leishmanias but not host macrophages by the antimicrotubulin herbicide trifluralin. Science, v. 249, p. 924–926, 1990.

[104] Bhattacharya, G., Salem, M. M., Werbovetz, K. A. Antileishmanialdinitroaniline sulfonamides with activity against parasite tubulin. Bioorganic and Medicinal Chemistry Letters. v.12, p. 2395–2398, 2002.

[105] Esteves, M.A., Fragiadaki, I., Lopes, R., Scoulica, E., Cruz. M.E.M. Synthesis and biological evaluation of trifluralin analogues as antileishmanial agents Bioorganic & Medicinal Chemistry. v. 18, p. 274–281. 2010.

[106] Bogitsh, B.J., Middleton, O.L., Ribeiro-Rodrigues, R. Effects of the antitubulin drug trifluralin on the proliferation and metacyclogenesis of Trypanosomacruziepimastigotes. Parasitology Research. v. 85, p. 475–480, 1999.

[107] Traub-Cseko, Y.M., Ramalho-Ortigao, J.M., Dantas, A.P., De Castro, S.L., Barbosa, H.S., Downing, K.H. Dinitroanilineherbicidesagainstprotozoan parasites: the case of Trypanosomacruzi. Trends in Parasitology. v. 17, p. 136–141 2001.

[108] Nath, J., Schneider, I., Antimalarial effects of the antitubulin herbicide trifluralin: studies in Plasmodium falciparum. Clinical Research. v. 40, p. A331, 1992.

[109] Lee, J.H., Arumuganathan, K.; Yen, Y., Kaeppler, S., Baenziger, P. S. Root tip cell cycle synchronization and metaphase-chromosome isolation suitable for flow sorting in common wheat (Triticumaestivum L.). Genome, Otawa v.40, p.633-638, 1997.

[110] Dow, G., Reynoldson, J., Thompson, A. Comparative efficacy of two tubulin inhibitors, aldabenzole and trifluralin, against Plasmodium berghei. Parasitol. Int., Tokio v. 47, p.133-281, 1998.

[111] Werbovetz, K.A., Brendle, J.J., Sackett, D.L. Purification, characterization, and drug susceptibility of tubulin from Leishmania. Mol. Bioch. Parasitol., Amsterdamv. 98, p. 53-65, 1999.

[112] Donna, A., Betta, P.G., Gagliardi, F., Ghiazza, G.F., Gallareto, M. and Gabutto,V. Preliminary experimental contribution to the study of possible carcinogenic activity of two herbicides containing atrazine-simazine and trrifluralin as active principle.Pathologica, Gênova v. 73, p. 707-721, 1981.

[113] Kang, D., Park, S.K., Beane-Freeman, L., Lynch, C.F., Knott, C.E. Sandler,D.P. Hoppin, J.A., Dosemeci, M., Coble, J.,Lubin, J., Blair, A., Alavanja., M. Cancer incidence

among pesticide applicators exposed to trifluralin in the Agricultural Health Study. Environmental Research. v. 107, p. 271–276, 2008.

[114] N.C.I. Institute Nacional Cancer. Biossay of trifluralin for possible carcinogenicity Bethesda, MD, 2000.

[115] Emmerson, J.L., Pierce, E.C., Mcgrath, J.P. The chronic toxicity of compound 36352 (trifluralin) given as a compound of the diet to the fischer 344 rats for two years. Studies r-87 and R-97 (unpublished study received September 18, 1980 under 1471-35; submitted by Elanco Products Co., Division of Eli Lilly and Co., Indianapolis, IN), 1980.

Chapter 5

Allelochemicals as Bioherbicides — Present and Perspectives

Dorota Soltys, Urszula Krasuska,
Renata Bogatek and Agnieszka Gniazdowska

Additional information is available at the end of the chapter

1. Introduction

Since the first implementation of synthetic herbicides in crop protection systems, weeds have continuously developed resistance. As a main reason of such evolution, long-lasting exploitation of herbicides with one target site in plants is considered. This has been the case with the first widely-used triazine herbicides, photosynthesis inhibitors, which have effectively eliminated a wide range of weeds. Unfortunately, inappropriate adjustment of herbicides to weed species occupying fields, application of herbicides at the incorrect developmental stage and in unsuitable weather conditions have contributed to the accumulation of active compounds in the soil, accumulation of weed species and acceleration evolution of resistant biotypes [1]. To date, there have been 211 species and 393 biotypes of herbicide resistant weeds identified [2]. Most of them are resistant to B, C1 and A groups of herbicides, inhibitors of: acetolactate synthase (ALS), photosystem II and acetyl CoA carboxylase, respectively. Ten species pose the biggest threat for crops due to causing yield losses, including the most important herbicide-resistant species which are characterized by multiple resistances: rigid ryegrass (*Lolium rigidum* Gaud.), wild oat (*Avena fatua* L.) and redroot pigweed (*Amaranthus retroflexus* L.).

Evolution of weeds resistant to herbicides demands new solutions to cope with the problem since economic losses generated by weeds can be higher than those caused by other pests. Due to the fact that abandoning chemical weed control is, with current agricultural practices, rather impossible, it is necessary to create new classes of herbicides with new mechanisms of action and target sites not previously exploited. Presently used synthetic herbicides are not approved for use in organic agriculture. Moreover, using crop protection chemicals also need public acceptance. [3]. The number of synthetic chemicals with new target sites are decreasing dramatically. Eco-friendly trends in weed management force scientists to reach for innovative sources and tools. Natural compounds pose a great field for the discovery of new environmentally safe herbicides, so called "bioherbicides", which are based on compounds produced by living organisms. According to the CAS (Chemical Abstracts Service) registry, among the 24 million organic compounds, a large group of secondary plant metabolites is represented. Some of these compounds take part in allelopathic interactions.

2. Allelopathic interactions and allelopathic compounds

Allelopathy is considered a multi-dimensional phenomenon occurring constantly in natural and anthropogenic ecosystems [4]. It is defined as the interaction between plants and microorganisms by a variety of compounds usually referred to as allelopathins, allelochemicals, or allelopathic compounds. This review is focused mainly on compounds taking part in complex allelopathic interactions between higher plants. However, determination of quality, quantity, direct or indirect effects of allelopathins on plant or microorganism communities in the natural environment is very difficult owing to the multi-dimensional character of those interactions. The development of analytical techniques allowing better specification of direct effects of allelopathins, have moved the exploration (or the research on) of this phenomenon from fields into laboratories. The term "allelopathy" refers rather to interactions occurring in the natural environment [5]. For studies with plant extracts, allelopathins isolated from plant tissue, collected from exudates or even synthetic compounds identical to natural ones, it was established the term "phytotoxicity" to distinguish allelopathy (as a phenomenon occurring in natural environment) from studies conducted in laboratory.

Allelopathins are products of the secondary metabolism and are non-nutritional primary metabolites [6,7]. These compounds belong to numerous chemical groups including: triketones, terpenes, benzoquinones, coumarins, flavonoids, terpenoids, strigolactones, phenolic acids, tannins lignin, fatty acids and nonprotein aminoacids. A wide range of these biochemicals are synthesized during the shikimate pathway [8] or, in the case of essential oils, from the soprenoid pathway. Allelochemicals can be classified into 10 categories [9] according to their different structures and properties:

1. water-soluble organic acids, straight-chain alcohols, aliphatic aldehydes, and ketones;

2. simple lactones;

3. long-chain fatty acids and polyacetylenes;

4. quinines (benzoquinone, anthraquinone and complex quinines);

5. phenolics;

6. cinnamic acid and its derivatives;

7. coumarins;

8. flavonoids;

9. tannins;

10. steroids and terpenoids (sesquiterpene lactones, diterpenes, and triterpenoids).

Allelochemicals are released into the environment by plant organs such as roots, rhizomes, leaves, stems, bark, flowers, fruits and seeds (Figure 1a). The huge number of allelopathic interactions is typically negative in character, with positive relations being rare. Allelopathic compounds affect germination and growth of neighboring plants by disruption of various physiological processes including photosynthesis, respiration, water and hormonal balance. The underlying cause of their action is mainly inhibition of enzyme activity. Ability of an allelochemical to inhibit or delay plant growth and/or seed germination is usually defined as its "allelopathic (or phytotoxic) potential". An excellent example of allelopathic interaction is seen in soil exhaustion due to the accumulation of allelopathins that can be prevented by using fertilizers and rotating crops. Plants producing allelopathins are considered as "donor" organisms while the plants which allelopathins are directed to are referred to as "target" plants or "acceptors". The after-effects and strength of allelopathic interactions are diverse due to modifications of the allelopathins taking place in soil (Fig 1b). Most of the allelochemicals penetrate the soil as already plant-active compounds, e.g. phenolic acids, cyanamide, momilactones, heliannuols etc. Some have to be modified into the active form by microorganisms or by specific environmental conditions (pH, moisture, temperature, light, oxygen etc.), e.g. juglone, benzoxazolin-2-one (BOA), 2-amino-3-H-phenoxazin-3-one (APO).

3. Advantages and disadvantages of allelopathins as bioherbicides

Mode of action of some allelochemicals is similar to synthetic herbicides. These features have allowed them to be considered for possible use in weed management as bioherbicides. However, the field of knowledge is poorly studied but it is a very attractive area to explore.

Allelochemicals are highly attractive as new classes of herbicides due to a variety of advantages. However, in the perspective of bioherbicides based on allelopathins, effects caused by these compounds on target plants are also classified as "phytotoxic".

Most of allelopathins are totally or partially water-soluble which makes them easier to apply without additional surfactants [3, 10]. Their chemical structure is more environmentally friendly than synthetic ones. They possess higher oxygen- and nitrogen-rich molecules with relatively few so called 'heavy atoms', a halogen substitute, and are characterized by the absence of 'unnatural' rings. These properties decrease a chemical's environmental half-life, prevent accumulation of the compound in soil and eventual influence on non-target organisms. On the other hand, these properties are an allelochemical's Achille's heel due to less than satisfactory duration of activity. Structure complexity generates more stereocenters making them more reactive and unstable. Therefore, rapid degradation of one of the chemical groups can significantly decrease bioactivity of the whole compound.

The diversity of allelopathins makes them promising tools possessing specific properties in discovering novel, specific target sites in acceptor plants. Even if they inhibit photosynthesis

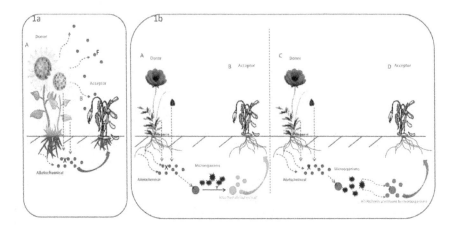

Figure 1. Multi-dimensional nature of allelopathic interacions. (1a) Plant A releases allelochemicals X and F which directly affect growth of plant B. (1b) left side; Plant A releases allelochemical X which is modified or activated by microorganisms to allelochemical Y that affects growth of plant B. (1b) right side; Plant A releases allelochemical X which stimulates microorganisms to produce allelochemical Z that affects growth of plant B.

or respiration, they may also bind to proteins at different sites than synthetic herbicides [11, 12]. This provides the opportunity to eliminate weeds that are already resistant to commercialized herbicides with the same mode of action. Allelochemicals are also characterized by multi-site action in plants without high specificity which is achieved in the case of synthetic herbicides. Therefore, this feature excludes the application of an allelopathic compound as a selective herbicide or totally prohibits its usage in weed management. On the other hand, effects of allelopathins in acceptor plants are highly dose-dependent [13]. This allows the opportunity to search out compounds exhibiting selectivity. Generally, monocotyledonous plants are more resistant to allelochemicals than dicotyledonous ones. Therefore, usage of a compound as a potential herbicide is possible but rather restricted to cultivation of exact crops with a defined weed composition.

The route of discovery is much more complicated with allelopathins. In contrast to synthetic herbicides where synthesis, bioassay, evaluation and quantitative structure-active relationship follow Quantitative Structure-Activity Relationship (QSAR), allelochemicals have to be first isolated from plant extracts [14]. The amount of recovered compounds is usually low in comparison to chemical synthesis. After extraction, purification and selection of the most attractive compound and determination of its mode of action in plants is done. At the end of the process, similar to synthetic herbicides, allellpathins are subjected to QSAR. The long discovery process is usually offset by a shorter, less expensive track of registration [15]. It is worth noting that before an allelochemical can become an herbicide, the following conditions have to be performed: phytotoxic activity at the range between 10^{-5} and 10^{-7} M, identified chemical structure, known mode of action in plants, time of residence in soil, possible influence on microbial ecology and non-target plants, possible toxic properties on human health and profitability of production on a commercial scale [16].

A high number of limitations does not exclude allelochemicals as possible herbicides. In particular, they can be alternatives in weed management strategy. Widely developed bioinformatics and cheminformatics support development of new herbicides [3, 15, 16]. Identified chemical structure of a particular allelochemical is a starting point to design a product with the compound-like properties using computer programs. Thanks to cheminformatics we are able to predict the potential structure of analogues and make several modifications, which make it more or less active, with higher environmental stability, as it was done for leptospermone. We may also predict the target site of compound action in plants due to comparison studies. Similar structure of a compound to a commercialized herbicide or other natural compound whose mode of action is well-known may allow us to predict the target site.

4. Allelopathic plant extracts as bioherbicides

Plant protection is effective but rather costly and problematic due to environmental pollution. Exploration of the allelopathic potential of some species allows the introduction of alternative techniques for weed management, e.g. extracts from allelopathic plants can be applied as foliar sprays. Apart from decreasing the costs of herbicide application, this method also improves crop production.

The best known examples of natural bioherbicides are phytotoxic water extracts from herbage of sorghum (*Sorghum bicolor* (L.) Moench.) (sorgaab) and sunflower (*Helianthus annuus* L.) (sunfaag) which can be effectively used in plant protection without yield losses.

Effects of sorgaab on weeds is time- and dose-depend but is typically used at 5% or 10% (w/v) concentration as double spray 20/30 and 40/60 days after sowing (DAS) or after seedling transplantation (AT) [17-19]. The best results to account for net profits have been elicited with a double spray of 10% extract in cotton (*Gossypim hirsutum* L.), soybean (*Glycine max* L.), wheat (*Triticum aestivum* L.) or rice (*Oryza sativa* L.). The highest efficacy of such extract applications has been verified in rice on reduction of barnyard grass (*Echinochloa cruss-galli* L.) biomass by 40%, without significant changes in weed density and accompanied yield increase by 18%.

Sunfaag has been widely used in wheat. The extract has been usually applied three times at 7-day intervals starting between 3-4 weeks post-emergence. This system of application has reduced biomass of the two most commonly occurring weeds, lambsquarters (*Chenopodium album* L.) and toothed dock (*Rumex dentatus* L.), by 70% and 97% respectively, although it has not eliminated all weed species in field. It has improved wheat biomass by 7-8% in comparison to weed free control without significant changes in number of tillers and total seed biomass. The herbicidal efficiency calculated as the effectiveness of sunfaag in comparison to synthetic herbicides showed a quite high value, 60% efficiency index. Weed management systems require high concentrations of sunfaag ranging up to 80% and can generate economic losses due to the necessity of cultivating higher amounts of sorghum or sunflower that also required an appropriate cultivation system [20, 21]. Therefore, sunfaag can be applied as a pre-emergence herbicide with much lower doses. The most promising application system has considered usage of 10% (w/v) extract at pre-emergence + 25 DAS + 35 DAS. Following the application, there has been noted a remarkably reduced population of wild oat, lesser swi-necress (*Coronopus didymus* L.) and littleseed canarygrass (*Phalaris minor* Retz.) without affecting germination of wheat and increased wheat yield in 7% [22]. However, the inhibitory effect on weed growth and crop yield is selective and highly dependent on duration or term of sorgaab and sunfaag application.

Aqueous extracts of sorghum and sunflower are effective on weed growth but unfortunately might not be profitable enough in crop production; however, crop allelopathy can be manip-ulated for achieving sustainable weed management. Combination of phytotoxic crop water extracts with lower rates of herbicides may provide reduced weed control levels with reduced herbicide usage. The interesting review of allelopathic crop plants in weed management strategy is presented in reference [23]. Two field studies were conducted utilizing water extracts of sorghum, sunflower and rapeseed (*Brassica napus* L.) with reduced glyphosate dosage for controlling purple nutsedge (*Cyperus rotundus* L) in cotton [24]. Sorghum and rapeseed water extracts were tank mixed (at 15 or 18 L ha^{-1}) in different combinations with reduced rates of glyphosate by 767 and 575 g active substance (a.s.) ha^{-1} and sprayed as directed post emergence at 21 DAS. Purple nutsedge density and dry weight were suppressed by 78% to 95% and 83% to 95%, respectively, when different crop water extracts were used in combi-nation with a reduced rate of glyphosate. Seed cotton yield was improved from 15-21% in sorgaab and rape water extract combinations with reduced rates of glyphosate (67-75%). Similar research has been conducted on water extracts of sorghum with sunflower in combi-nation with herbicides in wheat, soybean, rice, and canola (*Brassica* sp.) [25, 26]. Both extracts, in combination with herbicides, have the same or even better effect on inhibition of growth of the following weeds: littleseed canarygrass and lesser swinecress, compared to single synthetic herbicide applications [25, 26]. Spraying of wheat seedlings 30 DAS with sorgaab+sunfaag (18 L each ha^{-1}) with mesosulfuron+idosulfuron (4.32 g a.s. ha^{-1}) has the same effect on total weed density (reduction up to 90% in relation to control) as application of mesosulfuron+idosulfuron used alone, but with higher doses (120 g a.s. ha^{-1}). Herbicidal solution has also improved yield parameters, both in relation to control and in relation to single herbicide application: fertile tillers (10%), spikelets per spike (11%) and grains per spike (10%) [26]. In cotton, application of both extracts at 18 L ha^{-1} each with glyphosate (767 g a.s. ha^{-1}) 21 DAS has been the most effective in density reduction of the highly competitive weed purple nutsedge up to 93% [24]. However, the greatest benefit in wheat is the usage of a sorgaab/sunfaag combination which lowered by 70% doses of metribuzin and phenaxaprop (at 57 g a.s. ha^{-1}), applied at 18 L each ha^{-1}. In turn, in cotton, application of the same rates of extracts per ha with glyphosate (767 g a.s. ha^{-1}) seems to be the most economically reasonable costs of following weed management method [24, 25].

Selectivity of plant extracts on weeds without any negative implications on crop productivity is probably due to differences in the physiological stage of plants and following plant competition. Sunfaag has been applied when wheat seedlings were 3-4 weeks old while lambsquarters and toothed dock 1-week old at the stage of three to four leaf [20, 21].

High allelopathic potential conditioned by glucosinolates and isothiocyanates is present in Brassica sp. [27, 28]. Isothiocyanates have been strong suppressants of germination of spiny sowthistle (*Sonchus asper* L. Hill), scentless mayweed (*Matricaria inodora* L.), smooth pigweed (*Amaranthus hybridus* L.), barnyardgrass, blackgrass (*Alopecurus myosuroides* Huds.) and wheat [28]. Black mustard (*Brassica nigra* L.) extract of different plant parts like leaf, stem, flower and root have inhibited germination and radicle length of wild oat. Inhibitory effects on germination increased with increasing concentration of extract solution of the fresh plant parts [29]. Some experiments were conducted also using garden radish (*Raphanus sativus* L.) extract on germination of 25 weed and 32 crop species [30]. Garden radish extracts totally inhibited germination of 11 weeds such as Johnsongrass (*Sorghum halelense* L. Pers.), *Alhagi* spp., blackgrass (*Alopecurus myosuroides* Huds.), shepherd's-purse (*Capsella bursa-pastoris* L. Medik.), field bindweed (*Convolvulus arvensis* L.), dodder (*Cuscuta* sp.), carrot (*Daucus carota* L.), shortpod mustard (*Hirschfeldia incana* L.), *Ochtodium aegyptiacum* (L.), and shortfruit hedgemustard (*Sisymbrium polyceratium* L.), and 4 crop species namely lettuce (*Lactuca sativa* L.), tobacco (*Nicotiana tabacum* L.), bean (*Phaseolus vulgaris* L.), and clover (*Trifolium* sp.). Garden radish extracts at different rates (100, 66, 50 and 33% of pure extract) did not affect germination of wheat, cotton, and maize (*Zea mays* L.), but affected soybean germination at the 100% extract rate in vitro. Rhizome regeneration of Johnsongrass was inhibited by 54-99% depending on extract concentration. Regeneration of bermudagrass (*Cynodon dactylon* L. Pers.) rhizomes was inhibited to a lower extent at all concentrations; for instance, 54% inhibition occurred at the highest extract concentration. Lower extract rates stimulated redroot pigweed germination, while 66 and 100% extracts inhibited germination by 21 and 42%, respectively. Inhibition reached only 56 and 49% at the highest extract concentration for common purslane (*Portulaca oleracea* L.) and cocklebur (*Xanthium strumarium* L.), respectively. Garden radish residues which were cut into pieces and incorporated into the growing medium decreased weed intensity and increased maize yield [31].

Legumes crops may also be applied as a source of allelochemicals useful in weed suppression. Mulch of dead pea plants could be used to control growth of weeds. Pea cover crop has regulated germination and growth of lady's thumb (*Polygonum persicaria* L.), smooth pigweed, smallflower galinsoga, and common lambsquarters. Similarly, the aqueous leachates (1%) of all four legumes, velvetbean (*Mucuna deeringiana* (Bort.) Merr.), jackbean (*Canavalia ensiformis* (L.) DC.), jumbiebean (*Leucaena leucocephala* (Lam.) de Wit), and wild tamarind (*Lysiloma latisiliquum* (L.) Benth.), have been shown to suppress weeds [32]. These plants exhibited strong phytotoxic effects on the radicle growth of barnyardgrass, alegría (*Amaranthus* ssp.) and amaranth (*Amaranthus hypochondriacus* L.) [33]. Russian knapweed (*Acroptilon repens*) control is difficult in many crops. Allelopathic effects of extracts and plant parts of alfalfa (*Medicago sativa* L.) on Russian knapweed were reported both in Petri dishes and pot experiments [34]. Alfalfa has been recommended in fields with high mugwort (*Artemisia vulgaris* L.) infestation, as it decreased mugwort to 89% under field conditions, while extracts of alfalfa vegetative parts inhibited mugwort germination up to 83% in Petri dish assays.

Application of plant extracts as pre-emergence or as early post emergence herbicides resulted in reduction of doses of synthetic herbicide due to their synergistic or additive action. However, not all phytotoxic extracts are effective enough to inhibit weed growth or germination when applied as spray even when plants show high allelopathic potential as mulch, intercropping system or in rotation. This may be the result of masking the activity of one compound by another in water solution or other factors such as impossibility of extract penetration through the cuticle [12]. A new opportunity to enhance effectiveness of usage of bioherbicides based on natural extracts is associated with extraction of individual allelochemicals and/or its comparison with synthetic herbicides. The extraction of sesquiterpene lactone, dehydrozaluzanin C (DHZ) produced among Compositae family serves as an example [34]. Comparison studies of isolated DHZ (1 mM) and the commercial herbicide Logran® showed high inhibitory activity of DHZ on dicotyledonous plants while the synthetic herbicide showed no activity [34]. Also pure 2-benzoxazolinone (BOA) isolated from several graminaceous crops such as rye (*Secale cereale* L.), maize and wheat was active similarly as herbicide but its stability in the environment was much shorter than the synthetic herbicide [35].

5. Plant allelopathins as sources of bioherbicides

Plant phytotoxic extracts, after evaluation, can be successfully used in integrated weed management. However, as was aforementioned, not all systems of its application under field conditions are suitable and profitable enough. To circumvent masking effects of one allelopathin by another in plant extract, research is now focused on isolation and application of a single, specific compound for the purpose of weed elimination. The list of allelochemicals isolated from various plants that may act as inhibitors of weed seed germination and/or weed growth are summarized in Table 1. A purified allelopathic compound may act on target plants with much higher or much lower strength. Even in situations when an allelopathin is active at unprofitably high doses but has a favorable environmental profile, it still may be a source to explore due to several reasons such as biodegradability. Modifications of chemical structure can make a compound more active on target plants while preserving desire properties.

Herein, examples of purified allelopathins with possible roles as herbicides are described. Some herbicides based on modified allelopathins already launched on the market are also included.

5.1. Sorgoleone

The inhibitory effect of sorghum on various plant species has been known for many years. Accumulation of sorghum phytotoxins in soil affects crop growth and imposes the need for a crop rotation system. Besides crops, weeds are also vulnerable to its allelopathic influence [16, 36]. Sorghum toxicity is mainly determined by both hydrophilic phenols in herbage, as well as hydrophobic sorgoleone and its analogs exuded by the root hairs [37, 38]. Therefore, sorghum herbage reach can be successfully used against weeds as a foliar spray as it is discussed in detail in the previous chapter.

Compounds	Botanical source	Sensitive weeds
Glucosinolates, Isothiocyanates	mustard (*Brassica* sp.) garden radish (*Raphanus sativus*)	spiny sowthistle (*Sonchus asper* L. Hill), scentless mayweed (*Matricaria inodora* L.), smooth pigweed (*Amaranthus hybridus* L.), barnyardgrass (*Echinochloa cruss-galli* L. Beauv.), slender meadow foxtail or blackgrass (*Alopecurus myosuroides* Huds.), *Alhagi* spp., *Cachia maritime*, Shepherd's-purse(*Capsella bursa-pastoris* L.), morning glory (*Convolvulus arvensis* L.), dodders (*Cuscuta* spp.), wild carrot or bird's nest (*Daucus carota* L.), shortpod mustard, buchanweed or hoary mustard(*Hirschfeldia incana* L.), *Ochtodium aegyptiacum* (L.), shortfruit hedgemustard (*Sisymbrium polyceratium* L.)
Sorgoleone	sorghum (*Sorghum bicolor* L. Moench)	littleseed canarygrass (*Phalaris minor* Retz.), lesser swinecress (*Coronopus didymus* L.), purple nutsedge (*Cyperus rotundus* L.), black nightshade (*Solanum nigrum* L.), redroot pigweed (*Amaranthus retroflexus* L.), common ragweed (*Ambrosia atrtemisiflora* L.), sicklepod (*Cassia obtusifolia* L.)
Momilactone	rice (*Oryza sativa* L.), moss (*Hypnum plumaeform*)	barnyardgrass, (*Echinochloa colonum* L.), livid amaranth(*Amaranthus lividus* L.), hairy crabgrass (*Digitaria sanguinalis* L.), annual meadow grass, annual bluegrass or poa (*Poa annua* L.)
Artemisinin	annual wormwood (*Artemisia annua* L.)	redroot pigweed, pitted morning-glory (*Ipomoea lacunose* L.), common purslane (*Portulaca oleracea* L.), annual wormwood, duckweed (*Lemna minor* L.), algae (*Pseudokirchneriella subcapitata*)
Leptospermone	bottle brush (*Callistemon citrinus*), manuka (*Leptospermum scoparium* J.R., G. Forst)	barnyard grass, hairy crabgrass, yellow foxtail (*Setaria glauca* L.), california red oat (*Avena sativa* L.), Indian mustard (*Brassica juncea* L.), curly dock (*Rumex crispus* L.)
Essential oils	eucalyptus (*Eucalyptus* sp.)	barnyard grass, *Cassia occidentalis*, annual ryegrass (*Lolium rigidum*)
Sarmentine	pepper (*Piper* sp.)	barnyard grass, redroot pigweed, crabgrass, Sprangletop (*Leptochloa filiformis* Lam.), dandelion (*Taraxacum* sp.), lambsquarter or wild spinach (*Chenopodium album* L.), annual bluegrass or poa, morning glory or bindweed, wild mustard, curly dock

Table 1. Allelopathic compounds isolated from plants that exhibit inhibitory potential on seed germination and growth of weeds

However, allelochemical sorgoleone has enormous potential as an herbicide due to its high activity against various weed species. Studies conducted under laboratory conditions have shown that low doses of sorgoleone (100 µM) inhibit growth of the following weeds by 80%, black nightshade (*Solanum nigrum* L.), redroot pigweed, common ragweed (*Ambrosia atrtemisiflora* L.), and by 40% of sicklepod (*Cassia obtusifolia* L.), hairy crabgrass (*Digitaria sanguinalis* L.), velvetleaf (*Abutilon theophrasti* Medik.), barnyardgrass and tef (*Eragrostis tef* Zucc., Trotter) [11, 16].

Sorgoleone released into the soil may act as a pre-emergence herbicide. Its persistence in the soil during or after sorghum cultivation inhibits germination and growth of small-seeded weeds, e.g. hairy crabgrass and green bristlegrass (*Setaria viridis* (L.) Beauv.), due to its better absorption and translocation within the small seeds than in large seeds [39]. However, strength and final effect on seeds or seedling physiology is multifactor-dependent. Sorgoleone sorbs strongly to the organic matter. This allows an extended persistence in the soil but unfortunately, significantly reduces its bioavailability. Moreover, the dynamics of decomposition significantly influences sorgoleone bioactivity, e.g. the methoxy- group of the aromatic ring is decomposed by 26% 48 h after exudation; however, some amounts of sorgoleone are also extractable after 6 weeks [40, 41]. Nevertheless, constitutive production of the compound allows a continuous supply and accumulation in the soil around 1.5 cm of root zone [42].

Inhibition of H^+-ATPase in plant roots makes sorgoleone an effective growth inhibitor and potential post-emergence herbicide [43]. Decreased activity of that enzyme affects ion uptake and water balance by decreasing water uptake and affecting plant growth. Redroot pigweed, Jimson weed (*Datura stramonium* L.) and tef grown in hydroponic culture with 10 µM sorgoleone were characterized by lower H^+-ATPase activity in roots. Presence of sorgoleone in nutrient solution significantly suppressed growth and evoked brown coloration and necrosis [43, 44].

Sorgoleone may be taken up by roots but cannot be translocated acropetally by xylem due to high lipophilic properties. Therefore, its application as a post-emergence herbicide may be limited. However, as a spray (0.6 kg ha⁻¹), it has inhibited growth by 12% of green foxtail (*Setaria faberi* Herrm.), by 40-50% purslane, hairy crabgrass and velvetleaf, and up to 80-90% of common ragweed, redroot pigweed, and black nightshade [40].

Due to the structural similarity of sorgoleone to plastoquinon, it acts as a photosystem II (PSII) inhibitor [11, 43]. It binds to the niche of the D1 protein in PSII, gathers electrons and does not allow reoxidation of plastoquinon A by the secondary electron acceptor, plastoquinone B. Competition studies under sorgoleone *versus* synthetic herbicides such as atrazine, diuron, metribuzin and bentazon have shown that sorgoleone is an atrazine competitive inhibitor [11, 12]. Moreover, the I_{50} of sorgoleone is 0.1 µM and similar to other PSII inhibitors. It is worth mentioning that sorgoleone belongs to the His215 family of PSII inhibitors, while atrazine belongs to Ser264. Mutation in Ser264 of the D1 protein is responsible for resistance to triazines as well as other non-triazine herbicides, leading to cross-resistance. However, plants resistant to atrazine, with a QB binding site on PSII mutation (Ser264), are not resistant to sorgoleone. Application of sorgoleone is particularly justified in the case of triazine-resistant biotypes of redroot pigweed, due to the same physiological effects as applications of atrazine in redroot pigweed-susceptible biotypes [11]. These properties make sorgoleone a potential early post-emergence herbicide when applied as a spray with much less environmental implications than atrazine. Therefore, inhibition of photosynthesis is the main target site of sorgoleone action in young seedlings but its mode of action in older plants may be different [12]. Sorgoleone can be a useful inhibitor of p-hydroxyphenylpyruvate dioxygenase (HPPD), which takes part in α-tocopherol and plastoquinone synthesis. Inhibition of that enzyme leads to a decreased pool of available plastoquinone and indirectly affects activity of phytoene desaturase, a key enzyme in carotenoid synthesis. Such sequence of events causes declining carotenoid levels and affects photosynthesis [45]. Currently used triketone herbicides (e.g. sulcotrione, isoxaflutole) have the same mechanism of action on HPPD as sorgoleone, irreversible competitive inhibition, with I_{50} = 0.4 μM. Triketone herbicides are considered by the U.S. Environmental Protection Agency (EPA) to be a low environmental risk. They are usually utilized as selective herbicides to eliminate broadleaf weeds in corn [10]. It follows, due to similar action and chemical structure and environmental friendly profile, sorgoleone might also be useful as a selective herbicide; however, such comparison studies have yet to be conducted. Then, its mode of action also cannot explain whether it is more or less active on broadleaf or grass weeds species [44].

5.2. Momilactones

Extracts and residues of rice, the well-known cereal plant, also have allelopathic potential. Among isolated secondary metabolites, phenolic acids, hydroxamic acids, fatty acids, terpenes and indoles were identified [46]. The key role in rice allelopathy plays momilactone A and B isolated from root exudates. High allelopathic rice varieties release up to 2-3 μg of momilactone B per day [3]. These compounds inhibited the growth of typical weeds in rice, e.g. barnyard grass and awnless barnyard grass (*Echinochloa colona* (L.) Link.) at concentrations higher than 1 μM and 10 μM, respectively. Furthermore, phytotoxic abilities of momilacton A and B were also demonstrated on livid pigweed (*Amaranthus lividus* L.), hairy crabgrass and annual bluegrass (*Poa annua* L.) at concentrations higher than 60 μM and 12 μM, respectively [47]. The experiment has shown that momilactone B is secreted by rice roots into the rhizosphere over the entire life cycle [48]. Momilactone A and B belong to the diterpenoid phytoalexins which are known as antimicrobial secondary metabolites generated in response to signal molecules called elicitors (especially biotic elicitors) [49]. Both compounds thought to be unique to rice, recently have been found in the moss (*Hypnum plumaeforme* Wils.), a taxonomically distinct plant [49]. Despite the ability of momilactone A and B to inhibit plant growth, its mode of action in plants is still unknown.

5.3. Artemisinin

Artemisinin is a sesquiterpenoid lactone of annual wormwood (*Artemisia annua* L.). It is synthesized and sequestered in glandular trichomes located on the leaves and flowers [51]. It can also be excreted by the roots or root hairs, but only at the beginning of the growing season; therefore, dead leaves are the major source of artemisinin in soils [52]. Artemisinin is also lost from annual wormwood by rain runoff but to a minor degree (<0.5%),. This allelopathin is well known as a promising anti-malaric agent but also as a phytotoxin selective mainly to broadleaf weeds. Artemisinin (at 33 µM) significantly reduced shoot and root growth of lettuce, redroot pigweed, pitted morning-glory (*Ipomoea lacunose* L.) common purslane and annual wormwood [53]. However, the same treatment had no effect on sorghum or velvetleaf. Several studies have been aimed at identifying the molecular target site of this compound as well as the structural requirements for herbicidal activity [53-55].The effect of artemisinin is most evident on root growth and chlorophyll content. In onion root tips, artemisinin (10 - 100 µM) decreased the mitotic index, provoked abnormal mitotic figures and caused structural modifications of chromosomes [55]. However, no definite target site has yet been identified. The most recent studies on rice sprayed with 1.86 µM artemisinin indicated its inhibiting abilities on photo-synthetic electron transport [56]. Artemisinin site of action is probably plastoquinone B in photosystem II. Interestingly, as authors suggest, this effect is caused not directly by artemi-sinin itself, but rather by an unidentified artemisinin-metabolite occured in the plant after artemisinin application [56].

Other controversies around the phytotoxic potential of artemisinin arose when the dichloro-methane extracts of annual wormwood leaves containing artemisinin showed a stronger phytotoxic effect on redroot pigweed seed germination and seedling growth than pure artemisinin [57]. Moreover, aqueous extract with disposed artemisinin had equal inhibitory effects on both physiological processes as allelopathin alone. This experiment suggests a marginal role of artemisinin in plant extract and joint action of other allelochemicals. Although, most studies analyzing allelopathic weed–crop interferences using annual wormwood were conducted under laboratory and greenhouse conditions [58].

Toxic studies on duckweed (*Lemna minor* L.) and the fresh water algae *Pseudokirchneriella subcapitata* (Korshikov) had EC_{50} values 0.24 and 0.19 mg L^{-1} respectively, with growth rate as endpoint corresponding to those of the herbicide atrazine [59]. These profiles questioned environmental safety of artemisinin for the purpose as a bioherbicide. It may be a result of its complex chemical structure, but this compound may be used as the ba-sis for a new herbicide, based on artemisinin chemical structure. Such attempts have al-ready been made using artemisinin's analogues [55]. Four of the tested 12 analogues inhibited germination and root growth of lettuce, *Arabidopsis thaliana* (L.) and duckweed at extremely low concentrations (3 µM).

5.4. Leptospermone

Leptospermone (1-hydroxy-2-isovaloryl-4,4,6,6-tetramethyl cyclohexen-3,5-dione) is a natural triketone produced by the roots of the bottlebrush (*Callistemon citrinus* Curtis) [60]. In its pure form, it was tested both pre- and post-emergence on a range of plant species including: hairy crabgrass, yellow foxtail (*Setaria glauca* (L.) P. Beauv.), barnyard grass, California red oat (*Avena sativa* L.), redroot pigweed, Indian mustard (*Brassica juncea* L.) and curly dock (*Rumex crispus* L.). Leptospermone is a strong p- hydroxyphenylpyruvate dioxygenase (HPPD) inhibitor with I_{50} values 3 µg mL^{-1}[61]. Inhibition of this enzyme leads to disruption in carotenoid biosynthesis and loss of chlorophyll. Unfortunately, a pure compound rate of 9000 g a.s. ha^{-1} was required to give acceptable weed control. Such high doses excluded leptospermone from commercial development. The structure of this allelochemical was used as a basis for development of synthetic analogues including mesotrione (trade name Callisto), an herbicide produced by Syngenta AG. Mesotrione is applied for control of broadleaved weeds in maize. The rates of mesotrione are in the range from 75 to 225 g a.s. ha^{-1} (around 100 times more potent than leptospermone) [60].

However, leptospermone has lately been found as the main herbicidal component of manuka oil (*Leptospermum scoparium* J.R., G. Forst) [61]. Manuka oil (1%) applied as post-emergence spray, significantly decreased growth and dry weight of redroot pigweed, barnyardgrass, velvetleaf and hairy crabgrass. Though, hairy crabgrass seedlings that emerged after manuka oil application were totally blanched. Pre-emergence application of 0.17% manuka oil which corresponds to 0.2 L ha^{-1} of leptospermone inhibited hairy crabgrass growth by 50%. The pre-emergence effects are mainly dependent on its persistence in soil. Average time of leptospermone half-life in soil was calculated at 15 days while applied as a compound of manuka oil time extended by 3 days. This clearly shows that half-life of active compounds may be longer in mixture than applied alone due to additive or synergistic action. This type of leptospermone application poses another possibility of usage for this compound in its natural form without chemical modification of the structure [61].

5.5. Essential oils

Lately, there has been a growing interest for using essential oils as allelopathins with bioherbicide potential. Some of them have already been commercialized and successfully launched in organic agriculture. They disrupt the cuticle and contribute to desiccation or burn down young tissues. Examples of this are the commercially available bioherbicide with the trade name of GreenMatch EX which consists of lemongrass (*Cymbopogon* sp.) oils or Interceptor™ with 10% pine (*Pinus sylvestris* L.) oil [3]. Essential oils are complex mixtures of monoterpenes, sesquiterpenes, and aromatic phenols, oxides, ethers, alcohols, esters, aldehydes and ketones [62]. The main terpenoids of volatile essential oils are monoterpenes (C10) and sesquiterpenes (C15). It has been well documented that essential oils found in foliage of eucalyptus (*Eucalyptus* sp.) show phytotoxic potential. During field experiments it has been reported that common weeds such as coffee senna (*Cassia occidentalis* L.) and barnyardgrass sprayed with different concentrations of eucalyptus oil (from 5 % to 10 % v/v with 0.05 % v/v Tween-80) exhibited dose-dependent and species-dependent levels of injury. Coffee senna plants were more sensitive to the eucalyptus oil than barnyardgrass [62]. Phytotoxicity of eucalyptus oil is due to the components such as 1,8-cineole, citronellal, citronellol, citronellyl acetate, p-cymene, eucamalol, limonene, linalool, α-pinene, γ-terpinene, α-terpineol, alloocimene, and aromadendrene [62]. Pre-emergence herbicidal activity of 1,8-cineole 3, and 1,4-cineole 4 were tested against rigid ryegrass and garden radish var. Long Scarlet in laboratory-based bioassays. 1,8-cineole and its derivatives showed a dose-dependent herbicidal activity against both weed species [64]. Laboratory studies [64, 65] also have shown that soil-applied 1,8-cineole suppressed the growth of several weeds. However, field reports demonstrated that 1,8-cineole alone has poor herbicidal activity [67, 68]. The commercial herbicide cinmethylin is a 2-benzyl ether substituted analog of the monoterpene 1,4-cineole (1-methyl-4-(1-methylethyl)-7-oxabicyclo heptane). This compound was discovered and partially developed by Shell Chemicals as a derivative of the allelopathic natural monoterpene, 1,8-cineole [69]. The benzyl ether substitution appears to decrease the volatility of the cineole ring by several orders of magnitude thereby rendering it more suitable for herbicide use [70]. Cinmethylin is a moderately effective growth inhibitor used for monocot weed control [71]. Despite the fact that it has been used commercially in both Europe and Japan and has been studied experimentally for several decades, the mechanism of action of this herbicide is still unknown [54, 72]. Cinmethylin was commercialized outside the United States in 1982 under the trade names of Cinch and Argold. Cinmethylin is active on several important grasses in rice; Echinochloa sp., Cyperus sp. and heartshape false pickerelweed (*Monochoria viginalis* Burm.f.) at rates from 25 to 100 g a.s. ha^{-1} [73].

5.6. Sarmentine

Sarmentine was first isolated from long pepper (*Piper longum* L.) fruits [74] but is also present in varied organs of other *Piper* species (Huang and Asolkan patent). It has been known as a medicinal plant with many beneficial multidirectional properties on human health. However, methanol extract of long pepper dry fruits has been shown to be suppressive to lettuce [75]. Purification and fractioning of long pepper crude extract allows the dissection of the active compound – sarmentine, a molecule with a long unsaturated fatty acid chain and pyrrolidine. Due to the hydrophobic properties, sarmentine is suspended with surfactants, 0.2 % glycospere O-20, 2% ethanol and 0.1% sodium lauryl sulfate. As a foliar spray, it is active at 2.5 mg mL^{-1}, but its high phytotoxicity is manifested at 5 mg mL^{-1}. Higher concentrations of sarmentine caused almost 100% mortality of redroot pigweed, barnyardgrass, bindweed (*Convonvulus* sp.), hairy crabgrass, sprangletop (*Leptochloa* sp.), annual bluegrass, wild mustard (*Sinapis arvensis* L.), curly dock with impaired effects on horseweed (*Conyza canadensis* (L.) Cronquist) and sedge (*Carex* sp.) growth under laboratory conditions. First phytotoxic symptoms such as bent stems and contact necrosis, have been visible 30 minutes after application; however full-blown implications were seen 7 h after spraying. The most likely mechanism of sarmentine action on plants is disruption of the plant cuticle which leads to disruption of cell membranes and lipid peroxidation followed by formation of radicals [76, 77].

As an herbicide, sarmentine and its derivatives may be both obtained from fruits of long pepper and successfully chemically synthesized [75]. Despite the fact that the compound is active under laboratory conditions, its chemical and biological instability under field conditions may limit its application as an herbicide. However, it has been shown that crucial for sarmentine herbicidal activity is the presence of an amine bond with a secondary amine. Replacement of the acid moiety with structurally similar fatty acids has not changed its phytotoxic potential. Moreover, natural herbicides based on sarmentine may also contain other derivatives with similar modes of action on plants but higher environmental stability [75]. Sarmentine may be successfully applied in combination with synthetic herbicides, e.g. aryloxyphenoxypropionic, benzoic acid, dicarboximide, organophosphorus, triazine, sulfonamide herbicides and with many others. This gives an opportunity to further the structural modification that makes the compound more stable without any disadvantages on bioherbicide action in plants. It is worth noting that sarmentine has already been patented as an herbicide but not commercialized yet [75].

6. Biotechnology in bioherbicide investigation

A lot of effort has been done to explore the nature of allelopathic interactions. Studies on allelopathic compounds greatly increased thanks to chemical and biochemical techniques, which improved identification and knowledge about its mode of action. Since then, the crucial role of secondary metabolites synthesized and released by plants became better understood. It has been clearly demonstrated that allelopathins may take part in very complex inter- and intra-specific ecological interactions including soil microorganisms. However, despite the extensive research carried out under laboratory conditions, the higher level of such interactions at the ecosystem level has not been sufficiently explored. Structure, chemical properties, and mode of action in plants of multitude allelochemicals are already known but, unfortunately, only a part-per thousand of them have been successfully introduced in agricultural practices. This is mainly due to limitations of compounds as plant protection agents but also due to extended field experiments. A very important aspect that allows the introduction of allelopathy to natural weed management is knowledge about biology of donor and target plants and the exact chemicals responsible for the interaction [78]. All formerly described limitations of natural compounds as bioherbicides decreasing in case of plant extracts as herbicides due to simple and low cost of application. However, separation of one, specific compound that is the most interesting for us among hundreds synthesized by plants often required information about its synthesis *in vivo*.

One of the problems is to obtain adequate amounts of the compound, when its chemical synthesis is impossible or collection of plants, unprofitable. Increased synthesis of an allelopathin gives triple profits. First of all, enhanced allelopathic potential of a plant makes it more competitive against weeds. Second of all, increased concentration of a compound makes plant extract more active. Thirdly, this allows collection of the compound at a sufficient amount and makes it more profitable. However, it is much easier to obtain active compounds from the crop species than wild living ones. Difficulties in introducing plants to cultivation are due to the low ability to grow outside their natural ecosystem [79].

Cells and organ cultures provide opportunities to circumvent these limitations. Abilities of undifferentiated and differentiated cells to produce allelochemicals may be commercialized in bioreactors using cell suspension cultures [79]. Such attempts have been made on Artemisia suspension culture for artemisinin production; however, obtained amounts of that compound were insufficient. The addition of β-cyclodextrins to the growing medium has increased artemisinin synthesis up to 300% [80]. Allelochemicals produced by roots may be obtained from hairy root cultures, both *via* callogenesis or infection with *Agrobacterium tumefaciens*. Transgenic hairy roots are characterized by high genetic stability and facility to accumulate metabolites. The hairy root system already has been applied to increased production of phenolic compounds of nettleleaf goosefoot (*Chenopodium murale* Linn.) [81] and gossypol of cotton [82]. Active growth of roots and rapid colonization of the bioreactor allows rapidly reaching target weight, necessary to obtain an adequate quantity of the compound extracted from plants or growing medium.

The recombinant DNA technology can be useful to improve allelochemical production. Enhancing or suppression of gene expression, metabolic engineering and genetic transformation are promising new tools for allelochemical synthesis [79]. This approach is based on elucidation of the metabolic pathway, enzyme activities and identification of genes encoding crucial enzymes, associated with metabolite (allelochemical) synthesis.

Allelopathy is a quantitative trait. A genetic analysis of quantitative trait loci (QTL) is a promising approach to identify genes underlying this trait. Only a few crops are under genetic screening for its allelopathic properties including: rice, wheat, barley and oat [83, 84]. The first QTL map associated with allelopathic properties was developed in rice. A segregating population derived from a cross of two cultivars varying with allelopathic potential against barnyardgrass. The map contained 140 DNA markers with four main-effects QTL located on chromosome 2, 3 and 8 [85]. Proteomic studies on allelopathy of rice against barnyardgrass confirmed the crucial role of three enzymes: phenylalanine ammonia-lyse (PAL), thioredoxin and 3-hydroxy-3-methilglutarilcoenzyme A reductase 3 (HMGR) is highly involved in phenols biosynthesis [86]. Such a genetic approach may allow the location of the gene in the genome and better understanding of its function in plant allelopathy and create the chance of applying marker assisted selection (MAS) to enhance allelopathic abilities.

Just like breeding programs allow improved crop production, they may also improve production of allelopathic compounds increasing allelopathic potential.

Scopoletin has been known as allelopathic root exudates of oats (*Avena* sp.) that affects growth of neighboring plants. Screening of 3000 of Avena accessions has shown varying ability to scopoletin production. Twenty five of them have exuded higher amounts of scopoletin than control cultivar Garry, of which 4 were threefold more than the control [87]. Variation in allelopathin production was also discovered for sorgoleone of seven sorghum accessions [38] nomilacton A and B of 8 rice accessions [88] DIBOA and DIMBOA of 14 rye cultivars [88], gramine of 43 lines of modern cultivar of barley (*Hordeum vulgare* L.) and wild progenitor *H. spontaneum* (C. Koch) [90]. Enhanced production of active compounds from allelopathic plants can be developed by efficient breeding - selection of individuals with high allelopathic ability. Identification of a single gene, arranged in synthesis of allelopathin already has been performed for sorgoleone. *SOR1* (or compatible *SbDES3*) expression is specific for root hairs of two species of sorghum (*S. bicolor* and *S. halepense*) and associated with sorgoleone synthesis, while it is not expressed in other organs of sorghum *SOR1* encodes novel fatty acid desaturase (FAD), involved in the formation of a specific bond at $16:3\Delta^{9,12,15}$ pattern [91, 92]. Comparative studies of FAD derived from sorghum with other desaturases showed high similarity to omega-3 fatty acid desaturases (FAD3) [93]. However, none of the hitherto known desaturases can synthesize double bonding at this unique pattern along the aliphathic chain of the sorgoleone molecule. Characterization of this gene allows an overexpression of *SOR1* and increased sorgoleone synthesis and improved allelopathic potential of sorghum, as well as easier collection of the compound. Moreover, the well-known pathway of sorgoleone synthesis and characteristic of candidate genes may be a promising source of introducing sorgoleone production to grass crops [94].

The situation becomes more complicated when more than one gene encoding special enzymes is required to increase synthesis of a plant compound. Such difficulties have been encountered for DIBOA, synthesized by various grass species [95]. In maize, biosynthesis of this compound is determined by five genes (*Bx1* to *Bx5*) encoding three enzymes: tryptophan synthase α homolog, cytochrome P-450 dependent monooxygenase [95].

Monoterpenes are a large family of compounds produced by a varied family of aromatic plants. Some of the monoterpenes also take part in allelopathic interactions, e.g. linalool, cineole camphene, pinene, limonene, etc. Currently, metabolic engineering allows improved production of specific compounds in heterologous systems [96]. The most interesting are monoterpene synthases which catalyzed geranyl diphosphate (GPP) into output structure of numerous monoterpenes family, e.g. enhanced expression of limonene synthase in transgenic peppermint (*Mentha piperita* L.) has increased yield of monoterpenes. An alternative approach is to change the density of secretory structures by both plant hormone and transcriptional factors manipulation. Such attempts already have been made in annual wormwood and *A. thaliana*. It was recently found that the number of glandular trichomes increased in response to jasmonic acid. Spraying of annual wormwood with this hormone significantly increased density of these structures on leaves what was accompanied with higher artemisinin content [51]. This was an effect of enhanced expression of gene encoding enzymes taking part in artemisinin biosynthesis. On the other hand, in Arabidopsis, co-expression of two positive transcriptional factors (GL1, and R protein of maize) has significantly improved the number of trichomes [96].

However, we have to bear in mind that biosynthesis of natural compounds can be limited to organs, tissues or even cells. Specific locations of compound synthesis, accumulation or secretion often make that compound toxic to other tissues within the same plant organism. Moreover, even successful transformation of a plant does not guarantee successful and sufficient production of a desirable compound. The gene of (S)-linalool synthase (*Lis*) of fairy fans (*Clarkia breweri* Gray), constitutively expressed in transgenic petunia (*Petunia hybrida* Hook.), has produced linalool but in its glycosylated, non-volatile form [96].

All presented techniques provide greater knowledge on allelopathy. However, better understanding of such complex interactions among this phenomenon bring us one step forward to development of new strategies in weed management and finding new herbicides and new herbicidal target sites.

7. Conclusions

The phenomena of allelopathy and phytotoxic interactions between plants are strongly expanding branches of biological science. Allelochemicals, as a group of substances also called biocommunicators, seem to be a fruitful challenge for combining traditional agricultural practices and new approaches in pest management strategies. Allelochemicals have already been used to defend crops against pathogens, insects or nematodes, parallel to some attempts to use them for weed control. Crop rotation, cover crops, dead and living mulches are being employed in agriculture. Both in natural and agricultural ecosystems allelopathic interactions are involved in practically every aspect of plant growth, as they can play the role of stimulants and suppressants. Complex plant-plant and plant-microbe interactions in ecosystems and currently developing studies on molecular, cytological and physiological levels bring us to a better understanding of processes occurring around us. The ancient knowledge of well-known toxic properties of water extracts of a variety of allelopathic plants give us a basis that could be used in the creation of a novel approach in weed control.

Some allelochemicals, mainly these that are mentioned in the text above, may act as a starting point for production of new bioherbicides with novel target sites, not previously exploited, as the understanding of their mode of action is still growing. Creation of bio-herbicides based on allelochemicals generates the opportunity to exploit natural com-pounds in plant protection and shows the possibility to cope with evolved weed resistance to herbicides. Despite the fact that we have extensive knowledge about the chemical nature of natural compounds, we can synthesize its analogues, and we have ba-sically explored its phytotoxic potential, we still have insufficient data. Until recently, most studies on phytotoxicity have been conducted under laboratory conditions due to the ability to eliminate other environmental factors such us temperature, soil texture and its chemical and physical properties. Such approach allows the recognition of only direct effects of allelochemical action. There is still a great need to transfer laboratory data into field conditions. Such experiments are not willing to be taken on due to troublesome field experiments dependent on environmental conditions and a few year repetitions. New tools of molecular genetics, proteomics and metabolomics profiling as well as modern and sophisticated methods of chemistry and biochemistry will lead to the creation of sub-stances, maybe based on the structure of particular compounds occurring in nature, which could be used without any risks as selective and eco-friendly herbicides.

Author details

Dorota Soltys[1*], Urszula Krasuska[1], Renata Bogatek[2] and Agnieszka Gniazdowska[2]

*Address all correspondence to: d.soltys@ihar.edu.pl

1 Laboratory of Biotechnology, Plant Breeding and Acclimatization Institute - National Re-search Institute, Mlochow, Poland

2 Department of Plant Physiology, Warsaw University of Life Sciences – SGGW, Warsaw, Poland

References

[1] Rola, H, Marczewska, K, & Kucharski, M. Zjawisko odporności chwastów na herbicydy w uprawach rolniczych. Studia i Raporty IUNG-PIB (2007). , 8-29.

[2] International Survey of Herbicide Resistance Weedshttp://www.weedscience.org/In.aspaccessed 31 October (2012).

[3] Dayan, F. E, Cantrell, C. L, & Duke, S. O. Natural products in crop protection. Bioorganic & Medicinal Chemistry (2009). , 17(12), 4022-4034.

[4] Gniazdowska, A, & Bogatek, R. Alleopathic interaction between plants. Multiside action of allelochemicals. Acta Physiologiae Plantarum (2005). B) , 395-407.

[5] Soltys, D, Rudzinska-langwald, A, Gniazdowska, A, Wisniewska, A, & Bogatek, R. Inhibition of tomato (*Solanum lycopersicum* L.) root growth by cyanamide is due to altered cell division, phytohormone balance and expansin gene expression. Planta (2012). , 236(5), 1629-1638.

[6] Weir, T. L, Park, S-W, & Vivanco, J. M. Biochemical and physiological mechanisms mediated by allelochemicals. Current Opinion in Plant Biology (2004). , 7(4), 472-479.

[7] Iqbal, A, & Fry, S. C. Potent endogenous allelopathic compounds in *Lepidium sativum* seed exudate: effects on epidermal cell growth in *Amaranthus caudatus* seedlings. Journal of Experimental Botany (2012). , 63(7), 2595-2604.

[8] Hussain, M. I, & Reigosa, M. J. Allelochemical stress inhibits growth, leaf water relations, PSII photochemistry, non-photochemical fluorescence quenching, and heat energy dissipation in three C3 perennial species. Journal of Experimental Botany (2011). , 62(13), 4533-4545.

[9] Li, Z-H, Wang, Q, Ruan, X, Pan, C-D, & Jiang, D-A. Phenolics and Plant Allelopathy Molecules (2010). doi:10.3390/molecules15128933, 15(12), 8933-8952.

[10] Vyvyan, W. R. Allelochemicals as leads for new herbicides and agrochemicals. Tetrahedron (2002). , 58(9), 1632-1646.

[11] Nimbal, C. I, Yerkes, C. N, Weston, L. A, & Weller, S. C. Herbicidal activity and site of action of the natural product sorgoleone. Pesticide Biochemistry and Physiolology (1996). , 54(2), 73-83.

[12] Dayan, F. E, Howell, J, & Widenhamer, J. D. Dynamic root exudation of sorgoleone and its *in planta* mechanism of action. Journal of Experimental Botany (2009). , 60(7), 2107-2117.

[13] Belz, R. G, Hurle, K, & Duke, S. O. Dose-response- A challenge for allelopathy? Nonlinearity in Biology, Toxicology and Medicine (2005). , 3(2), 173-211.

[14] Duke, S. O. Natural pesticides from plants. In: J. Janick and J.E. Simon (eds.) Advances in new crops. Portland: Timber Press; (1990). , 511-517.

[15] Dayan, F. E, Owens, D. K, & Duke, S. O. Rationale for a natural products approach to herbicide discovery. Pest Management Science (2012). , 68(4), 519-528.

[16] Bhowmik, P. C. Inderjit. Challenges and opportunities in implementing allelopathy for natural weed management. Crop Protection (2003). , 22(4), 661-671.

[17] Irshad, A, & Cheema, Z. A. (2005). Comparative efficacy of sorghum allelopathic potential for controlling barnyardgrass in rice. Proceedings of the 4th World Congress on Allelopathy, Wagga Wagga, New South Wales, Australia. http://www.regional.org.au/au/allelopathy/2005/2/4/2220_irshada.htm

[18] Khaliq, A, Cheema, Z. A, Mukhtar, M. A, & Ahmad, S. M. Evaluation of sorghum (Sorghum bicolor) water extract for weed control in soybean. International Journal of Agriculture and Biology (1999). , 1(1), 23-26.

[19] Cheema, Z. A, & Khaliq, A. Use of sorghum allelopathic properties to control weeds in irrigated wheat in semi arid region of Punjab. Agriculture, Ecosystems & Environment. (2000).

[20] Anjum, T, & Bajwa, R. Field appraisal of herbicide potential of sunflower leaf extract against Rumex dentatus. Field Crops Research (2007).

[21] Anjum, T, & Bajwa, R. The effect of sunflower lear extracts on Chenopodium album in wheat fields in Pakistan. Crop Protection (2007). , 26(9), 1390-1394.

[22] Naseem, M, Aslam, M, Ansar, M, & Azhar, M. Allelopathic effects of sunflower water extract on weed control and wheat productivity. Pakistan Journal of Weed Science Research (2009). , 15(1), 107-116.

[23] Bhadoria PBSAllelopathy: a natural way towards weed management. American Journal of Experimental Agriculture (2011). , 1(1), 7-20.

[24] Iqbal, J, Cheema, Z. A, & Mushtaq, M. Allelopathic crop water extracts reduce the herbicide dose for weed control in cotton (Gossypium hirsutum). International Journal of Agriculture and Biology (2009). , 11(4), 360-366.

[25] Razzaq, Z. A, Cheema, K, Jabran, K, Farooq, M, Khaliq, A, & Haider, G. Basra SMA. Weed management in wheat through combination of allelopathic water extract with reduced doses of herbicides. Pakistan Journal of Weed Science Research (2010). , 16(3), 247-256.

[26] Razzaq, A, Cheema, Z. A, Jabran, K, Hussain, M, Farooq, M, & Zafar, M. Reduced herbicide doses used together with allelopathic sorghum and sunflower water extracts for weed control in wheat. Journal of Plant Protection Research (2012). , 52(2), 281-285.

[27] Fenwick, G. R, Heaney, R. K, & Mullin, W. J. Glucosinolates and their breakdown products in food and food plants. Critical Reviews in Food Science and Nutrition (1983). , 18-123.

[28] Petersen, J, Belz, R, Walker, F, & Hurle, K. (2001). Weed suppression by release of isothiocyanates from turnip-rape mulch. Agronomy Journal 2001;, 93(1), 37-43.

[29] Turk, M. A, & Tawaha, A. M. Allelopathic effect of black mustard (*Brassica nigra* L.) on germination and growth of wild oat (*Avena fatua* L.). Crop Protection (2003). , ??(4), 673-677,

[30] Uygur, F. N, Koseli, F, & Cinar, A. Die allelopathische Wirkung von *Raphanus sativus* L. Journal of Plant Diseases and Protection Zeitschrift für Pflanzenkrankheiten und Pflanzenschutz, Sonderheft (1990). XII , 259-264.

[31] Uludag, A, Uremis, I, Arslam, M, & Gozcu, D. Allelopathy studies in weed science in Turkey- a review Journal of Plant Diseases and Protection Zeitschrift für Pflanzenkrankheiten und Pflanzenschutz Sonderheft (2006). XX , 419-426.

[32] Akemo, M. C, Regnier, E. E, & Bennett, M. A. Weed suppression in spring-sown rye (*Secale cereale*)-pea (*Pisum sativum*) cover crop mixes. Weed Technology (2000). , 14(3), 545-549.

[33] Caamal-maldonado, J. A, Jiménez-osornio, J. J, Torres-barragán, A, & Anaya, A. L. The use of allelopathic legume cover and mulch species for weed control in cropping systems. Agronomy Journal (2001). , 93(1), 27-36.

[34] Macías, F. A. Galino JCG, Molinillo JMG, Castellano D. Dehydrozaluzanin C: a potent plant growth regulator with potential use as a natural herbicide template. Phytochemistry (2000). , 54(2), 165-171.

[35] Reigosa, M. J, Gonzalez, L, Sanches-moreiras, A, Duran, B, Puime, D, Fernandez, D. A, & Bolano, J. C. Comparison of physiological effects of allelochemicals and commercial herbicides. Allelopathy Journal (2001). , 8(2), 211-220.

[36] Alsaadawi, I. S, Al-ekeelie, M. H. S, & Al-hamzawi, M. K. Differential allelopathic potential of grain sorghum genotypes to weeds. Allelopathy Journal (2007). , 19(1), 153-160.

[37] Lehle, F. R, & Putman, A. R. Allelopathic potential of sorghum (Sorghum bicolor): isolation of seed germination inhibitors. Journal of Chemical Ecology (1983). , 9(8), 1223-1234.

[38] Czarnota, M. A, Rimando, A. M, & Weston, L. A. Evaluation of root exudates of seven sorghum accessions. Journal of Chemical Ecology (2003). , 29(9), 2073-2083.

[39] Netzly, D. H, & Butler, L. G. Roots of sorghum exude hydrophobic droplets containing biologically active components. Crop Science (1986). , 26(4), 775-778.

[40] Czarnota, M. A, Paul, R. N, Dayan, F. E, Nimbal, H. I, & Weston, L. A. Mode of action, localization of production, chemical nature, and activity of sorgoleone: a potent

PSII inhibitor in Sorghum spp. root exudates. Weed Technology (2001). , 15(4), 813-825.

[41] Weston, L. A, & Czarnota, M. A. Activity and persistence of sorgoleone, a long-chain hydroquinone produced by Sorghum bicolor. Journal of Crop Production (2001). , 4(2), 363-377.

[42] Trezzi, M. M, Vidal, R. A, & Dick, D. P. Peralba MCR, Kruse ND. Sorptive behavior of sorgoleone in soil in two solvent systems and determination of its lipophilicity. Journal of Environmental Science and Health (2006). , 41(4), 345-356.

[43] Hejl, A. M, & Koster, K. L. The allelochemical sorgoleone inhibits root H^+-ATPase and water uptake. Journal of Chemical Ecology (2004). , 30(11), 2181-2191.

[44] Einhellig, F. A, & Souza, I. F. Phytotoxicity of sorgoleone found in grain sorghum root exudates. Journal of Chemical Ecology (1992). , 18(1), 1-11.

[45] Meazza, G, Scheffler, B. E, Tellez, M. R, Rimando, A. M, Romagni, J. G, Duke, S. O, Nanayakkara, D, Khan, I. A, Abourashed, E. A, & Dayan, F. E. The inhibitory activity of natural products on plant p-hydroxyphenylpyruvate dioxygenase. Phytochemistry (2002). , 59(3), 281-288.

[46] Kato-noguchi, H, Hasegawa, M, Ino, T, Ota, K, & Kujime, H. Contribution of momilactone A and B to rice allelopathy. Journal of Plant Physiology (2010). , 167(10), 787-791.

[47] Chung, I-M, Hahn, S-J, & Ahmad, A. Confirmation of potential herbicidal agents in hulls of rice, Oryza sativa. Journal of Chemical Ecology (2005). , 31(6), 1339-52.

[48] Kato-noguchi, H, Ota, K, & Ino, T. Release of momilactone A and B from rice plants into the rhizosphere and its bioactivities. Allelopathy Journal (2008). , 22(2), 321-8.

[49] Okada, A, Okada, K, Miyamoto, K, Koga, J, Shibuya, N, Nojiri, H, & Yamane, H. OsTGAP1, a bZIP transcription factor, coordinately regulates the inductive production of diterpenoid phytoalexins in rice. The Journal of Biological Chemistry (2009). , 284(39), 26510-26518.

[50] Kato-noguchi, H. Convergent or parallel molecular evolution of momilactone A and B: Potent allelochemicals, momilactones have been found only in rice and the moss *Hypnum plumaeforme*. Journal of Plant Physiology (2011). , 168-1511.

[51] Nguyen, K. T, Arseault, P. R, & Wethers, P. J. Trichomes + roots + ROS = artemisinin: regulating artemisinin biosynthesis in *Artemisia annua* L. In Vitro Cellular and Developmental Biology- Plant (2011). , 47(3), 329-338.

[52] Jessing, K. K, Cedergreen, N, Mayer, P, Libous-bailey, L, Strobel, B. W, Rimando, A, & Duke, S. O. Loss of artemisinin produced by *Artemisia annua* L. to the soil environment. Industrial Crops and Products. (2013). , 43-132.

[53] Duke, S. O, Vaughn, K. C, Croom, E. M, & Elsohly, H. N. (1987). Artemisinin, a constituent of annual wormwood (*Artemisia annua*) is a selective phytotoxin. Weed Science 1987;, 35(4), 499-505.

[54] DiTomaso JMDuke SO. Is polyamine biosynthesis a possible site of action of cinmethylin and artemisinin? Pesticide Biochemistry and Physiology (1991). , 39(2), 158-167.

[55] Dayan, F. E, Hernandez, A, Allen, S. N, Moraces, R. M, Vroman, J. A, Avery, M. A, & Duke, S. O. Comparative phytotoxicity of artimisinin and several sesquiterpene analogues. Phytochemistry (1999). , 50(4), 607-614.

[56] Bharati, A, Kar, M, & Sabat, S. C. Artemisinin inhibits chloroplast electron transport activity: mode of action. PLOS ONE (2012). e38942. doi:10.1371/journal.pone.0038942http://www.plosone.org/article/info%3Adoi%2F10.1371%2Fjournal.pone.0038942

[57] Lydon, J, Teasdale, J. R, & Chen, P. K. Allelopathic activity of annual wormwood (*Artemisia annua*) and its role of artemisinin. Weed Science (1997). , 45-807.

[58] InderjitNilsen ET. Bioassays and field studies for allelopathy in terrestrial plants: progress and problems. Critical Reviews in Plant Sciences (2003).

[59] Jessing, K. K. Production of biomedicine under different climatic conditions- Artemisinin as study case IOP Conf. Series: Earth and Environmental Science 6 (2009). doi: 10.1088/http://iopscience.iop.org/1755-1315/6/34/342026/pdf/ 1755-1315_6_34_342026.pdf

[60] Cornes, D. (2006). Callisto: A very successful maize herbicide inspired by allelochemistry. Maize Association of Australia 6th Triennial Conference. http://www.regional.org.au/au/allelopathy/2005/2/7/2636_cornesd.htm

[61] Dayan, F. E, Howell, J. L, Marais, J. P, Ferreira, D, & Koivunen, M. Manuka oil, a natural herbicide with preemergence activity. Weed Science (2011). , 59(4), 464-469.

[62] Batish, D. R, Singh, H. P, Kohli, R. K, & Kaur, S. Eucalyptus essential oil as a natural pesticide. Forest Ecology and Management (2008). , 256(12), 2166-2174.

[63] Batish, D. R, Setia, N, Singh, H. P, & Kohli, R. K. Phytotoxicity of lemon-scented eucalypt oil and its potential use as a bioherbicide. Crop Protection (2004). , 23(12), 1209-1214.

[64] Barton, A. F, Dell, B, & Knight, A. R. Herbicidal activity of cineole derivatives. Journal of Agricultural and Food Chemistry 210;, 58(18), 10147-55.

[65] Vaughn, S. F, & Spencer, G. F. Volatile monoterpenes as potential parent structures for new herbicides. Weed Science (1993). , 41(1), 114-119.

[66] Romagni, J. G, Allen, S. N, & Dayan, F. E. Allelopathic effects of volatile cineoles on two weedy plant species. Journal of Chemical Ecology (2000). , 26(1), 303-313.

[67] Halligan, J. P. Toxic terpenes from *Artemisia californica*. Ecology (1975). , 56(4), 999-1003.

[68] Heisey, R. M, & Delwiche, C. C. Phytotoxic volatiles from *Trichostema lanceolatum*. American Journal of Botany (1984). , 71(6), 821-828.

[69] Grayson, B. T, Williams, K. S, Freehauf, P. A, Pease, R. R, Ziesel, W. T, Sereno, R. L, & Reinsfelder, R. E. The physical and chemical properties of the herbicide cinmethylin. Pesticide Science (1987). , 21(2), 143-153.

[70] Vaughn, S. F, & Spencer, G. F. Synthesis and herbicidal activity of modified monoter-penes structurally similar to cinmethylin. Weed Science (1996). , 44(1), 7-11.

[71] Russell, S. G, Monaco, T. J, & Weber, J. B. Influence of soil moisture on phytotoxicity of cinmethylin to various crops. Weed Science (1991). , 39(3), 402-407.

[72] Baum, S. F, Karanastasis, L, & Rost, T. L. Morphogenetic effect of the herbicide Cinch on Arabidopsis thaliana root development. Journal of Plant Growth Regulation (1998). , 17(2), 107-114.

[73] Duke, S. O. Allelopathy: Current status of research and the future of the discipline: A commentary. Allelopathy Journal (2010). , 25(1), 17-30.

[74] Huang, H, Morgan, C. M, Asolkar, N. R, Koivunen, M. E, & Marrone, P. G. Phytotox-icity of sarmentine isolated from long pepper (*Piper longum*) fruit. Journal of Agricul-tural and Food Chemistry (2010). , 58(18), 9994-10000.

[75] Huang, H, & Asolkar, N. R. (2011). Use of sarmentine and its analogs for controlling plant pests. Patent Patentdocs: http://www.faqs.org/patents/accessed 27 January 2011).(20110021358)

[76] Fukuda, M, Tsujino, Y, Fujimori, T, Wakabayashi, K, & Böger, P. Phytotoxic activity of middle-chain fatty acids I: effects on cell constituents. Pesticide Biochemistry and Physiolology (2004). , 80(3), 143-150.

[77] Lederer, B, Fujimori, T, Tsujino, Y, Wakabayashi, K, & Böger, P. Phytotoxic activity of middle-chain fatty acids II: peroxidation and membrane effects. Pesticide Bio-chemistry and Physiolology (2004). , 80(3), 151-156.

[78] Macías, F. A, Molinillo, J. M, Varela, R. M, & Galindo, J. C. Allelopathy--a natural al-ternative for weed control. *Pest Management Science* (2007). , 63(4), 327-48.

[79] Bourgaud, F, Gravot, A, Milesi, S, & Gontier, E. Production of plant secondary me-tabolites: a historical perspective. Plant Science (2001). , 161(5), 839-851.

[80] Durante, M, Caretto, S, Quarta, A, De Paolis, A, Nisi, R, & Mita, G. Cyclodextrins en-hance artemisinin production in *Artemisia annua* suspension cell cultures. Applied Microbiology and Biotechnology (2011). , 90(6), 1905-1913.

[81] Mitic, N, Damitrovic, S, Djordjevic, M, Zdravkovic-korac, S, Nikolic, R, Raspora, M, Djordjevic, T, Maksimovic, V, Živkovic, S, Krstic-miloševic, D, Stanišic, M, & Nin-

kovic, S. Use of *Chenopodium murale* L. transgenic hairy root in vitro culture system as a new tool for allelopathic assays. Journal of Plant Physiology (2012). , 169(12), 1203-1211.

[82] Triplett, B. A, Moss, S. C, Bland, J. M, & Dowd, M. K. Induction of hairy root cultures from *Gossypium hirsutum* and *Gossypium barbadense* to produce gossypol and related compounds. In Vitro Cellular & Developmental Biology- Plant (2008). , 44(6), 508-517.

[83] Olofsdotter, M, Jensen, L. B, & Courtois, B. Improving crop competitive ability using allelopathy- an example from rice. Plant Breeding (2002). , 121(1), 1-9.

[84] Belz, R. G. Allelopathy in crop/weed interactions- an update. Pest Management Science (2007). , 63(4), 308-326.

[85] Jensen, L. B, Courtois, B, Shen, L, Li, Z, Olofsdotter, M, & Mauleon, R. P. Locating genes controlling allelopathic effects against *Echinochloa crus-galli* (L.) in upland rice. Agricultural Journal (2001). , 93(1), 21-26.

[86] Lin, W-X, He, H-Q, Shen, L-H, Chen, X-X, Ke, Y, Guo, Y-C, & He, H-B. A proteomic approach to analysing rice allelopathy on barnyard grass (*Echinochloa crus-galli* L.). 4th International Crop Science Congress 26.(2004). Queensland, Australia. http://www.cropscience.org.au/icsc2004/poster/2/4/1/1414_xionglw.htm, 08-1.

[87] Fay, P. K, & Duke, W. B. An assessment of allelopathic potential in *Avena* germplasm. Weed Science (1977). , 25-224.

[88] Kato-noguchi, H. Allelopathic substance in rice root exudates: rediscovery of momilactone B as an allelochemical. Journal of Plant Physiology (2004). , 161(3), 271-276.

[89] Copaja, S. V, Villarroel, E, Bravo, H. R, Pizarro, L, & Argandona, V. H. Hydroxamic acids in *Secale cereale* L. and the relationship with their antifeedant and allelopathic properties. Zeitschrift fuer Naturforschung Section C Journal of Biosciences (2006). , 61-670.

[90] Lovett, J. V. Hoult AHC. (1992). Gramine: the occurrence of a self defence chemical in barley, *Hordeum vulgare* L. In: Hutchinson KJ, Vickery PJ. (eds) Looking Back- Planning Ahead conference proceedings, February Australian Agronomy Conference. "". Edited by Proceedings of the 6th Australian Agronomy Conference, 1992, The University of New England, Armidale, New South Wales. http://www.regional.org.au/au/asa/1992/concurrent/alternative-practices-plant-protection/p.htm#TopOfPage, 10-14.

[91] Pan, Z, Rimando, A. M, Baerson, S. R, Fishbein, M, & Duke, S. O. Functional characterization of desaturases involved in the formation of the terminal double bond of an unusual 16:3Δ 9,12,15 fatty acid isolated from *Sorghum bicolor* root hairs. Journal of Biological Chemistry (2007). , 282(7), 4326-4335.

[92] Yang, X, Scheffler, B. E, & Weston, L. A. SOR1, a gene associated with bioherbicide production in sorghum root hairs. Journal of Experimental Botany (2004). , 55-2251.

[93] Yang, X, Owens, T. G, Scheffler, B. E, & Weston, L. A. Manipulation of root hair development and sorgoleone production in sorghum seedlings. Journal of Chemical Ecology (2004). , 30(1), 199-213.

[94] Weston, L. A, & Duke, S. O. Weed and crop allelopathy. Critical Reviews in Plant Sciences (2003).

[95] Frey, M, Chomet, P, Glawischnig, E, Stettner, C, Grün, S, Winklmair, A, Wolfgang, E, Bacher, A, Meeley, R. B, Briggs, S. P, Simcox, K, & Gierl, A. Analysis of a chemical plant defense mechanism in grasses. Science (1997). , 277(3526), 696-699.

[96] Mahmoud, S. S, & Croteau, R. B. Strategies for transgenic manipulation of monoterpene biosynthesis in plants. Trends in Plant Science (2002). , 7(8), 366-373.

Chapter 6

Integrating Herbicides in a High-Residue Cover Crop Setting

Andrew J. Price and Jessica A. Kelton

Additional information is available at the end of the chapter

1. Introduction

Sustainable agriculture requires the use of multiple, integrated weed management practices to ensure long-term viability. A number of cultural, mechanical, and chemical weed control options can be utilized in a production system to reduce weed interference and safeguard crop yield. The dependence on one single weed control strategy may result in short-term success; however, long-term use can lead to multiple setbacks including poor soil health, reduced crop production, and increasing herbicide resistance. In turn, employing multiple weed control tactics simultaneously may prove difficult without previous knowledge as to how best to implement an integrated weed management system. To that end, this chapter is dedicated to illustrating successful herbicide use in conjunction with cover crops and their residues, practices proven not only to suppress weed germination and growth, but also to reduce soil erosion and water runoff and build soil organic matter and thus subseqent productivity.

Use of cover crops, particularly those producing high amounts of biomass (greater than 4,500 kg ha^{-1}), can provide numerous benefits for a cropping system [1]. However, care must be taken when choosing herbicides to apply to these cover crops both prior to and after primary crop planting. This chapter provides an overview of effective herbicide choices for use prior to and within cover crop as well as efficient application methods for use after planting the primary crop(s). We also discuss herbicide interception by cover crop residue and means to control reduced efficacy due to interception. It is hoped that this summary will aid in the adoption of sustainable farming practices to ensure successful agricultural productivity for future generations.

2. Conservation agriculture

As demands are placed on agriculture to produce increasing yields for a growing global population, the need to implement systems with high productivity and sound environmental standards is key to ensuring agricultural sustainability for future generations. To this end, conservation agriculture is a systems-based approach for food, feed, and fiber production that utilizes a number of practices aimed at maintaining yields while limiting energy and chemical inputs, minimizing soil degradation and erosion, and reducing long-term, detrimental impacts to the environment [2]. Conservation agriculture is comprised of many different management practices, particularly cultural techniques such as crop rotation, planting date, and seeding rate, that can reduce dependence on chemical inputs for successful yield production. More-over, limited tillage practices, or conservation-tillage, is essential to conservation agriculture systems to ensure soil quality, reduce runoff, and lessen energy consumption on agricultural lands.

2.1. Conservation tillage

Conservation-tillage, or reduced-tillage, has been proven to provide multiple benefits in agricultural settings. In addition to erosion and runoff control, soil health improvement, and reduced energy demands, reduced-tillage practices can produce crops yields similar to that of conventional systems [3-5]. The use of reduced-tillage, however, can alter weed communities. Seed production by annual weed species remains, in most part, on the soil surface where it is subject to increased decomposition and predation. With reduced competition and minimized soil disruption, perennial weed species can become established and dominate the weed community in conservation-tillage [6]. To aid in the control of both annual and perennial weeds, the use of cover crops for ground cover can reduce herbicide requirements in conservation-tillage settings.

2.2. Cover crops

A number of cereal and legume cover crops are utilized in various crop productions for several purposes. Currently, a large portion of cover crops are planted as a green manure which are turned under prior to sowing the primary crop [7,8]. In reduced-tillage, however, cover crops are grown as a ground cover and remain on the soil surface after cover crop termination. In addition to further reducing soil erosion, increasing soil organic matter, and improving water infiltration, cover crops can provide a level of weed suppression both prior to and during the primary growing season [9]. When compared to fallow conservation-tillage systems, cover crops offer increased weed control through direct resource competition while actively growing as well as through shading and/or allelopathy after termination. Covers grown to produce high levels of biomass, in particular, can increase shading of germinating weed species and provide greater ground cover for an extended period during the growing season. When employing cover crops, however, knowledge concerning herbicide use both during cover crop production and primary crop growth is essential.

2.3. Herbicide use

2.3.1. Cover crop establishment and termination

To produce substantial cover crop biomass, it is imperative to adequately manage cover crop production. Besides using correct seeding rates, early planting dates, and sufficient fertilizer applications, it is important to be aware of herbicide applications made prior to cover crop establishment. Often times, postemergent (POST) herbicides applied late season or post-harvest can have residual carryover than may be detrimental to cover crops. Rotation restrictions listed on herbicide labels should be referred to when planning POST applications and cover crop species.

To manage cover crops before cash crop planting, herbicides are typically utilized for cover crop termination. Most often, these herbicides, such as glyphosate and glufosinate, are non-selective with little to no carryover risk. However, consideration should be given to in-season chemical weed control regimes in order to limit repeated applications of a single herbicide mode of action. Moreover, care should be taken to avoid reduced herbicide rates applied for cover termination to reduce the risk of herbicide resistance [10]. Recent research has focused on mechanical termination with a roller or crimper which may reduce or eliminate the need of these herbicides for cover crop termination [11].

2.3.2. Cash crop establishment and management

Although use of in-season herbicides can be substantially reduced when using high-residue cover crops, some chemical applications are generally required to achieve the most effective weed suppression and minimize crop loss due to weed competition. While an ideal agricultural system would require no chemical inputs for sufficient weed control, practicality dictates the use of herbicides to guarantee crop yield since no system, as of yet, exists that can successfully suppress weed populations without intensive labor or mechanical requirements. To this end, cover crops are a means to minimize, rather than eliminate, herbicide inputs in crop systems. In recognizing the fact that the majority of agricultural systems will require chemical weed control measures for optimum crop production even when utilizing cover crops, it is essential to understand how cover crops affect herbicide selection and efficacy for each crop.

Primarily, the use of reduced-tillage and cover crops eliminates the ability to utilize preplant incorporated herbicides which offer residual soil activity [11]. Furthermore, cover crop residue can impede preemergent (PRE) herbicide applications from reaching the soil surface, reducing herbicide efficacy [12]. While postemergent chemical weed control can be effective alternatives in these settings, many weed species can prove to be difficult to control if not killed early in the season. Moreover, resistance concerns essentially necessitate the use of preemergent herbicides with differing mechanisms of action to avoid selection pressure for resistant weed biotypes [13].

Along with many cultural pracitces, production of crops under reduced-tillage with cover crops requires development of specific herbicide regimes to ensure minimal chemical inputs while achieving sufficient weed control to allow for successful crop production. The following sections review major crops produced globally, describe research conducted in respect to reduced-tillage production, as well as list available herbicides for use when using reduced-tillage and cover crops. These reviews are designed to provide information that can be beneficial for producers implementing conservation-tillage.

3. Wheat

Global production of wheat (*Triticum aestivum* L.) was estimated at approximately 217 million hectares in 2010 [14] representing the largest single crop, in area grown, and providing approximately 19% of the caloric intake of the world's diet [15]. In recent years, concerns have been noted over stagnant wheat yields due to drought and rising temperatures attributed to global warming [16]. Efforts to maintain current wheat production levels and identify potential measures to aid in increasing yield have led researchers to explore conservation practices in wheat systems. In addition to preserving high crop yields, long-term conservation systems are intended to protect environmental quality and reduce chemical and energy inputs necessary for crop production. Components of conservation systems such as reduced- or no-tillage can produce crop yields equal to or exceeding conventional tillage practices while reducing erosion, water runoff, and increasing water infiltration.

Much research has been conducted to evaluate wheat productivity in conservation-tillage practices. Reports reveal similar or increased grain yield for reduced-tillage compared to conventional tillage systems [17-19]. With little or no tillage operations, some chemical applications are required to achieve successful levels of weed control; however, with herbicide applications, weed species have been effectively controlled below levels that could reduce yield [20]. To offset the herbicide needs in conservation-tillage, evaluations of cover crops as ground cover have been conducted. Crops such as mustard (*Sinapis alba* L.), pea (*Pisum sativum* L.), and lentil (*Lens culinaris* Medik.) have proven to be good choices with little yield differences [21]. However, other reports show negative impacts on wheat production when implementing cover crops prior to wheat production for reasons such as increased weed competition, primarily *Bromus* spp., and reduced fertilizer uptake [22].

Like most crops produced in conservation-tillage, herbicide options may be limited to a degree whether utilizing a cover crop or not. With reduced-tillage, preplant incorporation of residual herbicides cannot be utilized. Moreover, when planting into cover crops, soil-applied pre-emergent herbicides may be less effective due to interception by crop residue. When planting wheat, preplant burndown herbicides may be necessary to control early weeds. POST herbicides are also necessary to control weeds that germinate after planting. Table 1 lists many of the herbicide options for use in conservation-tillage systems for wheat production.

4. Maize

Maize, or corn (*Zea mays* L.), is one of the most economically important grain crops worldwide with 162 million ha produced in 2010 [2]. In addition to being a staple in human and livestock diets in many countries, corn is also used for bioethanol production and the manufacturing of many non-food products. Consumption of corn and products derived from corn continues to increase. Given the demand, it is imperative for sustainable production systems that produce high yields while preserving long-term productivity of the land to be implemented.

Conservation-tillage practices have been researched and utilized for several decades in some regions such as the Midwest in the United States. As with many other crops, some variability has been noted for corn yield in no-tillage systems compared to conventional tillage methods. However, many reports show at least equal corn yields can be achieved when tillage practices are reduced [3]. Adequate yield potential, coupled with the reduction of on-farm expenses, have made conservation-tillage systems a good fit for corn production.

Herbicide			
Common Name	**Trade Name**[a]	**Application Timing**	**Weed Species Controlled**
Carfentrazone	Aim* [23]	Preplant Burndown	Non-selective control of emerged broadleaves and grasses
Glufosinate	Liberty* [24]		
Glyphosate	Roundup WeatherMax*[25]		
Paraquat	Gramoxone* [26]		
Chlorsulfuron + Metsulfuron	Finesse* [27]	PRE or POST[b]	*Bromus* species, annual ryegrass (*Lolium multiflorum*), kochia (*Kochia scoparia*)
Pyrasulfotole + Bromoxynil	Huskie™ [28]	Early POST	Emerged broadleaf seedlings such as dandelion (*Taraxacum officinale*); suppression of established dandelion and henbit (*Lamium amplexicaule*)
Thifensulfuron + Tribenuron	Harmony* Extra [29]	POST	Actively growing broadleaves, wild garlic (*Allium vineale*); suppression of Canada thistle (*Cirsium arvense*)
Clearfield wheat			
Imazamox	Beyond* [30]	POST	Broadleaves henbit and chickweed (*Stellaria media*), grasses barnyardgrass (*Echinochloa crus-galli*), jointed goatgrass (*Aegilops cylindrica*), volunteer cereals (non-Clearfield types)

[a]Trade names listed are representative of available herbicides. Inclusion of particular trade names does not suggest author endorsement.

[b]PRE, preemergence; POST, postemergence.

Table 1. Herbicides for use in reduced-tillage wheat production.

A major limiting factor to adopting reduced-tillage in corn production is the concern of less effective weed control. Tillage has long been used as a means for weed seed burial which reduces the number of seeds in the upper portion of the soil, the area most favorable for germination for most species. In addition to weed seed remaining in the upper layer of soil, shifts in weed species have also been noted. With the implementation of conservation-tillage, most crop systems experience a shift in weed species from annuals to perennials dominating the weed community.

Perennial weed species, largely controlled with tillage practices, can thrive on less distur-bed crop land. For effective weed control, producers implementing reduced-tillage have relied on increased herbicide applications. To curb herbicide use, cover crops have been adopted in conjunction with reduced-tillage corn systems. Research has shown that utiliz-ing a legume or grain cover crop can reduce weed density and growth while not affecting corn yield [31,32]. For corn in particular, cover crops offer a potential benefit in addition to weed suppression. Adequate nitrogen availability is essential for corn development. The use of legume cover crops, such as hairy vetch (*Vicia villosa* Roth), red clover (*Trifolium pratense* L.), or medics (*Medicago* spp.), may provide a portion of corn nitrogen requirements and reduce fertilizer inputs into the system [33]. Some research indicates that legume covers do not reduce fertilizer requirements but improves grain production with standard fertilizer applications [34]. Other research shows that legume covers can provide some nitrogen required for successful corn production[35,36]. Selecting the right legume cover crop for maximum nitrogen contribution with timely availability for corn uptake is key for utilizing these crops as nitrogen sources.

Use of burndown herbicides prior to corn planting is critical for early season weed control when using cover crops. A residual herbicide applied in conjunction with the herbicide used for cover crop termination can broaden weed species controlled as well as extend control into the season. A number of PRE herbicides are available that can be applied without incorporation into the soil and are effective even with plant residue on the soil surface. These herbicides and POST herbicide choices that can be successfully utilized in conservation-tillage corn with cover crops are listed in Table 2.

Herbicide			
Common Name	Trade Name[a]	Application Timing	Weed Species Controlled
Glufosinate	Liberty° [24]	Preplant burndown	Emerged weed species
Glyphosate	Roundup WeatherMax° [25]		
Paraquat	Gramoxone° [26]		
2,4-D	Agri Star° 2,4-D [37]		
Atrazine	Aatrex° [38]	Preplant or PRE[b]	Broadleaves such as kochia (*Kochia scoparia*); suppression of foxtail (*Setaria* spp.), velvetleaf (*Abutilon theophrasti*). Can also be applied POST

| Herbicide | | | |
Common Name	Trade Name[a]	Application Timing	Weed Species Controlled
Flumioxazin	Valor° [39]		Broadleaf species such as horseweed (*Conyza canadensis*); suppression of grass species such as panicum (*Panicum* spp.) and goosegrass (*Eleusine indica*)
Pendimethalin	Prowl° [40]		Germinating, small-seeded grass and broadleaf species such as crabgrass (*Digitaria* spp.) and common lambsquarters (*Chenopodium alba*)
S-metolachlor	Dual Magnum° [41]		Grass and broadleaf species such as foxtail and *Amaranthus* spp.
Carfentrazone	Aim° [23]	POST[c]	Certain broadleaf weed control; tank mix with atrazine or dicamba
Bromoxynil	Buctril° [42]		Broadleaf weeds such as burcucumber (*Sicyos angulatus*), giant ragweed (*Ambrosia trifida*)
Dicamba	Banvel° [43]		Annual broadleaf species as well as certain perennial species such as dock (*Rumex* spp.) and wild onion (*Allium* sp.)
Mesotrione	Callisto° [44]	POST	Broadleaf species such as wild mustard (*Sinapis arvensis*), nightshade (*Solanum* spp.), and Canada thistle (*Cirsium arvense*)
Tembotrione	Laudis° [45]		Broadleaf and grass species such as common chickweed, purple deadnettle (*Lamium purpureum*), *Amaranthus* spp., and large crabgrass (*Digitaria sanguinalis*)
Ametryn	Evik° [46]	POST-directed spray	Grass species such as Texas panicum, goosegrass, and foxtail
Linuron	Lorox° [47]		Broadleaf and grass species such as dog fennel, common ragweed (*Ambrosia artemisiifolia*), velvetleaf, and annual ryegrass (*Lolium multiflorum*)
Clearfield Corn			
Imazethapyr + Imazapyr	Lightning° [48]	POST	Broadleaves, grasses, and sedges such as kochia, ragweed, quackgrass (*Elytrigia repens*), and nutsedge (*Cyperus* spp.)
LibertyLink Corn			
Glufosinate	Liberty°	POST	Broadleaf and grass species; ragweed, horseweed, johnsongrass seedlings
Roundup Ready Corn			

Herbicide			
Common Name	Trade Name[a]	Application Timing	Weed Species Controlled
Glyphosate	Roundup Weathermax[c]	POST	Nonselective control of some broadleaf and grass species
Glyphosate + s-metolachlor + atrazine	Expert[c] [49]	PRE or POST	Annual broadleaves and grasses; perennials such as quackgrass, dandelion (*Taraxacum officinale*), and Canada thistle

[a]Trade names listed are representative of available herbicides. Inclusion of particular trade names does not suggest author endorsement.

[b]PRE, preemergence.

[c]POST, postemergence.

Table 2. Herbicides for use in reduced-tillage maize production.

5. Rice

Production of rice (*Oryza sativa* L.) in 2010 was near 154 million ha worldwide [2]. In many regions, rice provides nearly half or more of calories consumed by humans [50] and is the most important grain crop grown. Rice yield has steadily grown in the past several decades due to breeding and fertilizer advancements; however, it is necessary for rice yield to continue to improve in order to meet increased demands by a growing world population. Given that little land exists in rice-producing countries to expand production, it is necessary for methods to be established that can continue yield improvement without depleting future soil productivity.

Wetland, transplant rice production is the dominant and highest yielding rice system in most regions [50, 51]. However, the water and energy requirements may limit rice production as competition for resources increases [52]. To reduce strain on environmental and economic resources and to ensure sustainable rice systems in the future, dry-seeded rice production has been implemented in some areas [53]. Dry-seeded rice production can be initiated in conjunction with conservation-tillage with fewer water demands, lower energy and labor requirements, and reduced soil erosion. Research has reported that dry-seeded rice in no-tillage can be a successful alternative to conventional systems [52].

A limiting factor to widespread adoption of dry-seeded, reduced-tillage rice, however, is reduced weed control. For rice, transitioning from wetland, conventional systems to a dry system with reduced-tillage can affect weed compositions in multiple ways. Standing water can reduce germinating weed seeds while the transplanted rice becomes established; removing this water barrier can increase weed numbers [54]. Additionally, reduced-tillage practices can result in an increase of weed seed germination due to less seed burial.

In dry-seeded rice, mulches have been suggested as a means to combat weed increases [51]. Little research has been conducted to fully understand the benefits of cover crops for weed control in rice; however, legume covers have been associated with increased rice yield and reduced weed biomass in upland rice [55]. Future research needs include addressing the effects of cover crops on rice production in dry-seeded rice systems.

Due to challenging weed issues in rice systems, particularly dry-seeded rice, herbicide use will continue to be necessary for effective weed suppression in both conventional and reduced-tillage systems. The implementation of cover crops into these systems may lessen the herbicide requirements but will not eliminate the use of chemicals altogether. Currently there are a number of preemergent and postemergent herbicides available for use in rice production (Table 3); however, as dry-seeded, conservation-tillage rice systems increase in popularity, more herbicide options may become available.

Herbicide			
Common Name	Trade Name[a]	Application Timing	Weed Species Controlled
Clomazone	Command*[56]	PRE[b]	Grass species such as barnyardgrass (*Echinochloa crus-galli*), crabgrass (*Digitaria* spp.), and panicum (*Panicum* spp.)
Halosulfuron	Permit* [57]		Broadleaf species such as dayflower (*Commelina erecta*) and kochia (*Kochia scoparia*). Broadleaf and grass species may be controlled with a POST application.
Pendimethalin	Prowl* [40]		Germinating, small-seeded grass and broadleaf species such as crabgrass (*Digitaria* spp.), foxtail, and common lambsquarters (*Chenopodium alba*)
Quinclorac	Facet* [58]		Broadleaf and grass species such as morningglory (*Ipomoea* spp.), and barnyardgrass. Can also be applied POST
Thiobencarb	Bolero* [59]		Grass and broadleaf species such as barnyardgrass, dayflower (*Commelina communis*), and eclipta (*Eclipta alba*)
Acifluorfen	Ultra Blazer* [60]	POST[c]	Grasses and broadleaves such as foxtail (*Setaria* spp.), panicum, and eclipta
Bensulfuron	Londax* [61]		Broadleaf and sedge species, particularly aquatic weeds such as ducksalad (*Heteranthera limosa*) and ricefield bulrush (*Scirpus mucronatus*)
Bentazon	Basagran* [62]	POST	Broadleaf and sedge species such as dayflower, eclipta, and yellow nutsedge (*Cyperus esculentus*)
Carfentrazone	Aim* [23]		Broadleaf species such as common cocklebur (*Xanthium strumarium*), dayflower, and *Amaranthus* spp.

Herbicide			
Common Name	Trade Name[a]	Application Timing	Weed Species Controlled
Propanil	Stam° [63]		Grass, rush, and broadleaf species such as barnyardgrass, spikerush (*Eleocharis* spp.), and curly dock (*Rumex crispus*)
Cyhalofop	Clincher° [64]	After Flooding	Grass species such as barnyardgrass, broadleaf signalgrass (*Brachiaria platyphylla*), and junglerice (*Echnochloa colona*)
2,4-D	Agri Star® 2,4-D [37]		Annual and perennial weed species such as cocklebur, morningglory, and dock
Clearfield Rice			
Imazamox	Beyond° [30]	POST	Grass and broadleaf species such as morningglory, barnyardgrass, and panicum
Imazethapyr	Newpath° [65]		Grass, sedge, and broadleaf species such as barnyardgrass, morningglory, and nutsedge
Imazethapyr + Quinclorac	Clearpath° [66]		Grass, sedge, and broadleaf species such as junglerice, eclipta, morningglory, and nutsedge

[a]Trade names listed are representative of available herbicides. Inclusion of particular trade names does not suggest author endorsement.

[b]PRE, preemergence.

[c]POST, postemergence.

Table 3. Herbicides for use in reduced-tillage rice production.

6. Soybean

Production of soybean [*Glycine max* (L.) Merr.], estimated at 102 million ha in 2010 [2], meets a number of livestock and human food needs as well as industrial demands for use in products such as paints, lubricants, and biofuel. Due to its diversity of uses, the soybean is an important field crop for much of the world. In light of the value of soybeans, it is essential to establish sustainable growing practices to ensure global demand continues to be met.

Implementation of conservation practices, such as reduced-tillage, can be utilized as components of alternative management systems replacing conventional systems to provide erosion and runoff control while reducing labor and cost inputs. In the United States, in fact, approximately 80% of soybeans were produced with some form of conservation-tillage by 2006 [67]. This increase in conservation-tillage can be attributed to the environmental and economic benefits achieved with reduced-tillage as well as the commercial availability of herbicide-tolerant soybeans, which have made successful chemical weed control achievable with the use of fewer herbicides.

Early work in conservation-tillage soybean have reported equal or improved yield in soybean with reduced-tillage compared to conventional systems [68, 69]. Previous research has also examined soybean systems planted behind wheat or a cover crop such as rye with improved weed control being noted when compared to a fallow system [70] and greater yield with a cover crop than with just the previous crop's stubble [71]. The inclusion of plant residue, either from a cover crop or a previous crop, provides a level of weed control by acting as a physical barrier for germinating weed seed or through allelopathic inhibition released by some cover crop species. The weed control provided by ground cover is crucial in a no-till practice due to the loss of control from tillage reduction and the shift towards more difficult to control perennial weed species.

While cover crops and plant residue have been identified as means to reduce weed emergence when implemented in reduced-tillage practices further measures are required to keep the weed population below an acceptable level [70]. Many cultural practices, such as crop rotation, row spacing, and planting date, can be manipulated in such a way as to reduce weed populations; however, herbicide use is still necessary in many systems.

As with most field crops grown in conservation-tillage systems, soybean production with reduced-tillage has heavily relied on postemergent herbicide applications. Use of cover crops in these systems may also contribute to the tendency for fewer PRE herbicides due to interception concerns. However, the increase in herbicide-resistant weed species such as Palmer amaranth (*Amaranthus palmeri* S. Wats) and horseweed [*Conyza canadensis* (L.) Cronq.] in herbicide resistant crops, like soybean, necessitates the use of multiple herbicides to slow the development of weed resitance and safeguard the effectiveness of current herbicide options for the future. Table 4 provides a partial list of herbicides that can be utilized in reduced-tillage soybean with cover crops.

Herbicide			
Common Name	**Trade Name**[a]	**Application Timing**	**Weed Species Controlled**
Glufosinate	Liberty* [24]	Preplant Burndown	Emerged weed species
Glyphosate	Roundup WeatherMax* [25]		
Paraquat	Gramoxone* [26]		
2,4-D	Agri Star* 2,4-D [37]		
Clomazone	Command* [56]	PRE[b]	Grasses and broadleaves such as crabgrass (*Digitaria* spp.), panicum (*Panicum* spp.), velvetleaf (*Abutilon theophrasti*), and Florida beggarweed (*Desmodium tortuosum*)
Dimethenamid	Outlook* [72]		Grass and broadleaf species such as foxtail (*Setaria* spp.), panicum, and *Amaranthus* spp.
Flumioxazin	Valor* [39]		Broadleaf species such as horseweed (*Conyza canadensis*); suppression of grass species such as panicum and goosegrass (*Eleusine indica*)

Herbicide			
Common Name	**Trade Name**[a]	**Application Timing**	**Weed Species Controlled**
Imazaquin	Scepter* [73]		Broadleaf and grass species such as morningglory (*Ipomoea* spp.), velvetleaf, and foxtail
Metribuzin	Metribuzin [74]		Broadleaf and grass species such as *Amaranthus* spp.and broadleaf signalgrass (*Brachiaria platyphylla*)
Pendimethalin	Prowl* [40]		Grass and broadleaf species such as panicum and *Amaranthus* spp.
S-metolachlor	Dual Magnum* [41]		Grass and broadleaves such as barnyardgrass (*Echinochloa crus-galli*), crabgrass, and Florida pusley (*Richardia scabra*)
Bentazon	Basagran* [62]	POST[c]	Broadleaf weeds such as coffee senna (*Senna occidentalis*) and velvetleaf
Chlorimuron	Classic* [75]		Broadleaf weeds such as Florida beggarweed and morningglory
Cloransulam	FirstRate* [76]		Broadleaf weeds such as common cocklebur (*Xanthium strumarium*) and velvetleaf
Fluazifop	Fusilade* [77]		Annual and perennial grass species such as crabgrass and bermudagrass (*Cynadon dactylon*)
Imazethapyr	Pursuit* [78]		Broadleaf and grass species such as morningglory and crabgrass
Lactofen	Cobra* [79]		Broadleaf species such as croton (*Croton* spp.) and Florida beggarweed
Sethoxydim	Poast* [80]		Grass species such as foxtail, crabgrass, and panicum
LibertyLink Soybean			
Glufosinate	Liberty*	POST	Broadleaf and grass species such as *Amaranthus* spp., morningglory, and goosegrass
Roundup Ready Soybean			
Fomesafen + Glyphosate	Flexstar* [81]	POST	Broadleaf and grass species such as morningglory, velvetleaf, and broadleaf signalgrass
Glyphosate	Roundup WeatherMax®	POST	Grass and broadleaf species such as Florida beggarweed, crabgrass and groundcherry

[a]Trade names listed are representative of available herbicides. Inclusion of particular trade names does not suggest author endorsement.

[b]PRE, preemergence.

[c]POST, postemergence.

Table 4. Herbicides for use in reduced-tillage soybean production.

7. Cotton

Cotton production around the world is estimated at approximately 23 million tonnes (lint production) [2] with China, India, and the United States being the top producers [82]. Efforts to adopt sustainable cotton practices have led producers to utilize conservation-tillage systems in cotton production. Besides environmental benefits achieved with reduced-tillage, major economic advantages can be realized due to reduced time, labor, and fuel requirements when operating with less tillage. Prior to the introduction of herbicide-resistant crops, adoption of reduced-tillage was difficult due to control of weed species required multiple and costly herbicide inputs [13]. In some instances, effective herbicides were not available to control problematic weed species such as perennials that can thrive in reduced-tillage. When glyphosate-resistant cotton was made available, reduced-tillage became practical since a broad spectrum of weed species could be controlled with a single herbicide [83].

Extensive research has been carried out in conservation-tillage cotton with positive benefits seen for cotton yield [84-86]. Moreover, with herbicide-resistant cotton varieties, weed control has been as successful as conventional tillage cotton. Because of this success, conservation-tillage practices have been widely adopted in areas such as the southeastern United States. This dependence on a single herbicide, however, has led to the appearance of herbicide-resistant weed species and now threatens the feasibility of reduced-tillage cotton production. Currently, research efforts are focused on identifying ways to ensure the long-term viability of conservation-tillage while controlling established populations of herbicide-resistant weed species and reducing the risk of future development of resistant weeds.

Multiple weed management tactics are necessary to control weed resistance development with cover crops playing an important role in resistance management. The use of cover crops, particularly high-residue crops such as rye and black oat, can reduce herbicide inputs through shading and allelopathy. The use of high-residue crops allows for maximum shading of the soil surface during the beginning of the season while also providing a ground cover for a longer period into the growing season. Cover crops, along with multiple herbicide modes of action and rotation, have been shown to effectively control weeds in reduced-tillage cotton [87, 88].

A number of herbicide choices are available for use with conservation-tillage cotton (Table 5). PRE herbicides are especially important in early-season weed control to ensure management of weed species that are difficult to control later in the season. Although concerns have been raised as to whether cover crops reduce the efficacy of PRE herbicides, it has been suggested that any loss in weed control due to herbicide interception is offset by the control provided by cover crop residue [89-91].

Herbicide			
Common Name	Trade Name[a]	Application Timing	Weed Species Controlled
Dicamba	Banvel® [43]	Preplant Burndown	Emerged weed species
Flumioxazin	Valor® [39]		
Glufosinate	Liberty® [24]		

Herbicide			
Common Name	**Trade Name[a]**	**Application Timing**	**Weed Species Controlled**
Glyphosate	Roundup WeatherMax*[25]		
Paraquat	Gramoxone* [26]		
Clomazone	Command* [56]	Preplant or PRE[b]	Grasses and broadleaves such as crabgrass (*Digitaria* spp.), panicum (*Panicum*spp.), velvetleaf (*Abutilon theophrasti*), and Florida beggarweed (*Desmodium tortuosum*)
Fluometuron	Cotoran* [92]		Grasses and broadleaves such as signalgrass (*Brachiaria sp.),* horseweed (*Conyza canadensis*) and sicklepod (*Senna obtusifolia*)
Pendimethalin	Prowl* [40]		Grass and broadleaf species such as foxtail (*Setaria* spp.), panicum, and *Amaranthus* spp.
Prometryn	Caparol* [93]		Annual grass and broadleaves such as groundcherry (*Physalis* sp.), Florida pusley (*Richardia scabra*), and panicum
S-metolachlor	Dual Magnum*[41]		Grass and broadleaves such as barnyardgrass (*Echinochloa crus-galli*), crabgrass, and Florida pusley
Clethodim	Select* [94]	POST[c]	Grass species such as crabgrass, panicum, and foxtail
Herbicide			
Common Name	Trade Name	Application Timing	Weed Species Controlled
Quizalofop	Assure* [95]		Annual and perennial grasses such as foxtail, goosegrass (*Eleusine indica*), and bermudagrass (*Cynodon dactylon*)
Sethoxydim	Poast* [80]	POST	Grass species such as foxtail, crabgrass, and panicum
Trifloxysulfuron	Envoke* [96]		Broadleaf and grass species such as coffee senna (*Senna occidentalis*), barnyardgrass, and Florida beggarweed
Diuron	Direx* [97]	POST-directed spray	Broadleaf and grass species such as sicklepod, velvetleaf, and crabgrass
Linuron	Linex* [98]		Broadleaves and grasses such as morningglory, Florida pusley, and panicum
MSMA	MSMA [99]		Grass and broadleaf species such as crabgrass, Florida beggarweed, and *Amaranthus* spp.

Herbicide			
Common Name	Trade Name[a]	Application Timing	Weed Species Controlled
LibertyLink Cotton			
Glufosinate	Liberty®	POST	Broadleaf and grass species such as *Amaranthus* spp., morningglory, and goosegrass
Roundup Ready Cotton			
Glyphosate	Roundup WeatherMax®	POST	Grass and broadleaf species such as Florida beggarweed, crabgrass, foxtail, groundcherry, and velvetleaf

[a]Trade names listed are representative of available herbicides. Inclusion of particular trade names does not suggest author endorsement.

[b]PRE, preemergence.

[c]POST, postemergence.

Table 5. Herbicides for use in reduced-tillage cotton.

8. Peanut

Groundnut, or peanut (*Arachis hypogaea* L.), was planted on approximately 21 million ha between 2011 and 2012 wordwide with top production occurring in China, India, Indonesia, the United States, and some African countries such as Nigeria, Senegal, and Sudan [100]. Besides being a nutrient rich food source, the peanut is utilized for its oil in cooking and manufacturing as well as a livestock feed. In the United States, peanuts offer an exceptional rotational crop with cotton to replenish soil nitrogen. The benefits of peanuts to a cotton system, which have been shifting toward long-term, reduced-tillage practices, have necessitated the adoption of minimum tillage practices in peanut production as well.

The increased farming costs of conventional tillage systems have spurred producers to implement conservation-tillage to reduce expenses; however, peanut growers face unique difficulties when using these systems [101,102]. Particularly, concerns over peanut response to reduced-tillage due to peanut growth habits have required research in order to identify successful means of conservation-tillage integration into peanut production [102, 103].

Peanut yield variability under reduced-tillage compared to conventional tillage has been noted as one of the greatest concerns when adopting conservation-tillage practices [101,102]. Inconsistent yield response by peanut has been noted in previous studies investigating conservation-tillage. Research has reported yields of peanut to be reduced or equal to conventionally tilled peanut [101, 104]; other studies have shown reduced-tillage peanuts to produce equally or greater than conventional tillage peanuts [103,105]. Research efforts continue to recognize the contributing factors that affect peanut response to tillage systems.

Weed management in conservation-tillage peanut is also a concern for producers. Weed control in peanut, regardless of tillage system, can be problematic due to the extended growing season and unique growth habits [106,107]. Generally, peanut production requires an incorporated residual as well as a POST herbicide to provide effective weed control under the slow-closing canopy of peanut [107]. Moreover, in-season cultivation for weed management cannot be implemented due to the potential to damage developing peanut pods [106,108].

Weed control in reduced-tillage peanuts can be even more difficult than in conventional tillage due to the loss of weed control through seed burial and the inability to utilize preplant incorporated herbicides [109]. This results in increased dependence on post emergent herbicides which may or may not control the number of perennial weed species that may predominate in a reduced-tillage setting; the loss of effective weed management can reduce peanut productivity due to weed competition [102,107].

Utilization of cover crops in peanut systems may offer beneficial weed control while reducing the need for increased postemergent herbicide applications. Research has shown effective weed control with cover crops in strip-tillage peanut systems that use a dinitroaniline pre-emergent herbicide over cover crop residue [107]. Other effective herbicides used in conservation-tillage peanut systems are listed in Table 6.

9. Herbicide interception

Preemergent, residual herbicides must reach the soil surface to be effective. When spraying over cover crop residue, herbicide applications can be intercepted and absorbed prior to reaching the soil surface. Herbicides, such as acetochlor, chlorimuron, and oryzalin have been shown to be impeded by plant stubble [113,114]. While timely rainfall can move herbicides to the soil, some portion of herbicide can be retained in the residue.

Herbicide amounts intercepted by stubble can affect weed control achieved with the herbicide; efficacy can be reduced by cover crops either through physical interception preventing soil contact or through increased microbial activity in the residue speeding herbicide degradation [115]. Increases in soil organic matter from extended conservation-tillage practices may also increase herbicide adsorption within the soil [116]. Additionally, herbicide persistence and carryover risks may be increased when applied to residue [114]. Certain crops may be susceptible to herbicides at low doses that can persist in cover crop residue that would otherwise have dissipated in bare soil. However, little research has been done to determine the extent of persistence for most herbicides.

Methods to reduce herbicide interception are limited when using cover crops. Interception could potentially be managed, particularly in strip-till operations, through banded herbicide applications over the row allowing for in-row weed control while reducing herbicide inputs. Furthermore, a water-based, microencapsulated herbicide formulation, like Prowl H$_2$O® (pendimethalin), may allow more herbicide to reach the soil after a rain event or irrigation.

Herbicide			
Common Name	**Trade Name**[a]	**Application Timing**	**Weed Species Controlled**
Glyphosate	Roundup WeatherMax® [25]	Preplant Burndown	Emerged weed species
Paraquat	Gramoxone® [26]		
2,4-D	Agri Star® 2,4-D [37]		
Diclosulam	Strongarm® [110]	PRE[b]	Broadleaf species such as eclipta (*Eclipta prostrata*) and *Amaranthus* spp.
Flumioxazin	Valor® [39]		Broadleaf species such as horseweed (*Conyza canadensis*)
Pendimethalin	Prowl® [40]		Grass and broadleaf species such as foxtail (*Setaria* spp.) and *Amaranthus* spp.
Acifluorfen	Ultra Blazer® [60]	POST[c]	Broadleaf and grass species such as coffee senna (*Senna occidentalis*) and velvetleaf (*Abutilon theophrasti*)
Bentazon	Basagran® [62]		Broadleaf species such as morningglory (*Ipomoea* spp.) and velvetleaf
Chlorimuron	Classic® [75]		Broadleaf weeds such as Florida beggarweed (*Desmodium tortuosum*) and morningglory
Clethodim	Select® [94]		Grass species such as panicum, foxtail, and crabgrass (*Digitaria* spp.)
Imazapic	Cadre® [111]		Broadleaf and grass species such as morningglory, *Amaranthus* spp. and crabgrass
Imazethapyr	Pursuit® [78]		Broadleaf, grass, and sedge species such as Florida pusley (*Richardia scabra*), crabgrass, and nutsedge (*Cyperus* spp.)
Paraquat	Gramoxone®		Grass and broadleaf species
Sethoxydim	Poast® [80]		Grass species, foxtail and panicum
2,4-DB	Butyrac® [112]		Broadleaf species such as velvetleaf and prickly sida (*Sida spinosa*)

[a]Trade names listed are representative of available herbicides. Inclusion of particular trade names does not suggest author endorsement.

[b]PRE, preemergence.

[c]POST, postemergence.

Table 6. Herbicides for use in reduced-tillage peanut.

10. Conclusion

The ever increasing demands on global agriculture dictate the use of intensive, high-yielding production practices. However, the inability to sustain these systems long-term necessitates the implementation of more energy-efficient, environmentally-sound practices that can still produce successful yields. Conservation agriculture practices seek to achieve these goals in order to ensure current and future agricultural production. While components of conservation agriculture, such as reduced-tillage and cover crops, are fundamental practices in these systems, herbicides are still valuable and necessary weed management tools within conservation systems. Integrating these management practices can be challenging and continue to warrant research to identify the most successful means of utilizes herbicides in conjunction with reduced-tillage and cover crops.

Author details

Andrew J. Price[1*] and Jessica A. Kelton[2]

*Address all correspondence to: Andrew.price@ars.usda.gov

1 United States Department of Agriculture, Agricultural Research Service, National Soil Dynamics Laboratory, Auburn, Alabama, USA

2 Auburn University, Auburn, Alabama, USA

References

[1] Reiter, M.S., D.W. Reeves, C.H. Burmester, H.A. Torbert. Cotton nitrogen management in a high-residue conservation system: Cover crop fertilization. *Soil Science Society of America Journal*, 2008; 72, 1321-1329, ISSN 1435-0661.

[2] Food and Agriculture Organization of the United Nations (FAO). FAOSTAT 2010. Available online at http://www.fao.org/ag/ca/index.html (accessed 13 August 2012).

[3] DeFelice, M.S., P.R. Carter, and S.B. Mitchell. Influence of tillage on corn and soybean yield in the US and Canada. Online. Crop Management. 2006. http://www.plantmanagementnetwork.org/pub/cm/research/2006/tillage/ (accessed 12 August 2012).

[4] Reeves, D.W. 1997. The role of soil organic matter in maintaining soil quality in continuous cropping systems. *Soil and Tillage Research*, 43, 131-167, ISSN 0167-1987.

[5] Truman, C.C., D.W. Reeves, J.N. Shaw, A.C. Motta, C.H. Burmester, R.L. Raper, and E.B. Schwab. Tillage impacts on soil property, runoff, and soil loss variations of a

Rhodic Paleudult under simulated rainfall. Journal of Soil and Water Conservation, 2003; 58,258-267, ISSN 0022-4561.

[6] Swanton, C.J., K.J. Mahoney, K. Chandler, and R.H. Gulden. Integrated weed management: Knowledge-based weed management systems. *Weed Science*, 2008; 56, 168-172, ISSN 0043-1745.

[7] Norsworthy, J.K., M.S. Malik, P. Jha and M.B. Riley. Suppression of *Digitaria sanguinalis* and *Amaranthus palmeri* using autumn-sown glucosinolate-producing cover crops in organically grown bell pepper. *Weed Research*, 2007; 47, 425-432, ISSN 0043-1737.

[8] Treadwell, D.D, N.G. Creamer, J.R. Schultheis, and G.D. Hoyt. Cover crop management affects weeds and yield in organically managed sweetpotato systems. *Weed Technology* 2007; 21, 1039-1048, ISSN 0890-037X.

[9] Brennan, E.B. and R.F. Smith. Winter cover crop growth and weed suppression on the Central Coast of California. *Weed Technology* 2005; 19, 1017-1024, ISSN 0890-037X.

[10] Clark, A., editor. Managing Cover Crops Profitably. College Park, MD, USA: Sustainable Agricultural Research and Education (SARE); 2007.

[11] Price, A.J., J.A. Kelton. Weed control in conservation agriculture. In: Soloneski S. and M. Larramendy (ed.) Herbicides: Theory and Applications. Rijeka: InTech; 2010. p. 3-16.

[12] Gaston, L.A., D.J. Boquet, and M.A. Bosch. Pendimethalin wash-off from cover crop residues and degradation in a Loessial soil. *Communications in Soil Science and Plant Analysis* 2003; 34, 2515-2527, ISSN 0010-3624.

[13] Price, A.J., K.S. Balkcom, S.A. Culpepper, J.A. Kelton, R.L. Nichols, and H. Schomberg. Glyphosate-resistant Palmer amaranth: A threat to conservation tillage. *Journal of Soil and Water Convervation* 2011; 66(4), 265-275, ISSN 0022-4561.

[14] Mitchell, D.O. and M. Mielke. Wheat: The global market, policies, and priorities. In Aksoy M.A. and J.C. Beghin (eds.) Global Agricultural Trade and Developing Countries. Washington, DC, USA: World Bank; 2005. p. 195-214.

[15] Food and Agriculture Organization of the United Nations (FAO). Conservation agriculture 2011. Available online at http://www.fao.org/docrep/013/al977e/al977e00.pdf (accesssed 13 August 2012).

[16] Zhao, H., G. Gao, X. Yan, Q. Zhang, M. Hou, Y. Zhu, Z. Tian. Risk assessment of agricultural drought usning the CERES-Wheat model: a case study of Henan Plain, China. *Climate Research* 2011; 50, 247-256, ISSN 0936-577X.

[17] Bonfil, D.J., I. Mufradi, S. Klitman, and S. Asido. Wheat grain yield and soil profile water distribution in a no-till arid environment. *Agronomy Journal* 1999; 91, 368-373, ISSN 0002-1962.

[18] De Vita, P., E. Di Paolo, G. Fecondo, N. Di Fonzo, and M. Pisante. No-tillage and conventional tillage effects on durum wheat yield, grain quality and soil moisture content in southern Italy. *Soil and Tillage Research*, 2007; 92, 69-78, ISSN 0167-1987.

[19] Gruber, S., C. Pekrun, J. Mohring, and W. Claupein. Long-term yield and weed response to conservation and stubble tillage in SW Germany. *Soil and Tillage Research*, 2012; 121, 49-56, ISSN 0167-1987.

[20] Wilson, H.P., M.P. Masgianica, T.E. Hines, and R.F. Walden. Influence of tillage and herbicides on weed control in a wheat (*Triticum aestivum*)- soybean (*Glycine max*) rotation. *Weed Science*, 1986; 34, 590-594, ISSN 0043-1745.

[21] Guy, S.O. and R.M. Gareau. Crop rotation, residue durability, and nitrogen fertilizer effects on winter wheat production. *Journal of Production Agriculture*, 1998; 11, 457-461, ISSN 0890-8524.

[22] Dao, T.H. Crop residues and management of annual grass weeds in continuous no-till wheat (*Triticum aestivum*). *Weed Science*, 1987; 35, 395-400, ISSN 0043-1745.

[23] FMC Corporation. 2012. Aim® *Herbicide Label*. Philadelphia, PA, USA: FMC Corporation Agricultural Products Group. 15 p.

[24] Bayer CropScience. 2011. Liberty® *Herbicide Label*. Research Triangle Park, NC, USA: Bayer CropScience LP. 20 p.

[25] Monsanto Company. 2009. Roundup WeatherMax® *Herbicide Label*. St. Louis, MO, USA: Monsanto Company. 54 p.

[26] Syngenta Crop Protection. 2011. Gramoxone® *Herbicide Label*. Greensboro, NC, USA: Syngenta Crop Protection, LLC. 55 p.

[27] E. I. du Pont de Nemours and Company. 2009. DuPont™ Finesse® *Herbicide Label*. Wilmington, DE, USA: E.I. du Pont de Nemours and Company, Inc. 12 p.

[28] Bayer CropScience. 2011. Huskie™ *Herbicide Label*. Research Triangle Park, NC, USA: Bayer CropScience LP. 24 p.

[29] E. I. du Pont de Nemours and Company. 2010. DuPont™ Harmony® Extra *Herbicide Label*. Wilmington, DE, USA: E.I. du Pont de Nemours and Company, Inc. 13 p.

[30] BASF Corporation. 2011. Beyond® *Herbicide Label*. Research Triangle Park, NC, USA: BASF Corporation. 23 p.

[31] Yenish, J.P., A.D. Worsham, and A.C. York. Cover crops for herbicide replacement in no-tillage corn (*Zea mays*). *Weed Technology* 1996; 10, 815-821, ISSN 0890-037X.

[32] Clark, A.J., A.M. Decker, and J.J. Meisinger. Seeding rate and kill date effects on hairy vetch-cereal rye cover crop mixtures for corn production. *Agronomy Journal*, 1994; 86, 1065-1070, ISSN 0002-1962.

[33] Fisk, J.W., O.B. Hesterman, A. Shrestha, J.J. Kells, R.R. Harwood, H.M. Squire, and C.C. Sheaffer. Weed suppression by annual legume cover crops in no-tillage corn. *Agronomy Journal*, 2001; 93, 319-325, ISSN 0002-1962.

[34] Utomo, M., W.W. Frye, and R.L. Blevins. Sustaining soil nitrogen for corn using hairy vetch cover crop. *Agronomy Journal*, 1990; 82, 979-983, ISSN 0002-1962.

[35] Wagger, M.G. Cover crop management and nitrogen rate in relation to growth and yield of no-till corn. *Agronomy Journal*, 1989; 81, 533-538, ISSN 0002-1962.

[36] Decker, A.M., A.J. Clark, J.J. Meisinger, F. Ronald Mulford, and M.S. McIntosh. Legume cover crop contributions to no-tillage corn production. *Agronomy Journal*, 1994; 86, 126-135, ISSN 0002-1962.

[37] Albaugh. 2012. Agri Star® 2,4-D Amine *Herbicide Label*. Ankeny, IA, USA: Albaugh, Inc. 36 p.

[38] Syngenta Crop Protection. 2009. Aatrex® *Herbicide Label*. Greensboro, NC, USA: Syngenta Crop Protection, LLC. 24 p.

[39] Valent U.S.A. 2010. Valor® *Herbicide Label*. Walnut Creek, CA, USA: Valent U.S.A. Corporation. 27 p.

[40] BASF Corporation. 2008. Prowl® *Herbicide Label*. Research Triangle Park, NC, USA: BASF Corporation. 24 p.

[41] Syngenta Crop Protection. 2011. Dual Magnum® *Herbicide Label*. Greensboro, NC, USA: Syngenta Crop Protection, LLC. 54 p.

[42] Bayer CropScience. 2005. Buctril® *Herbicide Label*. Research Triangle Park, NC, USA: Bayer CropScience LP. 36 p.

[43] Arysta LifeScience. 2004. Banvel® *Herbicide Label*. Cary, NC, USA: Arysta LifeScience North America, LLC. 27 p.

[44] Syngenta Crop Protection. 2012. Callisto® *Herbicide Label*. Greensboro, NC, USA: Syngenta Crop Protection, LLC. 32 p.

[45] Bayer CropScience. 2010. Laudis® *Herbicide Label*. Research Triangle Park, NC, USA: Bayer CropScience LP. 19 p.

[46] Syngenta Crop Protection. 2011. Evik® *Herbicide Label*. Greensboro, NC, USA: Syngenta Crop Protection, LLC. 8 p.

[47] Tessenderlo. 2010. Lorox® *Herbicide Label*. Phoenix, AZ, USA: Tessenderlo Kerley, Inc. 14 p.

[48] BASF Corporation. 2008. Lightning® *Herbicide Label*. Research Triangle Park, NC, USA: BASF Corporation. 10 p.

[49] Syngenta Crop Protection. 2009. Expert® *Herbicide Label*. Greensboro, NC, USA: Syngenta Crop Protection, LLC. 31 p.

[50] Fairhurst, T.H. and A. Dobermann. Rice in the global food supply. *Better Crops International*, 2002; 16, 3-6. http://www.ipni.net/ppiweb/bcropint.nsf/$webindex/ 0E477FFC43BD62 DA85256BDC00722F62/$file/BCI-RICEpU3.pdf. (accessed 31 August 2012).

[51] Farooq, M., K.H.M. Siddique, H. Rehman, T. Aziz, D. Lee, and A. Wahid. Rice direct seeding: Experiences, challenges and opportunities. *Soil and Tillage Research*, 2011; 111, 87-98, ISSN 0167-1987.

[52] Mishra, J.S. and V.P. Singh. Tillage and weed control effects on productivity of a dry seeded rice-wheat system on a Vertisol in Central India. *Soil and Tillage Research*, 2012; 123, 11-20, ISSN 0167-1987.

[53] Chauhan, B.S. and D.E. Johnson. Influence of tillage systems on weed seedling emergence pattern in rainfed rice. *Soil and Tillage Research*, 2009; 106, 15-21, ISSN 0167-1987.

[54] Rao, A.N., D.E. Johnson, B. Sivaprasad, J.K. Ladha. and A.M. Mortimer. Weed management in direct-seeded rice. *Advances in Agronomy*, 2007; 93, 153–255, ISSN 0065-2113.

[55] Becker, M. and D.E. Johnson. Legumes as dry season fallow in upland rice-based systems of West Africa. *Biology and Fertility of Soils*, 1998; 27, 358-367, ISSN 0178-2762.

[56] FMC Corporation. 2011. Command® *Herbicide Label*. Philadelphia, PA, USA: FMC Corporation Agricultural Products Group. 19 p.

[57] Gowan Company. 2007. Permit® *Herbicide Label*. Yuma, AZ, USA: Gowan Company. 18 p.

[58] BASF Corporation. 2010. Facet® *Herbicide Label*. Research Triangle Park, NC, USA: BASF Corporation. 9 p.

[59] Valent U.S.A. 2001. Bolero® *Herbicide Label*. Walnut Creek, CA, USA: Valent U.S.A. Corporation. 4 p.

[60] United Phosphorus. 2009. Ultra Blazer® *Herbicide Label*. King of Prussia, PA, USA: United Phosphorus, Inc. 10 p.

[61] United Phosphorus. 2010. Londax® *Herbicide Label*. King of Prussia, PA, USA: United Phosphorus, Inc. 9 p.

[62] Arysta LifeScience. 2005. Basagran® *Herbicide Label*. Cary, NC, USA: Arysta LifeScience North America, LLC. 12 p.

[63] United Phosphorus. 2010. Stam® *Herbicide Label*. King of Prussia, PA, USA: United Phosphorus, Inc. 7 p.

[64] Dow AgroSciences. 2011. Clincher® *Herbicide Label*. Indianapolis, IN, USA: Dow AgroSciences LLC. 4 p.

[65] BASF Corporation. 2011. Newpath® *Herbicide Label*. Research Triangle Park, NC, USA: BASF Corporation. 12 p.

[66] BASF Corporation. 2011. Clearpath® *Herbicide Label*. Research Triangle Park, NC, USA: BASF Corporation. 10 p.

[67] Ebel, R. Soil management and conservation. In: Osteen, C., J. Gottlieb, and U. Vasavada (eds.) Agricultural Resources and Environmental Indicators, 2012. EIB-98, United States Department of Agriculture, Economic Research Service, August 2012. p 33-36. Available from http://www.ers.usda.gov/Publications/eib- economic-information-bulletin/eib98.aspx (accessed 5 September 2012).

[68] Campbell, R.B., R.E. Sojka, and D.L. Karlen. Conservation tillage for soybean in the U.S. Southeastern Coastal Plain. *Soil and Tillage Research*, 1984; 4, 531-541, ISSN 0167-1987.

[69] Robinson, E.L., G.W. Langdale, and J.A. Stuedemann. Effect of three weed control regimes on no-till and tilled soybeans (*Glycine max*). *Weed Science*, 1984; 32, 17-19, ISSN 0043-1745.

[70] Price, A.J., D.W. Reeves, and M.G. Patterson. Evaluation of weed control provided by three winter cereals in conservation-tillage soybean. *Renewable Agriculture and Food Systems*, 2005; 21, 159-164, ISSN 1742-1705.

[71] Liebl, R., F.W. Simmons, L.M. Wax, and E.W. Stoller. Effect of rye (*Secale cereale*) mulch on weed control and soil moisture in soybean (*Glycine max*). *Weed Technology*, 1992; 6, 838-846, ISSN 0890-037X.

[72] BASF Corporation. 2008. Outlook® *Herbicide Label*. Research Triangle Park, NC, USA: BASF Corporation. 17 p.

[73] BASF Corporation. 2009. Scepter® *Herbicide Label*. Research Triangle Park, NC, USA: BASF Corporation. 15 p.

[74] Loveland Products. 2008. Metribuzin *Herbicide Label*. Greeley, CO, USA: Loveland Products, Inc. 26 p.

[75] E. I. du Pont de Nemours and Company. 2010. DuPont™ Classic® *Herbicide Label*. Wilmington, DE, USA: E.I. du Pont de Nemours and Company, Inc. 15 p.

[76] Dow AgroSciences. 2011. FirstRate® *Herbicide Label*. Indianapolis, IN, USA: Dow AgroSciences LLC. 6 p.

[77] Syngenta Crop Protection. 2011. Fusilade® *Herbicide Label*. Greensboro, NC, USA: Syngenta Crop Protection, LLC. 39 p.

[78] BASF Corporation. 2011. Pursuit® *Herbicide Label*. Research Triangle Park, NC, USA: BASF Corporation. 27 p.

[79] Valent U.S.A. 2007. Cobra® *Herbicide Label*. Walnut Creek, CA, USA: Valent U.S.A. Corporation. 29 p.

[80] BASF Corporation. 2010. Poast® *Herbicide Label*. Research Triangle Park, NC, USA: BASF Corporation. 24 p.

[81] Syngenta Crop Protection. 2009. Flexstar® *Herbicide Label*. Greensboro, NC, USA: Syngenta Crop Protection, LLC. 26 p.

[82] National Cotton Council of America. Production Ranking 2012. http://www.cotton.org/econ/cropinfo/cropdata/rankings.cfm (accessed 5 September 2012).

[83] Carpenter, J. and L. Gianessi. Herbicide tolerant soybeans: Why growers are adopting Roundup Ready varieties. *AgBioForum*, 1999; 2, 65-72, ISSN 1522-936X.

[84] Schwab, E.B., D.W. Reeves, C.H. Burmester, and R.L. Raper. Conservation tillage systems for cotton in the Tennessee Valley. *Soil Science Society of America Journal*, 2002; 66, 569-577, ISSN 1435-0661.

[85] Nyakatawa, E.Z., K.C. Reddy, and D.A. Mays. Tillage, cover cropping, and poultry litter effects on cotton: II. Growth and yield parameters. *Agronomy Journal*, 2000; 92, 1000-1007, ISSN 1435-0645.

[86] Keeling, W., E. Segarra, and J.R. Abernathy. Evaluation of conservation tillage cropping systems for cotton on the Texas Southern High Plains. *Journal of Production Agriculture*, 1989; 2, 269-273, ISSN 0890-8524.

[87] Bauer, P.J. and D.W. Reeves. A comparison of winter cereal species and planting dates as residue cover for cotton growth with conservation tillage. *Crop Science*, 1999; 39, 1824-1830, ISSN 0002-1962.

[88] Reeves, D.W., A.J. Price, and M.G. Patterson. Evaluation of three winter cereals for weed control in conservation-tillage nontransgenic cotton. *Weed Technology*, 2005; 19, 731-736, ISSN 0890-037X.

[89] Johnson, M.D., D.L. Wyse, and W.E. Lueschen. The influence of herbicide formulation on weed control in four tillage systems. *Weed Science*, 1989; 37, 239-149, ISSN 0043-1745.

[90] Lindwall, C.W. Crop management in conservation tillage systems. In P. Unger (ed.) Managing Agricultural Residues. Boca Raton, FL, USA: Lewis Publishing, 1994. p. 185-210.

[91] Westerman, P.A., M. Liebman, F.D. Menalled, A.H. Heggenstaller, R.G. Hartzler, and P.M. Dixon. Are many little hammers effective? Velvetleaf (*Abutilon theophrasti*) population dynamics in two- and four-year crop rotation systems. *Weed Science*, 2005; 53, 382-392, ISSN 0043-1745.

[92] Makhteshim Agan. 2009. Cotoran® *Herbicide Label*. Raleigh, NC, USA: Makhteshim Agan of North America, Inc. 6 p.

[93] Syngenta Crop Protection. 2011. Caparol® *Herbicide Label*. Greensboro, NC, USA: Syngenta Crop Protection, LLC. 21 p.

[94] Valent U.S.A. 2007. Select® *Herbicide Label*. Walnut Creek, CA, USA: Valent U.S.A. Corporation. 30 p.

[95] E. I. du Pont de Nemours and Company. 2010. DuPont™ Assure® *Herbicide Label*. Wilmington, DE, USA: E.I. du Pont de Nemours and Company, Inc. 13 p.

[96] Syngenta Crop Protection. 2011. Envoke® *Herbicide Label*. Greensboro, NC, USA: Syngenta Crop Protection, LLC. 43 p.

[97] E. I. du Pont de Nemours and Company. 2011. DuPont™ Direx® Extra *Herbicide Label*. Wilmington, DE, USA: E.I. du Pont de Nemours and Company, Inc. 21 p.

[98] Tessenderlo. 2010. Linex® *Herbicide Label*. Phoenix, AZ, USA: Tessenderlo Kerley, Inc. 14 p.

[99] Drexel Chemical Company. 2009. MSMA *Herbicide Label*. Memphis, TN, USA: Drexel Chemical Company. 4 p.

[100] United States Department of Agriculture Foreign Agricultural Service (USDA-FAS). Peanut area, yield, and production. http://www.fas.usda.gov/psdonline/psdreport.aspx? hidReportRetrievalName=BVS&hidReportRetrievalID=918&hidReportRetrievalTemplateID=1#ancor (accessed 5 September 2012).

[101] Jordan, D.L., J.S. Barnes, C.R. Bogle, G.C. Naderman, G.T. Roberson, and P.D. Johnson. Peanut response to tillage and fertilization. *Agronomy Journal*, 2001; 95, 1125-1130, ISSN 0002-1962.

[102] Tubbs, R.S. and R.N. Gallaher. Conservation tillage and herbicide management for two peanut cultivars. *Agronomy Journal*, 2005; 97, 500-504, ISSN 0002-1962.

[103] Johnson, W.C. III, T.B. Brenneman, S.H. Baker, A.W. Johnson, D.R. Sumner, and B.G. Mullinix, Jr. Tillage and pest management considerations in a peanut-cotton rotation in the Southeastern coastal plain. *Agronomy Journal*, 2001; 93, 570-576, ISSN 0002-1962.

[104] Brandenberg, R.L., D.A. Herbert, Jr., G.A. Sullivan, G.C. Naderman, and S.F. Wright. The impact of tillage practices on thrips injury of peanut in North Carolina and Virginia. *Peanut Science*, 1998; 25, 27-31, ISSN 0095-3679.

[105] Marois, J.J. and D.L. Wright. Effect of tillage system, phorate, and cultivar on tomato spotted wilt of peanut. *Agronomy Journal*, 2003; 95, 386-389, ISSN 0002-1962.

[106] Wilcut, J.W., A.C. York, W.J. Grichar, and G.R. Wehtje. The biology and management of weeds in peanut (*Arachis hypogaea*). In H.E. Pattee and H.T. Stalker (eds.) Advan-

ces in Peanut Science. Stillwater, OK, USA: American Peanut Research Educational Society, 1995. p. 207-244.

[107] Grichar, W.J., B.A. Besler, R.G. Lemon, and K.D. Brewer. Weed management and net returns using soil-applied and postemergence herbicide programs in peanut (*Arachis hypogaea* L.). *Peanut Science*, 2005; 32, 25-31, ISSN 0095-3679.

[108] Rao, V.R. and U.R. Murty. Botany-morphology and anatomy. In J. Smartt (ed.) The Groundnut Crop: A Scientific Basis for Improvement. London: Chapman & Hall, 1994. p. 43-95.

[109] Price, A.J. and J.W. Wilcut. Weed management with diclosulam in strip-tillage peanut (*Arachis hypogaea*). *Weed Technology*, 2002; 16, 29-36, ISSN 0890-037X.

[110] Dow AgroSciences. 2010. Strongarm® *Herbicide Label*. Indianapolis, IN, USA: Dow AgroSciences LLC. 5 p.

[111] BASF Corporation. 2012. Cadre® *Herbicide Label*. Research Triangle Park, NC, USA: BASF Corporation. 9 p.

[112] Albaugh. 2010. Agri Star® Butyrac® *Herbicide Label*. Ankeny, IA, USA: Albaugh, Inc. 9 p.

[113] Banks, P.A. and E.L. Robinson. Soil reception and activity of acetochlor, alachlor, and metolachlor as affected by wheat (*Triticum aestivum*) straw and irrigation. *Weed Science*, 1986; 34, 607-611, ISSN 0043+1745.

[114] Schmitz, G.L., W.W. Witt, and T.C. Mueller. The effect of wheat (*Triticum aestivum*) straw levels on chlorimuron, imazaquin, and imazethapyr dissipation and interception. *Weed Technology*, 2001; 15, 129-136, ISSN 0890-037X.

[115] Reddy, K.N., M.A. Locke, and L.A. Gaston. Tillage and cover crop effects on cyanazine adsorption and desorption kinetics. *Soil Science*, 1997; 162, 501-509, ISSN 0038-075X.

[116] Locke, M.A., K.N. Reddy, and R.M. Zablotowicz. Weed management in conservation crop production systems. *Weed Biology and Management*, 2002; 2, 123-132, ISSN 1445-6664.

Chapter 7

Herbicide Safeners: Effective Tools to Improve Herbicide Selectivity

Istvan Jablonkai

Additional information is available at the end of the chapter

1. Introduction

Herbicide safeners, formerly referred to as herbicide antidotes, are chemical agents that increase the tolerance of monocotyledonous cereal plants to herbicides without affecting the weed control effectiveness. The use of safeners offer several benefits to agricultural weed control. Safeners may allow: (1) the selective chemical control of weeds in botanically related crops; (2) the use of nonselective herbicides for selective weed control; (3) the counteraction of residual activity of soil-applied persistent herbicides such as triazines in crop rotation systems; (4) an increase in the spectrum of herbicides available for weed control in "minor" crops; (5) an expansion and extension of the uses and marketability of generic herbicides; (6) the elucidation of sites and mechanism by serving as useful biochemical tools [1]. The commercial viability of safener concept is indicated by the growing number of herbicide-safener products available on the pesticide market. With the use of safeners, difficult weed control problems can be addressed and without safeners, many herbicidally active substances could have never been applied for weed control [2].

The concept to enhance crop tolerance to nonselective herbicide by using chemical agents was introduced by Otto Hoffman. In the late 1940s Hoffmann serendipitiously found that no herbicide injury symptons were developed in tomato plants previously treated with 2,4,6-T, an inactive analogue of herbicide 2,4-D when plant were exposed accidentally to vapors of 2,4-D due to the malfunction of the ventillation system of the greenhouse [3]. Following this observation Hoffmann reported later the antagonistic effects of 2,4-D against herbicidal injury by barban after foliar treatments of wheat plants [4]. Research and development in finding new safeners as well as subsequent commercialization proceeded very intensively in the 1970s. Since the patent application of the safening properties of 1,8-naphthalic anhydride (NA) intensive research on discovery of new safeners resulted in compounds with diverse chemistries (Table 1) successfully applied to alleviate injury symptoms by various classes of herbicides in cereal crops.

NA patented by Hoffmann [5] has been considered as the most versatile safener showing less botanical and chemical specificity than other safeners developed later. NA protected cereals as seed treatments against various herbicide chemistries [6]. NA was reported to be mildly phytotoxic to maize (chlorosis and growth inhibition) under some growing conditions. One problem in treating seeds with safeners prior to planting is that phytotoxicity can increase as the time the safener is exposed to the seed increases. With NA, the phytotoxicity to the crop increases with increased time the safener is in contact with the seed during storage. This problem has thus far prevented NA from being introduced to the commercial market [7].

The introduction of dichloroacetamide derivatives developed as safeners against thiocarbamates and chloroacetanilides was a breakthough in the history of the safeners since these compounds can be applied to the soil in preplant incorporated (PPI) or preemergence (PRE) technology in prepackaged tank mixture with the herbicide. Generally, prepackaged herbicide-safener mixtures offer several advantages over seed safeners. First of all, the manufacturer controls all components of the formulation secondly, the farmers buy and use a single and reliable product which allows a wider selection of crop cultivars. Dichlormid exhibited a remarkable degree of chemical and botanical specificity in protection of maize against thiocarbamates such as EPTC, butylate, vernolate but the safener was less protective to maize against chloroacetanilides. In addition to dichlormid, a number of dichloroacetylated amine derivatives were marketed. Among them AD-67, a spiro-oxazolidine compound was commercialized to protect maize plants against acetochlor while benoxaxor can be used to safen S-metolachlor or racemic metolachlor in maize. Furilazole, in addition to providing protection against acetochlor, has a very good safening effect on sulfonylureas particularly halosulfuron. The dichloromethyl-1,3-dioxolane MG-191 the most active member of dichloroacetal and ketal derivatives, protects maize against thiocarbamate and chloroacetanilide injuries. MG-191, similarly to dichlormid, is more effective against thiocarbamates than chloroacetanilides.

The oxime ethers such as cyometrinil, oxabetrinil, and fluxofenim were marketed as seed treatment safeners to protect sorghum plants against chloroacetanilides, in particular, metolachlor. Flurazole, a 2,4-disubstituted 5-thiazolecarboxylate is also a seed safener allowing the safe use of alachlor in sorghum. The phenylpyrimidine safener fenclorim was introduced against pretilachlor in rice and can be used in tank mixture formulated together with the chloroacetanilide herbicide.

The urea type dymron and the thiocarbamate dimepiperate are actually herbicidally active compounds that possess safening activity against pretilachlor [8] and bensulfuron [9] in rice.

Trends toward post-emergence herbicide treatments and the use of high-activity herbicide molecules have led to the development of safeners with post-emergence application in winter cereals. A new era in safener research began with the discovery of 1,2,4-triazolcarboxylates and fenchlorazole-ethyl was developed as a post-emergence safener against ACCase inhibitor fenoxaprop-ethyl in wheat in a tank mixture with the herbicide. Similarly, the dihydropyrazol dicarboxylate mefenpyr-diethyl was used against ACCase inhibitors including fenoxaprop-ethyl as well as mesosulfuron and iodosulfuron in a variety of cereals. The main application of 8-quinolinoxy-acetate cloquintocet-mexyl is against clodinafop-propargyl in wheat. Dihydroisoxazole-carboxylate isoxadifen-ethyl can safen herbicides of various mode of action. First, it was applied in maize in combination with foramsulfuron but mixture with foramsulfuron and iodosulfuron-methyl is also in use. In rice, it can be used with fenoxafop-P-ethyl and ethoxysulfuron. The arylsulfonyl-benzamide, cyprosulfamide is the latest achievement in safener research. It protects maize against isoxaflutole pre-emergence and can also be used in maize with isoxaflutole plus thiencarbazone in pre-emergence and early post-emergence applications [10].

Interestingly, no successful safeners have been developed for broad-leafed crops. Recently, the non-phytotoxic microbial inhibitor diethotate (O, O-diethyl-O-phenyl phosphorothioate) [11] used to inhibit soil microbes that degrade thiocarbamate herbicides was patented as a Table 1 safener for cotton plants against injuries by clomazone [12].

Despite large amount of information published on the activity, mode of action and uses of safeners during the 50-year history of these herbicide antagonists this overview will focus on several less addressed topics such as a) relationships between the molecular structure and the safening properties; b) basis for differential chemical selectivity; and c) safener effects on detoxifying enzymes in crop plants and weeds.

Chemical class	Name	Structure[a]	logP	Herbicide	Crop	Appl. method
Anhydride	1,8–Naphthalic anhydride (NA)		2.54	Thiocarbamates	Maize	Seed–treatment
Dichloro–acetamide	Dichlormid		1.84	Thiocarbamates Chloroacet-anilides	Maize	PPI, PRE
	Furilazole		2.12	Acetochlor Halosulfuron-methyl	Maize	PRE
	AD–67		2.32[b]	Acetochlor	Maize	PRE
	Benoxacor		2.69	Metolachlor	Maize	PRE

Chemical class	Name	Structure[a]	logP	Herbicide	Crop	Appl. method
Oxime ether	Cyometrinil		1.56	Chloroacet–anilides (metolachlor)	Sorghum	Seed–treatment
	Oxabetrinil		2.76	Chloroacet–anilides (metolachlor)	Sorghum	Seed-treatment
	Fluxofenim		2.90	Chloroacet-anilides (metolachlor)	Sorghum	Seed–treatment
Thiazole carboxylic acid	Flurazole		3.64[b]	Alachlor	Sorghum	Seed–treatment
Dichloromethyl-ketal	MG–191		1.35[b]	Thiocarbamates Chloroacet-anilides	Maize	PRE
Phenyl–pyrimidine	Fenclorim		4.17	Pretilachlor	Rice	PRE
Urea	Dymron		2.70	Pyributicarb Pretilachlor Pyrazosulfuron–ethyl	Rice	PRE, POST
Piperidine–1–carbothioate	Dimepiperate		4.02	Sulfonylureas	Rice	POST
8–Quinolinoxy–carboxylic esters	Cloquintocet–mexyl		5.03	Clodinafop–propargyl	Cereals	POST
1,2,4–Triazole–carboxylate	Fenchlorazole–ethyl		4.52	Fenoxaprop–ethyl	Cereals	POST

Chemical class	Name	Structure[a]	logP	Herbicide	Crop	Appl. method
Dihydropyrazole–dicarboxylate	Mefenpyr–diethyl		3.83	ACCase inhibitors Sulfonylureas	Wheat, Rye, Triticale, Barley	POST
Dihydroisoxazole–carboxylate	Isoxadifen ethyl		3.88[b]	ACCase inhibitors Sulfonylureas	Maize Rice	POST
Arylsulfonyl–benzamide	Cypro–sulfamide		2.09[b]	Isoxaflutole	Maize	PRE, POST

[a] Safeners used as racemic mixtures are indicated by *R/S* in their structures.

[b] Log P values unavailable were calculated by ALOGPS 2.1 program available online at www.vclab.org/articles/cite.html.

Table 1. Structure, logP and application of some important safeners.

2. Structure–safening activity relationships

Structure-activity correlations are very important in the search for biological activity because they provide useful information about chemical substituents that are necessary for the required bioactivity. Published structure-activity correlation studies with safeners and analogous compounds have been limited.

Hoffmann's original patent for NA against EPTC in maize claimed only a few NA analogs such as alkyl esters, barium and tin salts as well as *N,N'*-diallyl naphthalene-1,8-dicarboxylic acid, *N,N'*-diallyloxamide, *N,N'*-dipropynyloxamide, *N,N,N',N'*-tetrapropynyloxamide and dipropynylmalonamide [5]. In addition to the original patent, the effects of other structural analogs of NA were tested against EPTC in maize as seed dressing [13]. The presence of the dicarboxylic anhydride group and at least one aromatic ring attached directly to the anhydride appeared to be essential for the protective activity of NA structural analogues. Derivatives such as acenaphthylene-1,2-dione, benzoisoquinoline-1,3-dione, 4-amino-NA, naphthalic-dianhydride, phtalic anhydride as well as diphenic anhydride showed safening effects while chlorinated NA, 2-phenylglutaric anhydride and phenalene-1-one were toxic to maize.

Detailed structure-activity correlations were conducted mainly with various amide safeners that protect maize from thiocarbamate injury. Studies with several hundred of amides revealed that the most effective safeners were *N,N*-disubstituted acetamides [14] or substituted *N*-acetyl-1,3-oxazolidines [15, 16]. Structure-activity studies with dichloroacetamides revealed that *N,N*-disubstituted derivatives were more effective than monosubstituted amides. A variety of substituents on the nitrogen atom including alkyl, haloalkyl, alkenyl and heterocyclic groups impart various degrees of protective activity. Nevertheless, mono- and trichloroacetamides exhibited less safening activity than dichloro analogues [17, 18]. Based on these SAR studies similarities between the chemical structure of the herbicide and its safener, the possible competitive antagonism between the thiocarbamate and the safener molecules for a common target site has been postulated [19]. Computer-aided molecular modeling (CAMM) studies supported this theory [20]. Superimposing of the structures of dichlormid and EPTC revealed that the two chlorine atoms of the safener do not superimpose over any functional group of the EPTC. If structure of EPTC sulfoxide, the very phytotoxic EPTC metabolite, and the dichlormid were superimposed, the two compounds were similar with functional groups in the same location on both molecules. Comparative three-dimensional quantitative structure-activity relationship studies using comparative molecular field analysis (CoMFA) also supported the competitive antagonism theory and predicted a structure of N-allyl-N-methoxyethoxymethyl dichloroacetamide as a potent highly effective safener [21].

Structure-safening activity studies with oxime ether derivatives revealed that the safening activity is affected by the number of nucleophilic sites present in the molecule. An oxime ether with two nucleophilic sites was more effective than those with only one. In addition to cyometrinil, oxabetrinil and fluxofenim pyridin-2-aldoxime O-ethers such as benzyl and phenylethyl ethers were protective to grain sorghum in seed treatments against metolachlor. The oxime and aldehyde derivatives tested, in terms of decreasing safening effectiveness, were dimethyglyoxime > benzophenone oxime > pyridine-2-aldoxime > benzoin--oxime > methyl thioacetohydroxamate >pyridine-2-aldoxime methiodide > 5-nitro-furancarboxyaldehyde [22]. CAMM evaluations of the oxime ether analogues cyometrinil, oxabetrinil and fluxofenim revealed that as the effectiveness of the safener increases so does its molecular similarity to metolachlor [20].

Structure-safening activity relationships for thiazol-5-carboxylic acids against acetamide herbicides were described for 60 derivatives in the original patent [23]. Thiazolecarboxylates substituted by a trifluoromethyl in the 4-position are clearly superior to those substituted in the 4-position by methyl in reducing herbicidal injury to sorghum. Another preferred group of thiazolecarboxylates contained a halogen atom at position 2 preferably chlorine.

A structure-activity relationship study to safen maize against acetochlor was carried out with the herbicide safener MG-191 and its acetal and ketal analogues at preemergence application [24-26]. Open chain acetals formed from 1,1-dichloroacetaldehyde exhibited only marginal safening efficacy. Dialkyl ketals of 1,1-dichloroacetone showed increasing effectiveness up to 3 carbon length of the alkyl group with further increases in carbon atoms resulted in loss of activity. The 5-, 6- and 7-membered 1,3-dioxacycloalkanes prepared from dichloroacetaldehyde had hardly detectable safening activity. However, introducing alkyl or aryl substitution at the 2-position of the 1,3-dioxacycloalkane ring remarkably increased the safening activity. Regarding ring size the highest activity observed was for 2-dichloromethyl-2-methyl-1,3-dioxepane. Replacing an oxygen in the 1,3-dioxolane ring for nitrogen resulted in oxazolidines with reduced safening activities but alkyl or aryl substitution on the nitrogen increased the safening activity of compounds. Replacement of oxygens by sulfur atoms leads to less active derivatives among which 1,3-dithiolane derivative showed higher activity than the oxathio-lanes. Various 1,3-dioxolane-4-ones provided significant protection against the acetochlor. Benzo[1,3]-dioxoles were ineffective while benzo[1,3]dioxin-4-ones were protective in safening maize. 5-Dichloromethyl-3-substituted-isoxazoles were also active safeners.

Unfortunately, no publication has been reported for the other chemistry of safeners. However, no unifying structural motifs for compounds to be safeners can be predicted from these studies.

3. Chiral safeners

The importance of the chirality in the biological activity has long been recognized. Since biochemical processes in the cells take place in chiral environment and most enzymatic pathways are stereoselective, a high degree of enantiomeric and enantiotopic selectivity can be obtained when chiral or prochiral molecules are introduced into biological systems. About one fourth of the presently available pesticides are chiral, existing as two mirror images called enantiomers. These stereoisomers generally possess identical physico-chemical properties but widely different biological activities, such as toxicity, mutagenicity, and carciogenecity [27]. The active enantiomer of the chiral pesticide would have the desired effect on target species while the other may be inactive [28].

Among the commercially available safeners, four such as benoxacor, furilazole, cloquintocet-mexyl, and mefenpyr-dietyl are chiral compounds but used exclusively as racemic (R/S) mixtures in herbicidal compositions and no information accessible on the safening efficacy of the individual enantiomers. In one recent patent, the R enantiomer of furilazole is described in a herbicidal mixture as a safener [29].

Nevertheless, only a few molecules have been reported as safeners in enantiomerically pure form. The optical isomers of 4-(dichloromethylene)-2-[N-(α-methylbenzyl)imino]-1,3-dithio-lane hydrochloride were synthesized and were tested against triallate in wheat [30]. The R enantiomer exhibited high safening activity and its activity exceeded that of the S and the racemic compound. The monoterpene R-carvone was found more effective than the S enantiomer to safen maize against acetochlor injury [31]. 2-Dichloromethyl-2-methyl-[1,3]oxathiolane 3-oxide, a structural analogue of the MG-191 safener, was prepared and the enantiomers were separated by chiral HPLC [32]. The more polar diastereomeric pair was as effective as MG-191 while the other exhibited only marginal protection against acetochlor. Inducibility of ZmGSTF1-2 from roots was more enhanced by the stereoisomers with higher safening efficacy while only one of these enantiomers was effective in shoots. The findings indicated the importance of the stereochemistry in the protective effectiveness. The safener (S)-3-dichloroacetyl-2,2-dimethyl-4-ethyl-1,3-oxazolidine was found to induce the GSH content and GST activity in root and shoot of maize seedlings but the effect of the R form was not reported in these experiments [33]. As a future prospect, the needs for broad application of the green technology in the sustainable agriculture will probably induce a shift in the use and development of enantiomerically pure safeners.

4. Prosafeners and natural compounds with safening activity

The term prosafeners refers to molecules with safening activity undergoing biotransformation to the actual safening agent prior to exhibiting their safening effect. Substituted N-phenylmaleamic acids and their progenitors N-phenylmaleimides and N-phenylisomaleimides exhibited safening activity against alachlor in sorghum at preemergence application [34]. Simple hydrolytic ring-opening reaction of N-phenylmaleimides and N-phenylisomaleimides results in the N-phenylmaleamic acid derivatives with safening activity. Two thiazolidine derivative L-2-oxathiazolidine-4-carboxylic acid (OTC) [35] and thioproline (L-thiazolidine-4-carboxylic acid) [36] have been reported to safen sorghum against tridiphane injury. OTC is converted by 5-oxoprolinase to S-carboxy-L-cysteine which spontaneously decarboxylates to yield L-cysteine. The conversion of thioproline to cysteine takes place in two steps, first proline oxidase yields N-formyl-L-cysteine from which cysteine is forming by hydrolysis. Either source of cysteine elevates the glutathione level in plants and therefore enhance herbicide detoxication.

Safening activities of natural cyclic hydroxamic acids (DIMBOA, DIBOA, and MBOA) as well as synthetic analogues such as 1,4-benzoxazin-3-ones and 1,3-benzoxazolidin-2-ones were prepared and tested to safen maize against acetochlor and EPTC injuries [37]. Cyclic hydroxamic acids were supposed to act as safeners by catalyzing hydroxylation of herbicides containing reactive chlorine in their structure and they are ineffective against herbicides not possessing leaving groups. While no safening activities of natural hydroxamic acids were detected, the synthetic analogues exhibited low to moderate activity.

Metabolism of the herbicide safener, fenclorim resulting, in a semi-natural product with safening activity has recently been described in *Arabidopsis thaliana* cell cultures [38]. The metabolism of fenclorim mediated by GSTs yielded S-(fenclorim)-glutathione conjugate that was sequentially catabolized to S-(fenclorim)-cysteine then to 4-chloro-6-(methylthio)-phenylpyrimidine (CMTP). Although the fenclorim conjugates tested showed little GST inducing activity in *Arabidopsis*, the formation of CMTP resulted in metabolic reactivation, with the product showing enhancing activity similar to that of parent safener. In addition, CMTP safened rice plants and induced rice GSTs. The formation of CMTP by metabolic bioactivation can contribute to the longevity of safener action since it was found stable 8 – 24 h after application.

Oxylipins constitute a family of oxygenated natural products which are formed from fatty acids. Safeners and reactive electrophilic oxylipins (RES oxylipins) have a common biological activity in that they both strongly induce the expression of defence genes and activate detoxification responses in plants [39, 40]. Surprisingly, the application of oxylipin A has been found to reduce the herbicidal injury [41].

5. Interaction of safeners and herbicides on the absorption and translocation

Published results on how safeners affect the herbicide absorption are rather contradictory and, therefore, no general conlusion can be drawn. In an excellent summary the effect of 15 safeners toward various herbicides was reviewed [2]. Interestingly the majority of papers published report safener-enhanced herbicide uptake followed by no effect then reduced uptake results. According to a recent study mefenpyr-diethyl had no effect on the uptake of either mesosul-furon-methyl or iodosulfuron-methyl-sodium [10]. These results suggest that the influence of safeners on the herbicide uptake may not be a decisive factor in the protective action. However, the knowledge of absorbed amounts of safeners and herbicides by crops may help to determine the optimal herbicide/safener ratios applicable in the agricultural practice. In addition, determination of the site of safener and herbicide uptake can contribute to prepare the most selective herbicide-safener mixture. A suitable placement of soil-applied herbicides to roots or the emerging shoots is of great practical importance in achieving the most effective weed control and the least injury to crop plants.

Studies on how maize can differentiate in the absorbtion of herbicides and safeners were conducted with radiolabeled EPTC, acetochlor and MG-191 [42, 43]. Time-dependent uptake of root-applied [14C]EPTC reached a maximum after 6h and decreased up to 3 days (Figure 1). The first measurable shoot growth inhibition appeared just after 1-day-exposition to the herbicide and 38% shoot length inhibition was observed 3 DAT. In general, the MG-191 safener had no influence on the herbicide uptake except for 1 DAT when the safener enhanced the herbicide uptake by 1.5-fold as compared to that in the unsafened plants. Nevertheless, the safener conferred a complete protection to maize throughout the study. The highest amount of herbicide uptake was 65 µg/g fresh weight.

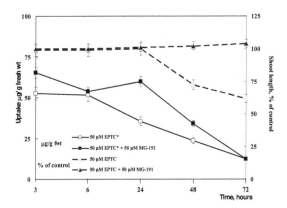

Figure 1. Influence of MG-191 safener on uptake and shoot length inhibition of root-applied [14C]EPTC.

As a comparision, the amount of root-absorbed [14C]acetochlor was continuously increased up to 3 days (Figure 2). As a result of increasing uptake the first detectable shoot length inhibition occurred 6h after treatment. At 3 DAT 28% shoot and 52% root (data not shown) growth inhibition by the herbicide occurred. Addition of the MG-191 safener did not affect the acetochlor absorption by maize seedlings but completely antagonized the herbicide shoot growth inhibition. The maize seedlings absorbed much higher amounts of acetochlor (377 µg/ g fresh weight).

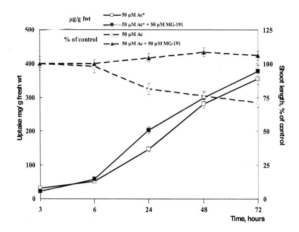

Figure 2. Influence of MG-191 safener on uptake and shoot length inhibition of root-applied [^{14}C]acetochlor.

All previous efforts to elucidate modes of action of safeners focused on the fate of various herbicides as affected by the safener treatments while no studies were conducted on how uptake, translocation and metabolism of safeners were influenced by herbicides. For a better understanding of the herbicide-safener interaction, absorption of [^{14}C]MG-191 by maize seedlings was studied as influenced by EPTC. Absorbed amount of the labeled safener following application to the roots of 5-day-old maize plants increased over the time and no influence of EPTC on uptake was observed (Figure 3). At a higher safener concentration (50 µM), plants absorbed higher amounts of radiolabel than at a lower concentration (10 µM) but plants contained low levels (3% and 1%) of the safener applied. The highest value for the safener content in the maize seedlings was less than 8 µg/g fresh weight.

These data clearly suggest that even this small amount of safener offer protection to maize. The absorbed herbicide/safener ratio (µg/µg) at 3 DAT accounted for 50 with acetochlor and 1.7 with EPTC at same concentrations of the herbicide. These results may partly explain why safening efficacy of MG-191 toward EPTC is higher than toward acetochlor under field conditions. Site of uptake can also affect the MG-191 effectiveness. In experiments using a charcoal barrier to separate shoot and root zones of maize, the influence of site of safener placement on acetochlor phytotoxicity was studied [44]. MG-191 was the most protective when both the safener and the herbicide were applied simultaneously to shoots and roots but also satisfactory protection was achieved when the safener was applied in the root zone and the herbicide to the emerging shoots. This also indicates the main site of uptake for acetochlor absorption is the coleoptile while the root-uptake is very significant in the safener performance. Under field conditions the more water-soluble MG-191 (log P, 1.35) can be more easily leached

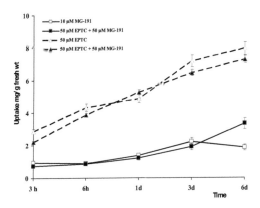

Figure 3. Time-course uptake of root-applied [^{14}C]MG-191 by 5-day-old maize seedlings and the influence of EPTC.

to the roots of maize plants than the less water-soluble acetochlor (log P 4.14). The higher logP of acetochlor also supports its higher uptake as compared to MG-191.

It is also difficult to put the results of safener affected translocation of absorbed herbicides in perspective. Reduction of translocation of herbicides such as acetochlor, methazachlor, and imidazolinones from roots of maize to the shoots following treatments with dichlormid, BAS 145,138 and NA is likely a consequence of the safener-enhanced herbicide metabolism to more polar and less mobile products [45-49]. On the other hand, no effect of MG-191 on EPTC and acetochlor translocation has been observed [42, 44]. It is interesting to note that safener MG-191 and the herbicide acetochlor exhibit different translocation patterns (Figure 4). While the majority of the absorbed radiolabel from [^{14}C] acetochlor was found in the roots and coleoptiles of maize seedlings (Figure 4a), the root-applied [^{14}C]MG-191 distributed evenly within the plants (Figure 4b) showing similar mobility and distribution as EPTC (data not shown). This may be further evidence for the higher protective efficacy of this safener against EPTC as compared to acetochlor. The similar translocation pattern of the herbicide and the safener may be a prerequisite for the high level of safening activity.

6. Action of safeners on the glutathione-mediated detoxification of herbicides

Various chemistries of safeners were found to enhance the herbicide detoxification in the safened plants by elevating the activity of the mediating enzymes such as glutathione S-transferases (GSTs), cytochrome P450 mixed function oxidases (CYPs), glycosyltransferases (UGTs) and ATP-binding casette (ABC) transporter proteins as well as a cofactor endogenous glutathione (GSH) involved in detoxification of herbicides [2, 50-52]. The best studied group

a)

Autoradiograph of [¹⁴C]acetochlor-
treated (2.5 kg/ha) maize seedling
11 DAT

b)

Autoradiograph of [¹⁴C]MG-191-treated (50 μM)
maize seedling 3 DAT

Figure 4. Distribution of root- and shoot-applied [¹⁴C]acetochlor and root-applied [¹⁴C]MG-191 in maize seedlings.

of plant enzymes involved in herbicide metabolism is the GSTs that mediate the conjugation of the major cellular thiol tripeptide, GSH with herbicide substrates. GSTs are multifunctional enzymes, each composed of two subunits which catalyze conjugation of a broad range of electrophilic substrates with GSH [53]. Herbicides known to conjugate with GSH include thiocarbamates, chloro-s-triazines, triazinone sulfoxides, chloroacetanilides, diphenylethers, some sulfonylureas, aryloxyphenoxypropionates, thiazolidines, and sulfonamides [54, 55]. Plant GSTs comprise a large and diverse group, with 54 GST genes encoded by the *Arabidopsis* genome, and have been classified on sequence similarity, genomic organization and functions into several distinct subclasses [56]. In plants, phi (F) and tau (U) classes are the most prominent GSTs involved in herbicide detoxification [57-59]. In addition to up-regulating GST expression, safeners also enhance the activity of enzymes involved in sulfate assimilation and GSH biosynthesis thereby elevating the level of GSH [50, 60].

Only two studies are available in the literature on how the safener structure affects the expression of GST isoforms. The herbicide safener MG-191 (2-dichloromethyl-2-methyl-1,3-dioxolane) and its less effective structural analogue dichloromethyl-dioxolanone (NO-17; 2-dichloromethyl-2,5-dimethyl-1,3-dioxolane-4-one) were reported to differentially enhance the expression of members of the GSTs in maize [61].

None of these safener molecules had influence on the expression of ZmGSTF1-2 (Figure 5a and b). However, MG-191 and, to a lesser extent NO-17 selectively enhanced the expression of tau class ZmGSTU1 in both root and shoot tissues after 1 day of treatment (Figure 5c and d). Addition of cycloheximide to the treatment solutions suppressed the enhancement of expres-

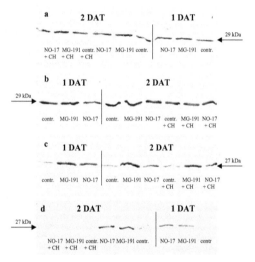

NO–17

sion of *ZmGSTU1* only in the roots. *ZmGSTU1* has previously been shown to play a key role in metabolism of nitrodiphenyl ether type herbicides [54].

Figure 5. Western blots of crude GST extracts from maize roots and shoots (a) Analysis of GSTs from maize shoots using the anti-*ZmGSTF1-2* serum.(b) Analysis of GSTs from maize roots using the anti-*ZmGSTF1-2* serum.(c) Analysis of GSTs from maize shoots using the anti-*ZmGSTU1-2* serum.(d) Analysis of GSTs from maize roots using the anti-*ZmGSTU1-2* serum.

Analysis of isoenzyme profile of maize GSTs revealed that phi class of GSTs predominate, with ZmGSTF1 as the major subunit which is present constitutively and shows high specificity to 1-chloro-2,4-dinitrobenzene (CDNB) substrate [62]. A second phi type GST termed *ZmGSTF2* accumulates following treatments with herbicide safeners. These subunits can dimerise together to form *ZmGSTF1-1* and *ZmGSTF2-2* homodimers as well as *ZmGSTF1-2* heterodimer. In addition to these three phi GST isoenzymes a phi type GST *ZmGSTF3* and three tau class GSTs *ZmGSTU1*, *ZmGSTU2* and *ZmGSTU3* are present in lower amounts [63, 64]. While the expression of *ZmGSTF2* was enhanced by auxins, herbicides, the herbicide safener dichlormid and glutathione, the *ZmGSTU1* subunit was induced more selectively, only accumulating significantly in response to dichlormid treatment [63]. Although *ZmGSTF2* has been considered more active in detoxifying metolachlor and alachlor than *Zm*GSTF1 it is far less abundant [65]. The importance of *de novo* synthesis of the isoenzyme *Zm*GSTU1 in its safening action is difficult to explain. Nevertheless, these results indicate that dichloromethyl-dioxolane type MG-191 is a more specific inducer of maize GSTs than other compounds commonly used to safen thiocarbamate or chloroacetanilide herbicides in maize.

						Protection[a]	GSH[b]	GST(CDNB)[c]	GST (Ac)[d]
Code	R	X	Y	R^1	R^2	(%)		treated/control	
Ac	-	-	-	-	-	-	1.11	1.48	3.74
1a	-	H	Cl	Et	Et	24	1.49	0.69	2.03
1b	-	H	Br	Et	Et	60	1.53	0.94	3.76
1c	-	Cl	Cl	Et	Et	8	0.69	1.42	1.83
1d	-	Cl	Cl	Pr	Pr	0	0.80	0.95	1.38
1e	-	Cl	Cl	Bu	Bu	-6	1.22	0.96	0.91
1f	-	Cl	Cl	i-Bu	i-Bu	-2	1.20	0.90	1.61
1g	-	Cl	Cl	-(CH$_2$)$_2$-		18	0.93	0.88	1.23
1h	-	Cl	Cl	-(CH$_2$)$_3$-		14	0.60	0.89	0.90
1i	-	Cl	Cl	-CH$_2$C(CH$_3$)$_2$CH$_2$-		-3	0.91	0.88	1.33
1j	-	Cl	Cl	-(CH$_2$)$_4$-		11	0.95	1.03	1.33
1k	-	Cl	Cl	-(CH$_2$)$_5$-		0	0.98	1.24	0.58
1l	-	Cl	Cl	-(CH$_2$)$_6$-		3	0.82	1.32	0.83
2a	Me	Cl	Cl	Et	Et	62	1.15	1.22	0.65
2b	Me	Cl	Cl	Pr	Pr	63	0.98	1.18	0.85
2c	Me	Cl	Cl	Bu	Bu	38	0.78	0.88	3.93
2d	Me	Cl	Cl	i-Bu	i-Bu	14	0.85	1.07	4.72
2e	Ph	Cl	Cl	-(CH$_2$)$_2$-		41	2.00	1.94	2.23
2f	Me	Cl	Cl	-(CH$_2$)$_2$-		64	1.18	1.83	3.93
2g	Me	Cl	Cl	-(CH$_2$)$_3$-		68	1.31	1.49	1.96
2h	Me	Cl	Cl	-CH$_2$C(CH$_3$)$_2$CH$_2$-		66	1.62	1.77	1.44
2i	Me	Cl	Cl	-(CH$_2$)$_4$-		70	1.71	1.48	0.92
2j	Me	Cl	Cl	-(CH$_2$)$_5$-		50	1.24	1.27	1.19

Structures: 1a-l, 2a-k, 3a-d

Code	R	X	Y	R¹	R²	Protection[a] (%)	GSH[b]	GST(CDNB)[c]	GST (Ac)[d]
								treated/control	
2k	Me	Cl	Cl	-(CH₂)₆-		60	1.38	1.39	1.14
3a	-	Cl	Cl	allyl	allyl	81	1.78	1.24	4.69
3b	-	H	Cl	H	allyl	48	2.25	1.25	3.60
3c	-	H	Cl	allyl	allyl	2	1.45	1.16	2.39
3d	-	H	Br	allyl	allyl	22	0.98	0.90	2.98

[a] based on shoot length; protection (%) = 100 x [(herbicide + safener)] / [control - herbicide]; shoot lengths 14 DAT: control, 27.9+5.3 cm, acetochlor, 3.1±0.3 cm;

[b] GSH content relative to that of untreated control; $GSH_{contr.}$: 0.55±0.09 µmol/g fresh weight;

[c] GST(CDNB) activity as compared to that of untreated control; $GST_{contr.}$: 3.87±0.33 nkat/mg protein;

[d] GST(Ac) activity as compared to that of untreated control; $GST_{contr.}$: 8.26±1.68 pkat/mg protein

Table 2. Safening activity and inducibility of shoot GSH content and GST activities by acetals, ketals and amides in maize

In other, structure and GST isoform expressing ability studies with acetal and ketal analogues of MG-191 as well as mono-and dichloroacetamides (Table 2) demonstrated that the safener structure affects the specific expression of GSTs mediating the detoxication of acetochlor (Matola et al., 2003). Nevertheless, no correlation was found between the degree of induction of GSH and GSTs and the safening activity as related to the structure. A higher inducibility of these GST isoforms was observed in root tissues (Figure 6a and c). In shoots, when the heterodimer ZmGSTF1-2 was used the expression of the constitutive ZmGSTF1 and inducible ZmGSTF2 was enhanced only by 2f (MG-191) and its analogue 2g having a 6-membered ring (Figure 6b). These molecules and also 2h were the most potent inducers of the expression of tau class ZmGSTU1 in shoot tissues (Figure 6c). ZmGSTU1 has previously been shown to play a key role in metabolism of nitrodiphenyl ether herbicides [54]. These results confirm previous findings that dichloromethyl-ketal safeners are more specific inducers ZmGSTU1-2 than other compounds commonly used to safen thiocarbamate and chloroacetanilide herbicides in maize [61].

The exact mechanism of the safener-mediated enhancement of GST activity is not completely understood. GSTs are induced by a diverse range of chemicals and accompanied by the production of active oxygen species. Thus the connection between safener-mediated protection of crops and oxidative stress tolerance has been suggested [66]. Many GSTs are effective not only in conjugating electrophilic substrates but also function as glutathione peroxidases. Safeners may induce GST expression by mimicking oxidative insult [67]. Our results indicate that safener structure plays a decisive role in specific expression of GSTs mediating the detoxication of chloroacetamide herbicides. Since no correlation between the degree of induction of levels of GSH and GST isoforms and the safener activity was found, the mode of action of safeners is a more complex process than simply promoting the metabolism of herbicides.

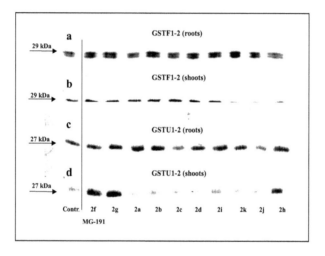

Figure 6. Western blots of crude GST extracts from maize roots and shoots; (a) and (b) analysis of GSTs using the anti-*Zm*GSTF1-2 serum from maize roots and shoots; (c) and (d) analysis of GSTs using the anti-*Zm*GSTU1-2 serum from maize roots and shoots.

7. Effect of safeners on herbicide detoxification enyzmes in weeds

Studies on the mechanism of action of safeners revealed that herbicide safeners improve crop tolerance to herbicides by regulating the expression of genes involved in herbicide metabolism [68]. It is widely accepted that safeners selectively protect crop plants against herbicide injury by stimulating the plant detoxifying mechanism at herbicide rates required for effective weed control. Nevertheless, only a few papers were published on the safener effect of GSTs and cytochrome P450 monooxygenases of various weed species. To a better understanding on why safeners do not provide protection to weeds it is essential to explore the safener action on detoxification enzymes of weeds.

7.1. Effect of safeners on weed glutathione (GSH) content and glutathione S-transferase enzyme (GSTs) activities

Safeners such as MG-191, dichlormid, AD-67, BAS-145138, and flurazol were reported to reduce phytotoxicity of EPTC in grassy weeds [69]. MG-191, BAS-145138 and flurazole offered moderate safening to *Bromus secalinus* (bromegrass) and flurazole was also moderately protective in *Setaria glauca* (yellow foxtail) at sublethal rate of EPTC. Safener-induced elevation of GSH contents and GST activities is widely considered as key element for increased tolerance to thiocarbamates and chloroacetanilides of safened plants [50]. Tolerance of plant species such as maize, soybean and several weeds to acetochlor has been correlated with their glutathione and homoglutathione content [70]. It was also apparent that a relationship exists between the relative GST activities toward alachlor and metolachlor in maize and various weed species ([71]. GST activities toward metolachlor were found to correlate well with the selectivity of the herbicide toward the broadleaf weeds but not toward the grass weeds [72]. However, there was no correlation between total activity of cysteine biosynthesis from serine (CBS) and susceptibility to metolachlor of sorghum, maize, and various grassy weeds [73]. GST isozymes involved in herbicide metabolism is cell suspension culture of a grass weed *Setaria faberi* (giant foxtail) exhibited a similar level of complexity to those from maize cell cultures [74].

Nevertheless, much less is known about GSH or other non-protein thiol contents and GST activities of different weed species following treatments by herbicides and safeners. In order to explain differential physiological and biochemical responses of monocot and dicot weeds to these herbicides, non-protein thiol levels and GST activities were studied in selected mono- and dicot weeds species [75]. The most sensitive *Echinochloa crus-galli* (ECHCR, barnyard-grass) contained higher level of non-protein thiols than less sensitive dicot seedlings (Figure 5). Nevertheless, thiol contents in the most tolerant maize and in the least sensitive monocotyledonous *Bromus secalinus* (BROSE, cheatgrass) were comparable. In general, either herbicide or safener pretreatments did not alter thiol contents substantially. *Abuthilon theophrasti* (ABUTH, velvetleaf) was the only exception because 1 µM acetochlor and 10 µM AD-67 resulted in remarkable increases (73% and 87%, respectively) in the levels of non-protein thiols.

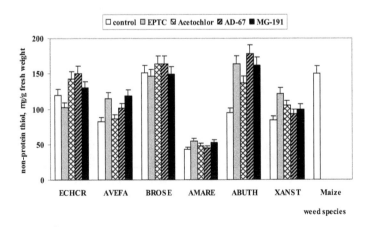

Figure 7. Effect of treatments on non-protein thiol contents of mono- and dicot weed species.

Glutathione S-transferase (GST) activities using CDNB substrate were not correlated with herbicide susceptibility of the selected weed species (Figure 6a). The GSTs extracted from monocot seedlings exhibited much higher activities than from dicot seedlings. GST$_{CDNB}$ activity detected in *Avena fatua* (AVEFA, wild oats) exceeded that in maize. In general, elevation of GST$_{CDNB}$ activities following pretreatments with both herbicides and safeners were more pronounced (2- to 10-fold of controls) in the highly sensitive *Echinochloa crus-galli* and *Amaranthus retroflexus* (AMARE, redroot pigweed) compared to less sensitive species.

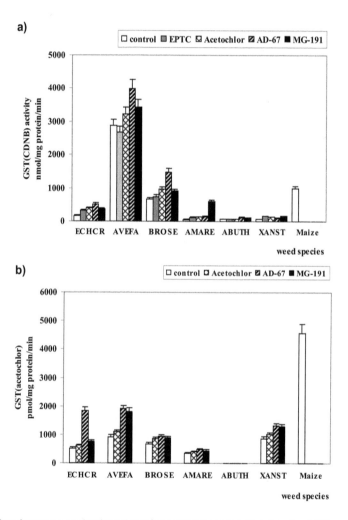

Figure 8. Effect of treatments on glutathione S-transferase activities of selected weed species. a) GST$_{CDNB}$ activities; b) GST$_{acetochlor}$ activities of untreated and treated 6-day-old etiolated seedlings.

With [^{14}C]acetochlor substrate, GST$_{acetochlor}$ activities of both mono- and dicot seedlings were in the same range except for velvetleaf (ABUTH) (Figure 6b). Regardless of treatment, extractable GSTs from velvetleaf did not show specificity for acetochlor. Nevertheless, GST$_{acetochlor}$ activities in all weed species were less expressed than in maize. No correlation was found between enzyme activity and acetochlor susceptibilities of these weed species. In monocot seedlings higher enzyme inductions (up to 2-fold increase) were observed as compared to those in dicots following safener treatment. Nevertheless, GST$_{acetochlor}$ activity of the maize seedlings exceeded those of weed species which may indicate that the higher detoxication capability of crop plant is closely related to the herbicide tolerance. It is also noteworthy that both GSH and cysteine conjugates of chloroacetamides were found inhibitory to GSTs from maize, *Avena fatua*, and *Echinochloa crus-galli* suggesting that GSH conjugation in crops and weeds takes place in a complex manner [76].

Interestingly, *Arabidopsis* plant cultures were more responsive to induction by safeners than either maize or wheat [77]. Enhancement of GST$_{CDNB}$ activity was greatest with fenclorim however treatment with flurazole, CMPI and benoxacor also offered significant increases. O-Glucosyltransferase and N-glucosyltransferase activities were also stimulated but to a lesser extents. Safeners mefenpyr diethyl and fenchlorazole-ethyl enhanced fenoxaprop-ethyl tolerance of weed *Alopecurus myosuroides* (black-grass) [78]. In black-grass, these detoxification pathways were only slightly enhanced by safeners, suggesting that metabolism alone was unlikely to account for increased herbicide tolerance. Instead, it was determined that safening was associated with an accumulation of glutathione and hydroxymethylglutathione and enzymes with antioxidant functions including phi and lambda glutathione transferases, active as glutathione peroxidases and thiol transferases respectively. In addition to enhanced glutathione metabolism safener treatment resulted in elevated levels of flavonoids in the foliage of black-grass plants, notably flavone-C-glycosides and anthocyanins. Safening of grass weeds was concluded as a mechanism associated with an inducible activation of antioxidant and secondary metabolism. The ability of safeners to induce GSTs of grassy weeds can be exploited in phytoremediating herbicide-contaminated soils. In recent studies safener benoxacor was used to enhance GSTs of the perennial grass *Festuca arundinancea* to establish a basis for preventing environmental herbicide pollution [79]. Further studies revealed that in addition to benoxacor cloquintocet-ethyl, fenchlorazol-ethyl, fenclorim, fluxofenim and oxabetrinil were also able to enhance GST activity in *Festuca* [80]. These results indicate that herbicide diffusion following the runoff of surface waters can be prevented or significantly reduced by vegetating buffer strips with *Festuca* and by the combination of herbicide and a suitable safener. By this way, the application of safeners can be extended by using non crop-species in phytoremediating contaminated soils.

7.2. Interaction of safeners on weed cytochrome P450 monooxygenases

The involvement of cytochrome P450 monooxygenases in herbicide detoxication and selectivity has been well demonstrated [81, 82]. The role of cytochrome P450 monooxygenases in enhanced metabolism of resistant weed species has also been documented [83, 84]. Nevertheless, only a few examples can be found in the literature as to cytochrome P450-dependent monooxygenase system in weed species [85].

Monocotyledonous (*Avena fatua, Bromus inermis, Echinochloa crus-galli*) and dicotyledonous (*Amaranthus retroflexus, Abuthilon threophrasti, Xantium strumarium*) weeds were used to study the interaction of safeners, herbicides metabolized by cytochrome P450 enzymes, and P450 inhibitors on herbicide phytotoxicity and P450 levels of weeds and maize [86]. The safener NA was slightly protective to all monocots at the reduced rate (50 g/ha) of nicosulfuron and also exhibited safening effects on dicots against all herbicides. MG-191 reduced growth inhibition of EPTC in *A. fatua* and *E. crus-galli*.

Species	Cytochrome P450, pmol/mg protein		
	Control	NA[a]	ABT[b]
A. fatua[c]	41±11	49±12	36±17
B. inermis	ND[d]	ND	ND
E. crus-galli	17±8	14±9	ND
A. retroflexus	10±4	21±8	ND
A. theophrasti	51±24	89±32	54±27
X. strumarium	ND	ND	ND
Maize[e]	67±14	73±15	96±18

[a] 0.5 %w/v; [b] 1 µM; [c] 7-day-old etiolated weed seedlings; [d] ND not detectable; [e] 4-day-old etiolated maize seedlings.

Table 3. Cytochrome P450 contents of mono- and dicot weeds and influence of treatment with the safener NA and P450 inhibitor ABT.

Weed microsomal cytochrome P450 enzymes were found less stable than those from maize. Carbon-monoxide difference spectra for *B. inermis* and *X. stumarium* could not be recorded probably due to dark colors of microsomal preparations and difficulties in resuspending the microsomes. Cytochrome P450 content in the microsomal membrane fraction of *A. fatua* was 2.4-fold greater than in *E. crus-galli* (Table 3). Among dicotyledonous plants, *A. theophrasti* contained 5.1-fold higher level of the enzyme as compared to that of *A. retroflexus*. However, the P450 level was higher in maize than in weeds.

It is difficult to evaluate changes in the enzyme contents of weed species pretreated with the safener NA or the P450 inhibitor ABT due to the high values of standard deviation of the data. Following treatments with NA, a stimulating tendency could be observed for weeds except *E. crus-galli*. With maize the NA treatment had no enhancing effect on the enzyme content. However, a significant increase (43%) was found when maize seedlings treated with ABT but the P450 inhibitor was uneffective on weed P450s.

For further characterization of *in vivo* interaction of the combination of the herbicides with safeners and inhibitors microsomes isolated from etiolated maize seedlings were used (Figure 7). Treatment of maize seedlings with nicosulfuron resulted in 30% elevation in P450 level while no effect of EPTC was found. The combination of NA with either bentazon or nicosulfuron decreased P450 levels by about 50% as compared to the untreated control. Interestingly, without herbicide pretreatment with NA had no influence on maize P450. The inhibitory effect of NA *in vitro* on maize P450 was reported by the formation of an enzyme-NA Type I complex [87]. Pretreatments with the combination of MG-191 and all herbicides yielded slight increases in the enzyme concentration. It is interesting to note that no binding of MG-191 to P450 was detected [88] which may indicate why MG-191 was not inhibitory to P450. The P450 inhibitor PBO simultaneously applied with bentazon and nicosulfuron substantially reduced P450 levels while the ABT was less inhibitory.

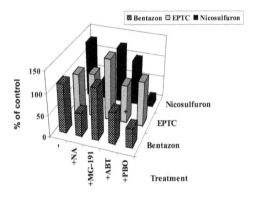

Figure 9. Interaction of herbicides with safeners and cytochrome P450 inhibitors on P450 enzymes extracted from 4-day-old etiolated maize seedlings. Treatments were as follows: bentazon, 10 µM; EPTC, 10 µM; nicosulfuron, 10 µM; NA 0.5%w/v; MG-191, 10 µM; ABT, 1 µM; PBO, 10 µM.

These results demonstrate that safeners can marginally protect weed species by stimulating the herbicide detoxifying enzymes but the lower level of these enzymes in weeds as compared to those in crops provide a basis for the botanical selectivity of safeners.

8. Mechanism of safener action

The mechanism by which safeners act is currently unknown despite the widespread agricultural use and the substantial experimental evidence accumulated on the biochemical basis of action. Safeners appear to induce a set of genes that encode enzymes and biosynthesis of cofactors involved in the herbicide detoxication [50, 52, 89, 90].

The exact mechanism of safener-mediated enhancement of GST activity is not completely understood. GSTs are induced by a diverse range of chemicals and accompanied by the production of active oxygen species. Thus the connection between safener-mediated protection of crops and oxidative stress tolerance has been suggested [66]. Many GSTs are effective not only in conjugating electrophilic substrates but also function as glutathione peroxidases. Safeners may induce GST expression by mimicking oxidative insult [67]. Herbicide safeners increase herbicide tolerance in cereals but not in dicotyledonous crops. The reason(s) for this difference in safening is unknown. Treatment of *Arabidopsis* seedlings with various safeners resulted in enhanced GST activities and expression of GSH-conjugate transporters such as *At*MRP1-4 [91]. Safeners also increased GSH content of *Arabidopsis* seedlings. However, treatment of *Arabidopsis* plants with safeners had no effect on the tolerance of seedlings to chloroacetanilide herbicides. Immunoblot analysis confirmed that *At*GSTU19 was induced in response to several safeners. These results indicate that, although *Arabidopsis* may not be protected from herbicide injury by safeners, at least one component of their detoxification systems is responsive to these compounds.

Concerning the location of safener binding site(s) of plants few studies have been conducted. A high-affinity cytosolic-binding site for the dichloroacetamide safener (R,S)-3-dichloroace-tyl-2,2,5-trimethyl-1,3-oxazolidine was found in etiolated maize seedlings ([92]. The binding was highest in the coleoptiles and lowest in the leaves. A good correlation was shown between the safener effectiveness. Chloroacetanilide and thiocarbamate herbicides were effective inhibitors of safener binding at low concentrations. The inhibition by alachlor and EPTC was shown to be competitive. The safener binding protein (SafBP) was purified to homogeneity having a molecular mass of 39 kDa [93]. Based on the peptides obtained from proteolytic digests of SafBP a cDNA encoding SafBP was cloned and expressed in *E. coli*. The predicted primary structure of SafBP was related to a phenolic *O*-methyltransferase but SafBP did not catalyze O-methylation of catechol or caffeic acid. It was concluded that SafBP may not be the primary site of action of the dichloroacetamide safeners. Supporting the participation of *O*-methyltransferases in the safener action, treatment of wheats (*Triticum aestivum* L.) with cloquintocet-mexyl resulted in an accelerated depletion of flavone *C*-glycosides and a selective shift in the metabolism of endogenous phenolics [94]. Changes in phenolic content were associated with an increase in *O*-methyltransferase and *C*-glucosyltransferase activity toward flavonoid substrates.

Proteomic methods were used to identify herbicide safener-induced proteins in the coleop-tile of *Triticum tauschii* [95]. The herbicide safener, fluxofenim, dramatically increased protein abundance in the molecular range in the molecular weight range of 24 to 30 kDa as well as a few higher molecular weight protein and overall 20 proteins were identified. Among the eighteen inducible proteins 15 were glutathione *S*-transferase subunits that fall into three subclasses: eight proteins were from the tau subclass, six proteins were from phi subclass, and one was from the lambda class. Another three safener inducible proteins showed homology to the aldo/keto reductase family with proteins that have roles in glycolysis and the Krebs cycle. One of the two constitutively expressed proteins showed the highest homology to the dehydroascorbate reductase subclass of GSTs while the other to an ascorbate peroxidase. Results indicated that the induced proteins were associated with herbicide detoxication and with general stress response. In another study with cloquintocet-mexyl safener and dimethenamid herbicide 29 safener-induced and 10 herbicide-regulated proteins were identified in *Triticum tauschii* seedlings [39]. Surprisingly, mutually exclusive sets of proteins were identified following herbicide or safener treatment suggesting a different signaling pathway for each chemical. Safener-responsive proteins were mostly involved in xenobiotic detoxication whereas herbicide-regulated proteins belonged to several classes involved in general stress responses. Quantitative RT-PCR revealed that multidrug resistance-associated protein (MRP) transcripts were highly induced by safeners and two MRP genes were differently expressed.

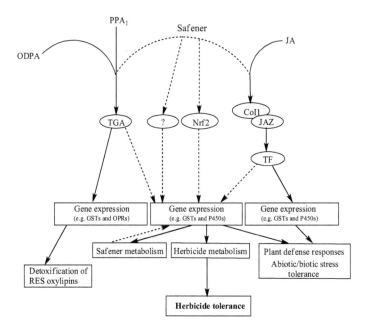

Figure 10. Suggested safener-mediated signalling pathway for regulation of defense genes and activation of detoxification pathways in plants by Riechers et al. [52]. Dashed lines indicate possible but unproven signaling pathways while solid lines indicate known signaling pathways. ODPA: 12-oxo-phytodienoic acid; OPRs: ODPA-reductases; PPA$_1$: A$_1$-type phytoprostanes; JA: jasmonic acid; TGA: TGA transcription factor; Nrf2: nuclear factor (erythroid-derived 2)-like 2; Col1: coronative insensitive protein 1; JAZ: transcriptional repressor protein; TF: transcription factor/activator.

Safeners were suggested to trigger an unidentified, preexisting signaling pathway for detoxification of endogenous toxins or xenobiotics [96]. According to a new hypothesis, safeners may be utilizing an oxidized lipid-mediated (oxylipins) or cyclopentenone-mediated signaling pathway which subsequently leads to the expression of GSTs and other proteins involved in detoxification and plant defense [52]. Some possible safener-mediated signaling pathways for the regulation of defense genes and activation of detoxification pathways have been suggested (Figure 8). Safeners may tap into a RES oxylipin-mediated signaling pathway and up-regulate TGA transcription factors, an Nrf2-Keap1-mediated as well as jasmonic acid-mediated signaling pathways. Safeners and oxylipins as reactive electrophilic species (RES oxylipins) have a common biological activity since both strongly induce the expression of defense genes and activate detoxification responses in plants [39, 40].

9. Conclusions

Fifty-year of herbicide safeners resesearch and use confirms that these molecules offered new ways to improve herbicide selectivity. Although this technology now competes with herbicide-tolerant, genetically-modified or naturally-selelected crops, safeners still comprise an important part of the herbicide market in maize, cereals and rice [10]. Many of the commercial safeners are in off-patent status offering a chance for the generic manufacturers to enter the market together with off-patent herbicides. In contrast, recent herbicide mixture patents with new herbicides still allow their exclusive usage by the patent holder [10].

Although safeners do not improve herbicide tolerance in dicot plants, but the utilization of biotechnology tools may help in extending the safener response from monocot to dicots. It was found, however that *Arabidopsis* transgenic plants did not respond to safeners at whole-plant level despite the increase of the expression of tau class protein in the roots [91]. Additionally, knowledge of critical regulatory elements in the promoters or untranslated regions of genes encoding detoxification enzymes, or a comprehensive understanding how gene expression is up-regulated by safeners might lead to the precise manipulation of transgene expression of plants [52].

The use of safeners to enhance tolerance of plants to organic pollutants such as herbicides, heavy metals or oils in the environment (soil, water) could also be a promising application of these chemicals. Phytoremediation studies with soils contaminated with oils and heavy metals and safener-treated wheat seeds have recently been reported [97]. While untreated seeds were unable to germinate on the contaminated soil, safener treatments resulted in seedlings briefly growing before succumbing to the pollutants.

Author details

Istvan Jablonkai

Institute of Organic Chemistry, Research Centre for Natural Sciences, Hungarian Academy of Sciences, Budapest, Hungary

References

[1] Hatzios K. K. (1989). Development of herbicide safeners: Industrial and university perspectives. In: *Crops safeners for herbicides*. Hatzios K. K., Hoagland R. E. (Eds). pp 3-45, Academic Press, San Diego, USA.)

[2] Davies J., Caseley J. C. (1999). Herbicide safeners: a review. *Pestic. Sci.*, 55, 1043–1058.

[3] Hoffmann O. L. (1978). Herbicide antidotes: From concept to practice. In: *Chemistry and action of herbicide antidotes*. Pallos F. M., Casida J. E. (Eds). pp 35-61, Academic Press, New York, NY, USA.

[4] Hoffmann O. L., Gull P. W., Zeising H. C., Epperly J. R. (1960). Factors influencing wild oat control with barban. *Proc. North Cent. Weed Control Conf.*, 17, 20.

[5] Hoffman O. L. (1971). Coated corn seed. *US Patent* 3,564,768.

[6] Abu-Hare A. Q., Duncan H. J. (2002). Herbicide safeners: uses, limitations, metabolism, and mechanism of action. *Chemosphere*, 48, 965-974.

[7] Monaco T. J., Weller S. C., Ashton F. M. (Eds). (2002). Herbicides and the plants, In: *Weed science: Principles and practices*. pp 98-126, Wiley, New York, NY, USA.

[8] Miyauchi N., Kobayashi K., Usui K. (2002). Differential safening activity of dymron and fenclorim on pretilachlor injury in rice seedlings in soil. *Weed Biol. Manag.* 2, 46-51.

[9] Matsunaka S., Wakabayashi K.(1989). In: *Crop Safeners for Herbicides*. Hatzios K. K., Hoagland R. E. (Eds). pp 47-61, Academic Press, San Diego, USA.

[10] Rosinger C., Bartsch K., Schulte W. (2012). Safener for Herbicides. In: *Modern Crop Protection Compounds*. Krämer W., Schirmer U., Jeschke P., Witschel M. (Eds) Vol. 1. pp 371-398. Wiley-VCH, Weinheim, Germany.

[11] Tam A. C., Behki R. M., Khan S. U. (1988). Effect of dietholate (R-33865) on the degradation of thiocarbamate herbicide by an EPTC-degrading bacterium. *J. Agric. Food. Chem.*, 36, 654-657.

[12] Keifer D. W. (2005). Method for safening crop from the phytotoxic effect of herbicide by use of phosphorated esters. US Patent 6,855,667.

[13] Hatzios K. K., Zama P. (1986). Physiological interactions between the herbicide EPTC and selected analogues of the antidote naphthalic anhydride on two hybrids of maize. *Pestic. Sci.*, 17, 25-32.

[14] Pallos F. M., Brokke M. E., Arneklev D. R. (1975). Antidotes protect corn from thiocarbamate herbicide injury. *J. Agric. Food Chem.*, 23, 821-822.

[15] Dutka F., Komives T., Marton A. F., Hulesch A., Fodor-Csorba K., Karpati M. (1979). Structure-activity relationships of herbicide antidotes. *Proc. Hung. Annu. Meet. Biochem.*, 19, 1-4.

[16] Görög K., Muschinek G., Mustardy L. A., Faludi-Daniel A. (1982). Comparative studies of safeners for prevention of EPTC injury in maize. *Weed Res.*, 22, 27-33.

[17] Pallos F. M., Reed A. G., Arneklev D. R., Brokke M. E. (1978). Antidotes protect corn from thiocarbamate herbicide injury. In: *Chemistry and action of herbicide antidotes*, Pallos F. M., Casida J. E., (Eds), pp 15-20, Academic Press, New York, NY, USA.

[18] Stephenson G. R., Chang F. Y. (1978). Comparative activity and selectivity of herbicide antidotes. In: *Chemistry and action of herbicide antidotes*, Pallos F. M., Casida J. E., (Eds), pp 35-61, Academic Press, New York, NY, USA.

[19] Stephenson G. R., Bunce J. J., Makowski R. I., Curry J. C. (1978). Structure-activity relationships for *S*-ethyl-*N*,*N*-dipropyl thiocarbamateantidotes in corn. *J. Agric. Food Chem.*, 26, 137-140.

[20] Yenne S. P., Hatzios K. K. (1990). Molecular comparisons of selected herbicides and their safeners by computer-aided molecular modeling. *J. Agric. Food Chem.*, 38, 1950-1956.

[21] Bordas B., Komives T., Lopata A. (2000). Comparative three-dimensional quantitative structure-activity relationship study of safeners and herbicides. *J. Agric. Food Chem.*, 48, 926-931.

[22] Chang T. S., Merkle M. G. (1982). Oximes as seed safeners for grain sorghum (*Sorghum bicolor*). *Weed Sci.* 30, 70-73.

[23] Howe R. K., Lee L. F. (1980). 2,4-Disubstituted-5-thiazolecarboxylic acids and derivatives. *US Patent* 4,199,506.

[24] Jablonkai I., Matola T. (2002). Structure–activity relationships of 2-dichloromethyl-1,3-dioxacycloalkanes and heteroanalogues in safening maize against chloroacetanilide herbicides. 10th IUPAC International Congress on Chemistry of Crop Protection, Basel, Switzerland, August 4-9, 2002, Book of Abstracts, p. 132.

[25] Matola T., Jablonkai I., Dixon D., Cummins I., Edwards R. (2003). Structure of dichloromethyl-ketal safeners affects the expression of glutathione S-transferase isoforms. *Proceedings of BCPC - Weeds*, vol 2, 527-532.

[26] Matola T., Jablonkai I. (2007) Safening efficacy of halogenated acetals, ketals and amides and relationships between the structure and effect of glutathion and glutathione S-transferases in maize. *Crop Prot.*, 26, 278-284.

[27] Lewis D. L., Garrison A. W., Wommack K. E., Whittemore A., Steudler P., Melillo J. (1999). Influence of environmental changes on degradation of chiral pollutants in soils. *Nature*, 401, 898-901.

[28] Garrison A. W. (2006). Probing the enantioselectivity of chiral pesticides. *Environ. Sci. Technol.*, 40, 16-23.

[29] Wittingham W. G. (2012). 6-Amino-4-pyridine carboxylate derivatives and their preparation, agrochemical compositions and use as herbicidal safener. *UK Patent Application*, GB 2484982.

[30] Bollinger F. G., Hemmerly D. M., Mahoney M. D., Freeman J. J. (1989). Optical isomers of the herbicidal antidote 4-(dichloromethylene)-2-[*N*-(α-methylbenzyl)imino]-1,3-dithiolane hydrochloride. *J. Agric. Food Chem.*, 37, 484-485.

[31] Jablonkai I., Matola T., Cummins I., Dixon D., Edwards R. (2010). Safening activity of carvone stereoisomers against acetochlor herbicide in maize. Royal Australian Chemical Institute's 13th National Convention in conjunction with the 12th IUPAC International Congress of Pesticide Chemistry, Melbourne, Australia, July 4-8, 2010, Abstr. No. 624.

[32] Jablonkai I., Visy J., Matola T., Cummins I., Dixon D., Edwards R. (2010). Diastereomers of a chiral safener 1,2-dichloromethyl-[1,3]oxathiolane 3-oxide exhibit differential safening activity against acetochlor in maize. Royal Australian Chemical Institute's 13th National Convention in conjunction with the 12th IUPAC International Congress of Pesticide Chemistry, Melbourne, Australia, July 4-8, 2010, Abstr. No. 623.

[33] Zhao L. X., Liu C. G., Fu Y., Xing Z. Y., Gao S. (2012). Induction of maize glutathione S-transferase by herbicide safeners and their effect on enzyme activity against chlorsulfuron. *Advanced Materials Research*, vol. 518-523, 5480-5483.

[34] Rubin B., Kirino O. (1989). Herbicide prosafeners: Chemistry, safening activity, and mode of action. In: *Crop Safeners for Herbicides*. Hatzios K. K., Hoagland R. E. (Eds). pp 317-351, Academic Press, San Diego, USA.

[35] Hilton J. L., Pillai P. (1986). L-2-oxathiazolidine-4-carboxylic acid protection against tridiphane toxicity. *Weed Sci.*, 34, 669-675.

[36] Hilton J. L., Pillai P. (1988). Thioproline protection against herbicide toxicity. *Weed Technol.*, 2, 72-76.

[37] Jablonkai I., Dutka F. (1996). Safening activity of natural hydroxamic acids and analogous compounds against herbicide injury to maize. *J. Environ. Sci. Health. Part B. - Pesticides, Food Contaminants, and Agricultural Wastes*, 31, 555-559.

[38] Brazier-Hicks M., Evans K. M., Cunningham O. D., Hodgson D. R. W., Steel P. G., Edwards R. (2008). Catabolism of glutathione conjugates in *Arabidopsis thaliana*: Role in metabolic reactivation of the herbicide safener fenclorim. *J. Biol. Chem.*, 283, 21102-21112.

[39] Zhang Q., Xu F. X., Lambert K. N., Riechers D. E. (2007). Safeners coordinately induce multiple proteins and MRP transcripts involved in herbicide metabolism and detoxication in *Triticum tauschii* seedling tissues. *Proteomics*, 7, 1261-1278.

[40] Mueller M. J., Berger S. (2009). Reactive electrophilic oxylipins: pattern recognition and signalling. *Phytochemistry*, 70, 1511-1521.

[41] Kreuz K., Riechers D. E., Zhang Q. (2010). The use of oxylipins as safeners and safening herbicidal compositions comprising oxylipins. WO 2011/134539.

[42] Jablonkai I. (1991). Basis for differential chemical selectivity of MG-191 safener against acetochlor and EPTC injury to maize Z. *Naturforsch.*, 46c, 828-835.

[43] Jablonkai I., Dutka F. (1995). Uptake, translocation and metabolism of MG-191safener in corn (Zea Mays L.), *Weed Sci.*, 43, 169-174.

[44] Jablonkai I., Repasi J., Dutka F. (1991). Effect of the site of MG-191 application on acetochlor herbicide uptake, distribution and phytotoxicity. *Pestic. Sci.*, 31, 91-93.

[45] Jablonkai I., Dutka F. (1985). Effect of R-25788 antidote on the uptake, translocation and metabolism of acetochlor herbicide by corn. *J. Radioanal. Nucl. Chem., Letters*, 96, 419-426.

[46] Barrett M. (1989) Protection of corn (*Zea mays*) and sorghum (*Sorghum bicolor)* from imazethapyr toxicity with antidotes. *Weed Sci.* 37, 296-301.

[47] Fuerst E. P., Lamoureux G. L. (1992). Mode of action of the dichloroacetamide antidote BAS 145138 in corn. II. Effects on metabolism, absorption and mobility on metazachlor. *Pestic. Biochem. Physiol.*, 42, 78-87.

[48] Little D. L., Ladner D. W., Shaner D. L. (1994). Modeling root absorption and translocation of 5-substituted analogs of the imidazolinone herbicide, imazzapyr. *Pestic Sci.*, 41, 171-185.

[49] Davies J., Caseley J. C., Jones O. T. G., Barrett M., Polge N. D. (1998). Mode of action naphthalic anhydride as a safener for herbicide AC 263,222 in maize. *Pestic. Sci.*, 52, 29-38.

[50] Hatzios K. K., Burgos N. (2004). Metabolism-based herbicide resistance: regulation by safeners. *Weed Sci.*, 52, 454-467.

[51] Coleman J. O. D., Blake-Kalff M. M. A., Emyr Davies T. G. (1997). Detoxication of xenobiotics by plants: chemical modification and vacuolar compartmentation. *Trends Plant Sci.*, 2, 144-151.

[52] Riechers D. E., Kreuz K., Zhang Q. (2010). Detoxification without intoxication: Herbicide safeners activate plant defence gene expression. *Plant Physiol.*, 153, 3-13.

[53] Marrs K. A. (1996). The function and regulation of glutathione S-transferases in plants. *Ann. Rev. Plant Physiol. Plant Mol. Biol.*, 47, 127-158.

[54] Cole D. J., Cummins I., Hatton P. J., Dixon D., Edwards R. (1997). Glutathione transferases in crops and weeds. In: *Regulation of enzymatic systems detoxifying xenobiotics.* Hatzios K. K. (Ed) pp 107-154, Wiley, Chichester, UK.

[55] Hatzios, K. K. (2001). Functions and regulation of plant glutathione S-transferases. In: *Pesticide biotransformation in plants and microorganisms: Similarities and divergencies.* Hall J. C, Hoagland R. E., Zablotowicz R. M. (Eds). pp 218-239, ACS Symposium Series 777, American Chemical Society, Washington, DC, USA.

[56] Dixon D. P., Hawkins T., Hussey P. J., Edwards R. (2009). Enzyme activities and subcellular localization of members of the *Arabidopsis* glutathione transferase superfamily. *J. Exp. Bot.*, 60, 1207-1218.

[57] Edwards R., Dixon D. P., Walbot V. (2000). Plant glutathione S-transferases: enzymes with multiple functions in sickness and in health. *Trend Plant Sci.*, 5, 193-198.

[58] Dixon D. P., Lapthorn A., Edwards R. (2002). Plant glutathione transferases. *Genome Biol.*, 3, 3004.1-3004.10.

[59] Dixon D. P., Skipsey M., Edwards R. (2010). Roles for glutathione transferases in secondary plant metabolism. *Phytochemistry*, 71, 338-350.

[60] Farago S., Brunold C., Kreuz K. (1994). Herbicide safeners and glutathione metabolism. *Physiol Plant.*, 91, 537-542.

[61] Jablonkai I., Hulesch A., Cummins I., Dixon D. P., Edwards R., (2001). The herbicide safener MG-191 enhances the expression of specific glutathione S-transferases in maize. *Proceedings of BCPC - Weeds*, vol 2, 527-532.

[62] Dixon D. P., Cole D. J. Edwards R (1997). Characterisation of multiple glutathione transferases containing the GST I subunit with activities toward herbicide substrates in maize (*Zea mays*). *Pestic. Sci.* 50, 72-82.

[63] Dixon D. P., Cole J., Edwards R. (1998). Purification, regulation and cloning of glutathione transferase (GST) from maize resembling the auxin-inducible type-III GSTs. *Plant Mol. Biol.* 36, 75-87.

[64] Dixon D. P., Cole D. J., Edwards R. (1999). Dimerization of maize glutathione transferases in recombinant bacteria. *Plant Mol. Biol.* 40: 997-1008.

[65] Rossini L., Jepson I., Greenland A. J., Sari Gorla M. (1996). Characterization of glutathione S-transferase isoforms in three inbred lines exhibiting differential sensitivity to alachlor. *Plant Physiol.* 112: 1595-1600.

[66] Theodoulou F. L., Clark I. M., He X. L., Pallett K. E., Cole D. J., Hallahan D. L. (2003). Co-induction of glutathione S-transferases and multidrug resistant protein by xenobiotics in wheat. *Pestic. Manag. Sci.*, 59, 202-214.

[67] Dixon D. P., Cummins I., Cole D. J., Edwards R. (1998b). Glutathione-mediated detoxication system in plants. *Curr. Opin. Plant Biol.* 1, 258-266.

[68] Davies J. (2001). Herbicide safeners – commercial products and tools for agrochemical reseseach. *Pesticide Outlook*, February 2001, 10-15.

[69] Hulesch A., Dutka F. (1993). Investigation of the safening of EPTC on several grassy crops and weeds by various safeners. *Proceedings of Brighton Crop Protection Conference – Weeds*, vol 1, 207-212.

[70] Breaux E. J., Patanella J. E., Sanders E. F. (1987). Chloroacetanilide herbicide selectivity: analysis of glutathione and homoglutathione in tolerant, susceptible, and safened seedlings. *J. Agric. Food Chem.*, 35 474–478.

[71] Hatton P. J., Dixon D., Cole D. J., Edwards R. (1996). Glutathione transferase activities and herbicide selectivity in maize and associated weed species. *Pestic. Sci.*, 46, 267-275.

[72] Andrews C. J., Skipsey M., Townson J. K., Morris C., Jepson I., Edwards R. (1997). Glutathione transferase activities toward herbicides used selectively in soybean. *Pestic. Sci.*, 51, 213-222.

[73] Hirase K., Molin W. T. (2002). Measuring cysteine biosynthesis activity from serine in extracts from sorghum, corn and grass weeds, and their metolachlor susceptibility. *Weed Biol. Manag.*, 2, 52-59.

[74] Hatton P. J., Cummins I., Price L. J., Cole D. J., Edwards R. (1998). Glutathione transferases and herbicide detoxification in suspension-cultured cells of giant foxtail (*Setaria faberi*). *Pestic Sci.* 53, 209-216.

[75] Jablonkai I., Hulesch A., Dutka F. (1995). Influence of herbicides and safeners on glutathione content and glutathione S-transferase activities of monocot and dicot weeds. Proceedings of the International Symposium on Weed and Crop Resistance to Herbicides -Cordoba (Spain), DePrado R., Jorrin J., Garcia-Torres L., Marshall G. (Eds), p. 89-91.

[76] Jablonkai I., Hulesch A, Barta I. C. (1997). Glutathione and cysteine conjugates inhibit glutathione S-transferase enzymes mediating GSH conjugation of the herbicide acetochlor. *Proceedings of the Brighton Crop Protection Conference - Weeds*, vol 2, p.801-806.

[77] Edwards R., Del Buono D., Fordham M., Skipsey M., Brazier M., Dixon D. P., Cummins I. (2005). Differential induction of glutathione transferases and glucosyltransferases in wheat, maize and *Arabidopsis thaliana* by herbicide safeners. *Z. Naturforsch.*, 60c, 307-316.

[78] Cummins I., Bryant D. N., Edwards R. (2009). Safener responsiveness and multiple herbicide resistance in the weed black-grass (Alopecurus myosuroides). *Plant Biotech. J.*, 7, 807-820.

[79] Del Buono D., Scarponi L., Espen L. (2007). Glutathione S-transferases in *Festuca arundinacea*: Identification, characterization and inducibility by safener benoxacor. *Phytochemistry*, 68, 2614-2624.

[80] Scarponi L., Del Buono D., Quagliarini E., D'Amato R. (2009). *Festuca arudinacea* grass and herbicide safeners to prevent herbicide pollution. *Agron. Sustain. Dev.*, 29, 313-319.

[81] Scalla R. (1991). Interaction of herbicides with safeners and synergist. In: *Pesticide chemistry: Advances in international research, development, and legislation*. Frehse H. (Ed). pp 141-150, Wiley-VCH, Weinheim, Germany.

[82] Durst F., Benveniste I., Lesot A., Salaün J.-P., Werck-Reichhart D. (1997). Induction of plant cytochrome P450. In: *Regulation of enzymatic systems detoxifying xenobiotics in*

plants. Hatzios K. K. (Ed). pp 19-34, Kluwer Academic Publishers, Dordrecht, Netherlands.

[83] Burnett M. W. M., Loveys B. R., Holtum J. A. M., Powles S. B. (1993). A mechanism of chlortoluron resistance in *Lolium rigidum. Planta,* 190, 182-189.

[84] Burnett M. W. M., Loveys B. R., Holtum J. A. M., Powles S. B. (1994). Identification of two mechanisms of sulfonylurea resistance within one population of rigid ryegrass (*Lolium rigidum*) using a selective germination medium, *Weed Sci.,* 42, 153-157.

[85] Burton J. D., Maness E. P. (1992) Constitutive and inducible bentazon hydroxylation in shuttercane (*Sorghum bicolor*) and Johnsongrass (*Sorghum halapense*). *Pestic. Biochem. Biochem. Physiol.,* 44, 40-49.

[86] Jablonkai I., Hulesch A. (1996). Cytochrome P450 levels of monocot and dicot weeds and influence of herbicides, safeners and P450 inhibitors on enzyme contents. Proceedings of the 2nd International Weed Control Congress - Copenhagen (Denmark), Brown H., Cussans G. W., Devine M. D., Duke S.O., Fernandez-Quintanilla C., Helweg A., Labrada R. E., Landes M., Kudsk P., Streibig J. C. (Eds), Vol 3, p. 789-794.

[87] Barta I. C., Dutka F. (1991) Interaction of maize cytochrome P450 with safeners and 1-aminobenzotriazole. *Proceedings of Brighton Crop Protection Conference – Weeds,* vol 3, 1127-1132.

[88] Jablonkai I., Hatzios K. K. (1994). Microsomal oxidation of the herbicides EPTC and acetochlor and of the safener MG-191 in maize. *Pestic. Biochem. Physiol.,* 48, 98-109.

[89] Gatz C. (1997). Chemical control of gene expression. *Ann. Rev. Plant Physiol. Plant Mol. Biol.,* 48, 89–108.

[90] Padidam M. (2003). Chemically regulated gene expression in plants. *Curr. Opin. Plant Biol.,* 6, 169-177.

[91] DeRidder B. P., Goldsbrough P. B. (2006). Organ-specific expression of glutathione S-transferases and the efficacy of herbicide safeners in Arabidopsis. *Plant Physiol.,* 140, 167-175.

[92] Walton J. D., Casida J. E. (1995). Specific binding of a dichloroacetamide herbicide safener in maize at a site that also binds thiocarbamate and chloroacetanilide herbicides. *Plant Physiol.,* 109, 213-219.

[93] Scott-Craig J. S., Casida J. E., Poduje L., Walton J. D. (1998). Herbicide safener-binding protein of maize. *Plant. Physiol.,* 116, 1083-1089.

[94] Cummins I., Brazier-Hicks M., Stobiecki M., Franski R., Edwards R. (2006). Selective disruption of wheat secondary metabolism by herbicide safeners. *Phytochemistry, 67,* 1722-1730.

[95] Zhang Q., Riechers D. E. (2004). Proteomic characterization of herbicide safener-induced proteins in the coleoptile of *Triticum tauschii* seedlings. *Proteomics*, 4, 2058-2071.

[96] Riechers D. E., Vaughn K. C., Molin W. T. (2005). The role of plant glutathione *S*-transferases in herbicide metabolism. In: *Environmental fate and safety management of agrochemicals*. Clark J. M., Ohkawa H. (Eds), pp 216-232, ACS Symposium Series 899, American Chemical Society, Washington, DC, USA.

[97] Taylor V. L., Cummins I., Brazier-Hicks M., Edwards R. (2012). Protective responses induced by herbicide safeners in wheat. *Environ. Exp. Botany*, doi: 10.1016/jenvexpbot.2011.12.030.

Chapter 8

Managing Commelina Species: Prospects and Limitations

Wendy-Ann Isaac, Zongjun Gao and Mei Li

Additional information is available at the end of the chapter

1. Introduction

Commelina species, notably *C. communis* L, *C. diffusa* Burm, *C. elegans* Kunth. and *C. benghalensis* L. as well as their biotypes, are perennial herbs of Neotropical origin which now have a pantropical distribution. Members of this family (Commelindeae: Commelinaceae) are common throughout the Caribbean, North and Latin America, Africa, Asia, the Middle East and parts of Oceania [18, 27, 28, 63, 64]. There are 500 - 600 species reported in the family Commelinaceae [50]. Recent data indicates that the Commelinaceae family contains 23 genera and at least 225 species native to or naturalized in the New World and 23 genera and about 200 species in the Neotropics [41] and also website reports of 50 genera and 700 species [16, 31]. There are 170 species of Commelina in the warmer regions of the world and 50 species of Murdannia occurring in the tropics and warm temperate regions worldwide with Tropical Asia having the greatest diversity [17].

Wilson [84] presented a comprehensive review on Commelina species and its management with emphasis on chemical weed control in 1981. Since Wilson's review much has been written about the weedy members of this family, notably Commelina species [84]. Indeed, the CAB ABSTRACTS Database contains well over 1200 references on Commelinaceae from 1981 to the present. *Commelina benghalensis* in particular has been the most reported species with several reports of research conducted on its control in southern states of the United States of America (USA) including Alabama, Florida, Georgia, Louisiana and North Carolina [18, 74, 75, 78-81]. Many of these studies should be consulted for basic details of the biology and ecology. The National American Plant Protection Organization (NAPPO) offers a comprehensive global distribution list of this weed species [47].

The current review is an attempt to provide an update on the status of the weedy Commelina species in agricultural production systems. This review is based on world literature over the last 45 years and considers major Commelina species found in the tropics and warm temperate regions in relation to their status, distribution, biology and spread and management.

2. Weed Status

Commelina benghalensis (Tropical spiderwort or Benghal dayflower) has become increasingly important, gaining pest significance in agronomic production systems in the southeastern coastal plain of the United States of America (USA) in crops such as cotton (*Gossypium* spp.) and peanut (*Arachis hypogea*) [70, 71] and in the North China Plain in crops such as potato (*Solanum tuberosum*) and summer corn (*Zea mays*) [37, 71, 72, Li et al. unpublished data 2007). It is commonly associated with wet locations. This weed was in fact listed as a Federal Noxious weed in Florida and Georgia where it is the most troublesome weed in cotton and a pest in peanut, corn (*Zea mays*), soybean (*Glycine max*), nursery stock and orchards [81]. This species which was first observed in USA in 1928 [18] gained noxious weed status in 1983 [81]. Between 1998 to 2001 and then to 2004 this weed which was ranked among the top 39 most troublesome weeds across all crops by Georgia extension agents (in 1998) moved to the 9th most trouble-some (in 2001) to the most troublesome cotton weed in Georgia (in 2003) [77] and Florida (2004) and the 3rd most troublesome weed of peanut in several south Georgia counties [54, 80]. In Georgia alone the weed is estimated to infest more than 80,000 ha [80-82] with a confirmed presence in 29 Georgia counties [54]. It is also observed throughout the panhandle and central Florida and listed by the United States Department of Agriculture (USDA) as appearing in more than 12 Florida counties [82].

Commelina communis has become one of the three most troublesome weeds in soybean fields in the Northeast China, and has caused significant reduction in production and quality of soybean [42]. Commelina species, namely *C. diffusa* and *elegans,* were reported as the 3rd most troublesome weed in the Caribbean where they are a serious problem of banana and other crops in the Windward Islands of Dominica, Grenada, St. Lucia and St. Vincent and the Grenadines [24]. Presently, Commelina species, commonly called watergrass, caner grass, pond grass, spiderwort, spreading dayflower, wandering Jew or French weed in these Islands, are by far the most serious in these countries. *Commelina diffusa* was once encouraged as a ground cover to reduce soil erosion [13] and has been identified as the host of the reniformis nematode *Rotylenchulus reniformis* [57], the banana lesion nematode *Pratylenchus goodeyi* [87] and recent data have confirmed its association with the burrowing nematode *Radopholus similis* [55]. These nematodes all contribute to significant reductions in banana production particularly *R. similis*, which may reduce banana production by more than 50 % and decrease the production duration of banana fields [55].

3. Biology and spread

Commelina species are C-3, monocotyledonous plants and therefore have a high efficiency of CO_2 uptake at low irradiance [34]; therefore, they tolerate shade very well and could become persistent. They are both annuals and perennials and therefore dominate the fallow vegetation because they are most competitive due to their growth and regeneration characteristics [72].

The plant is propagated mainly by seeds, stem cuttings and rooting from nodes and pieces [19, 46, 74, 75]. Plants may arise asexually when buds grow into autonomous, adventitiously erect leafy shoots, which later become separated from each other [12]. Occasionally the buds may sprout and grow into erect shoots directly without undergoing a period of inactivity [12]. The plant produces roots readily at the nodes of the creeping stems and will do so especially when broken or cut [27, 28]. Farmers in the Windward Islands report that Commelina species may be intensified when cut with a weed whacker as stolons spread more extensively.

The stems of Commelina species have a high moisture content and once it is well rooted the plant can survive for long periods without moisture [84]. This fact is evident in young banana plantations in the Windward Islands where stems become dried and shrivelled due to the direct contact with solar radiation particularly in the dry season. However, at the onset of rains and when the canopy of the banana closes, stems regain moisture, re-establish and rapidly begin to spread by runners which root at the nodes.

The mature aerial seeds of *C. benghalensis* are produced within 14 to 22 days after flower opening [74] and in some instances, e.g., the rice paddies of the Philippines, can produce in excess of 1,600 seeds/plant [53] or even 12,000 seeds/m^2 [74], whereas seeds grown from underground seeds are capable of producing 8,000 seeds/m^2 [74]. In cultivated areas the plant is spread by irrigation water and waterways. Animals may also spread the seeds.

Commelina species has gained noxious weed status in the Windward Islands because of several factors. Firstly, the fact that the weed was encouraged as a groundcover was compounded by inappropriate agricultural practices, notably irrational herbicide use which farmers have relied on for decades. The non-judicious use of herbicides has created imbalances and disturbances within the ecosystem in these Islands causing resistant biotypes. Secondly, the move within recent years by banana growers to adopt a Fairtrade system which uses no herbicides has catapulted the spread to an all-time high in the Windward Islands. Farmers have been forced to rely on the use of the cutlass or weed whacker as the only alternative strategies which have further intensified the problem by spreading plant propagules [30]. Most importantly these Islands which are characterized by hilly landscapes have ideal moist conditions for the proliferation of Commelina species. Finally, many of the banana plantations have been farmed for several years with virtually no crop rotations or tillage practices and this has further contributed to the stabilization of Commelina species populations.

In the USA, its sudden emergence as a noxious weed is attributed to crop production practices which are well suited for prolific weed growth such as minimum – tillage production (which is undertaken in conjunction with the use of glyphosate – resistant crops) and extreme tolerance to glyphosate [79-81]. The weed appears to be well-suited for high input agricultural production where high levels of fertilizers, irrigation and herbicides are used [79, 80]. The spread of *C. benghalensis* is attributed, in part, to the adoption of weed management programmes that lack the use of residual herbicides along with the adoption of reduced-tillage production practices [54]. Additionally, after introduction, invasive species often go long periods of time (lag period) during which the pest increases in distribution or density without being noticed as an obvious pest [54].

4. Economic impact in crop production

Three species of the Commelinaceae family are considered to be major problem weeds in cropping systems where they have become persistent and difficult to manage [27]. *Commelina benghalensis* is the most important of the three and it occurs as a weed in 25 different crops in 28 countries [27]. This weed has gained high importance in peanut and cotton in the southern United States [78, 79]. *Commelina diffusa* occurs as a weed in 17 crops in 26 countries and *Murdannia nudiflora* occurs as a weed in 16 crops in 23 countries [27].

Commelina diffusa thrives on cultivated soils of cocoa (*Theobroma cacao*), citrus, root crops such as dasheen (*Colocasia esculenta*) that tolerate water, and it is also a major weed in sugarcane (*Saccharum officinarum*), upland rice (*Oryza sativa*), soybean (*Glycine max*), cassava (*Manihot esculenta*), corn (*Zea mays*), banana and plantain (*Musa* spp.) [27]. *Commelina benghalensis* has been reported as a principal weed in upland rice in India and the Philippines, tea (*Camellia sinensis*) in India, coffee (*Coffee arabica*) in Tanzania and Kenya, soybean in the Philippines and cotton and maize in Kenya [27, 47]. This species is common in rice in Sri Lanka, sugarcane in India, the Philippines and Mozambique, cassava in Taiwan and maize in Zimbabwe [9]. *Commelina benghalensis* was reported as a weed of jute (*Corchorus olitorius*), sisal (*Agave sisalana*), beans (*Phaseolus* spp.), pastures, sweet potatoes (*Ipomoea batatas*), vineyards and barley (*Hordeum vulgare*) and other cereals in many countries [7].

Because of Commelina's vigorous growth habit, which allows the plant to form dense pure stands, they may compete easily with low growing crops such as vegetables, pulses and cereals as well as pasture grasses and legumes by smothering them [27]. Because Commelina species is a broadleaved weed it is generally not considered highly competitive for nutrients however this fact is not well researched and its allelopathic potential also needs to be ascertained. Invasive species such as *C. benghalensis* had higher plant growth rate at high nutrient availability and across water availability compared to a related non – invasive, but alien, congener, *C. bracteosa* Hassk. [6]. Interestingly, severe stunting has been reported in *C. diffusa* caused by high nitrogen [59] and altered growth and physiological characteristics for different *C. erecta* clones with increased phosphorus supply [71]. Results from systematic studies on the influence of *C. benghalensis* populations on crop yield are limited [54]. Increased reduction in aboveground and root dry matter as well as a 100% reduction in the number of leaves in lettuce (*Lactuca sativa*) plants were recorded with 1% and 3% hydro – alcoholic extracts of *C.benghalensis* suggesting its allelopathic potential [68].

Studies on the critical periods of interference in Commelina species are limited. Generally crops are affected most severely during the first 2 – 5 weeds of crop growth although mature plants can also be affected [7]. *Commelina benghalensis* in particular may affect crop growth and yield but this varies with environmental conditions [47]. Research aimed at evaluating the periods of interference of *C. benghalensis* in the initial growth of coffee seedlings reported prevention periods of 15 to 88 and 22 to 38 days after coffee seedling sowing under winter and summer conditions, respectively [11]. In cotton it was found that yield loss from *C. benghalensis* can be minimized by planting cotton early in the growing season, prior to substantial emergence of the weed [81].

5. Pests and diseases associated with commelina species

Commelina diffusa is an alternate host plant for the nematodes *Rotylenchulus reniformis*, *Helicotylenchus* spp., *Pratylenchus* spp., *Meloidogyne* sp. and *Radopholus similis* in banana [13, 27, 29, 44, 55, 57, 60, 87] and coffee [58]. The plant is also a collateral host of *Helicotylenchus dihystera* infecting guava fields [35]. *Commelina benghalensis* has also been identified as an alternate host of the southern root-knot nematode (*Meloidogyne incognita*) [55]. The southern root-knot nematode is widely distributed across cotton regions in Georgia [54]. Snails and slugs feed on *C. diffusa* plants and these affect crops such as pineapple and soybean [84].

Five viruses have been found naturally infecting species of Commelinaceae. Aneilema a potyvirus has also been found infecting 15 species of the Commelinaceae family including 4 of Commelina. There have been reports of *Commelina diffusa* potyvirus, which causes a mosaic in *Commelina diffusa* and *C. benghalensis* [2]. The virus is transmitted by two insect vectors, *Aphis gossypi* and *Myzus persicae*; Aphididae. It is transmitted in a non – persistent manner. The virus is transmitted by mechanical inoculation and not by grafting or contact between plants or by seeds. The isolate for cucumber mosaic virus (CMV) is originally from *Commelina elegans* but it is transmitted by Aphis gossypi, and not *Myzus persicae*. *Commelina diffusa* is susceptible to Commelina X potexvirus, Commelina yellow mottle badnavirus, Spring beauty latent bromovirus, Tradescantia – *Zebrina potyvirus*, spotted wilt and Cherry leaf roll nepovirus [2]. However, *Commelina elegans* is insusceptible to Tradescantia – *Zebrina potyvirus*. U2- tobacco mosaic virus has also been found infecting *C. communis* and *Z. pendula*. Brome mosaic virus isolates have been identified [70] infecting *C. diffusa* and *C. communis* in Fayetteveille, Arkansas, USA.

6. Methods of management in selected crops

Wilson's review on the control of these weed species was directed towards finding suitable chemicals for their control in the early stages of growth, summarizing results of trials from difference parts of the world [84]. However, he suggested that since dense mats of plant material make chemical weed control of older plants difficult, removal by hand is the only effective control at that stage [84].

Currently, chemical control is still generally considered the only practical means of controlling large infestations of Commelina species [78-82]. However, no single method of control seems to be effective for control of Commelina spp. in any crop. The difficulty lies in its ability for regeneration after attempted management even by cultural, mechanical or chemical control.

An Integrated Management Strategy (IWM) is therefore suggested for the best control of this weed species. A multi-component approach including an effective herbicide for successful management has been suggested [80-82].

7. Chemical management

Herbicides are not usually very effective against most Commelina species. The first verified resistance was registered in 1957, when C. diffusa biotypes were identified in the United States [26]. Commelina elegans has shown resistance to growth – regulator type herbicides [32]. Control using herbicides is, however, variable depending on the herbicide, accuracy of leaf coverage and environmental conditions [7]. Spraying with a selective or non – selective herbicide may work but repeated treatments are required for regrowth. Plants should not be under moisture stress when sprayed. Surfactants will improve penetration into the waxy-coated leaves.

Many standard herbicides have relatively low activity on species of Commelina [84]. These include 2,4-D, propanil, butachlor, trifluralin and pendimethalin. Treatment with 2,4-D or MCPA at the pre-emergent stage has been shown to be ineffective and although a reasonable kill of very young seedlings can be obtained, the plants develop a rapid resistance with age [32]. Particular biotypes are resistant to 2,4-D and they may be cross resistant to other Group O / 4 herbicides [83]. It has been found that one biotype of C. diffusa could withstand five times the dosage of a susceptible species [83].

In rice, bentazone, molinate, oxyfluorfen and bifenox are herbicides with good activity [7]. Post-emergent sequential treatments of propanil followed by nitrogen or of molinate followed by KN3 controlled C. diffusa in rice [61]. In soybean, bentazone and metribuzin are effective [7]. In corn, combination of bromoxynil and 2,4-D butylate produced a synergistic effect in post-emergent control of 3-4 leaf stage C. communis [85]. In plantation crops such as banana, paraquat is not always effective but mixture with diuron is recommended [7]. Dinoseb has been found to kill seedlings as well as dalapon but paraquat is reported to be relatively ineffective [32]. Prodiamine has been reported to be effective in ornamental fern beds [62]. Extreme tolerance to glyphosate has been documented [54]. Glyphosate has been shown to be effective but additives or mixtures may be needed for good results at moderate doses [7]. However, C. diffusa has been reported to have larger possibilities of recovery after glyphosate application because of its larger starch reservation [71].

Resistance to residual herbicides has also been reported and relatively high doses of simazine and diuron appear to be necessary to achieve control [32]. Recent studies on use of residual herbicides have identified Dual Magnum® (s-metolachlor) (applied as a preplant incorporated, pre-emergent and post-emergent) as providing excellent residual control (>80%) of C. benghalensis in peanut [54]. Atrazine and Dual Magnum®, two commonly used corn herbicides used in the USA, also gave good to excellent residual activity on C. benghalensis [3]. The most effective herbicide control strategies for C. benghalensis involve combinations of both pre-emergence and postemergence conventional herbicides [54]. These include preemergence herbicides with residual activity such as Axiom® (flufenacet + metribuzin), Dual Magnum®

Canopy SP® (metribuzin + chlorimuron) and Sencor® (metribuzin) and postemergence herbicides with fair to good activity such as Basagran®, Classic® (acetochlor) and Pursuit® (Imazethapyr). Gramoxone Max® and Aim® (acetochlor) can be used post-directed. In evaluating the effectiveness of several pre-emergence herbicides in suppressing *C. benghalensis* emergence, it was reported that s-metolachlor (at 1.07 and 1.60 kg a.i./ha), clomazone (at 0.42 and 1.05 kg a.i./ha) and flumetron (at 1.68 kg a.i./ha) provided ≥ 80% control at 6 weeks after treatment (WAT) in cotton [80]. It was stressed that the application of herbicides with soil residual activity will be crucial for the management of *C. benghalensis* [80].

In the Windward Islands, farmers started using paraquat around 1989 and noticed that it was ineffective. In an interview on August 10, 2002, Paddy Thomas, an experienced banana grower and pesticide salesman in St. Vincent and the Grenadines revealed that farmers started using gramocil (paraquat + diuron) at high doses for example and this too was not effective and resistance in Commelina spp. began to show. He also stated that Reglone, Round – up and Talent (paraquat + asulam) have also been used with little success for the control of Commelina species in the Windward Islands. Glufosinate has since been promoted as an environmentally-friendly option for the control of broad-leaved weeds including Commelina species.

Studies were conducted into the efficacy of glufosinate for weed control in coffee plantations and it was found that it did not effectively control Commelina spp. at a rate of 0.3 – 0.6 kg a.i. / ha, however, paracol and gardoprim suppressed this perennial weed better [50]. Fomasefen and lactofen have shown good potential for control of this broadleaf weed [10]. Glufosinate (240 g a.i./ha) and fomasefen (WIP 276 g a.i./ha) were used in St. Vincent and the Grenadines in Fairtrade banana fields to compare their efficacy in controlling *C. diffusa* [30]. They were both applied at the early post-emergence, 3-5 leaf stage with a backpack sprayer using a TJ-8002 fan-nozzle. Regrowth of *C. diffusa* and other weeds were observed 6 weeks after application with glufosinate, however, no regrowth was observed for up to 3 months with fomasefen. Fomasefen, however, caused damage by burning banana suckers and leaves (about 30%) of established banana plants [30]. Studies were conducted to evaluate the efficacy of several post-emergence herbicides in controlling *C. communis* in soybean, the results showed that imazethapyr (150 g a.i./ha), cloransulam-methyl (31.5 g a.i./ha), fomasefen (375 g a.i./ha) and mixture (756 g a.i./ha) of fomasefen plus imazethapyr with clomazone provided > 80% control of this weed at 30 days after treatment (DAT) [36, 37, 65, 67]. The efficacy of imazethapyr (90 g a.i./ha) in controlling *C. communis* reduced with increased leaf stage, and the control levels at 15 DAT were 100% (at 1 leaf stage), 89.17% (at 2 leaf stage), 56.45% (at 3 leaf stage) and 52.71% (at 4 leaf stage), respectively [41]. Therefore, the optimal application time of imazethapyr was 1-2 leaf stage of *C. communis* [41].

To screen more suitable herbicides for control of *C. benghalensis* and *C. communis* and determine the level of weed control provided by a single application of selected post-emergence herbicides, greenhouse studies on the laboratory toxicity of 23 herbicides to these weeds were conducted in 2010 [21]. The results indicated that, as for *C. benghalensis*, mesotrione, lactofen, oxyfluorfen, clomazone and flumioxazin provide complete control (100%), oxadiazon, fomasefen, metribuzin, acifluorfen, isoproturon, MCPA-sodium, carfentrazone-ethyl, fluroxypyr, fluoroglycofen-ethyl and bentazone are herbicides with excellent activity (90.0 - 100%

control), paraquat, 2,4-D butylate, rimsulfuron and thifensulfuron-methyl are herbicides with good activity (80.0 - 90.0% control), and nicosulfuron, bensulfuron-methyl, dicamba and glyphosate-isopropylammonium are relatively ineffective (< 80.0% control) at their own recommended dose, respectively. As for *C. communis*, mesotrione and thifensulfuron-methyl provide complete control (100%); metribuzin, paraquat, carfentrazone-ethyl, 2,4-D butylate, nicosulfuron, MCPA-sodium, fluroxypyr, flumioxazin and acifluorfen are herbicides with excellent activity (90.0 - 100% control); rimsulfuron, lactofen and fomesafen are herbicides with good activity (80.0 - 90.0% control); and glyphosate-isopropylammonium, bensulfuron-methyl, fluoroglycofen-ethyl, bentazone, clomazone, oxadiazon, oxyfluorfen, isoproturon and dicamba are relatively ineffective (< 80.0% control) at their own recommended dose, respectively. There are 19 and 14 herbicides which provided good to excellent control (> 80%) to *C. benghalensis* and *C. communis* under greenhouse conditions, respectively. However, the performance of those herbicides applied in different crops to control *C. benghalensis* and *C. communis* also needs to be ascertained.

8. Cultural management

This method depends on the crop infested, land size, level of technology available, value of crop, labour availability and costs, availability of draft power and the associated equipment and availability of herbicides [47]. The document further indicates that the methods currently used include proper land preparation, hand hoeing and pulling, removing the plants from the fields and drying, use of ox-drawn and tractor drawn cultivation, slashing and herbicide application. *Commelina diffusa* is very difficult to control manually as the stolons are cut into small pieces which can easily regenerate. Hand weeding and rolling the weed up like a carpet is considered suitable for removal of small infestations [30], if care is taken to remove every last piece. In Uganda, it was reported that heaping of stubborn weeds of Commelina plants is practical during the rainy season to speed up rotting and reduce the frequency of weeding [48]. In the dry season, heaps are then scattered as the dry conditions desiccate Commelina stems rapidly. A small percent of Ugandan farmers (5.9%) dig ditches and bury Commelina species, turning it into manure. Some farmers in St. Vincent have also tried this technique in the field with varying success.

A potential solution to overcoming Commelina weed infestations in banana is by intercropping with a fast, low – growing shade tolerant cover crop. This can be done by intercropping with melons, *Mucuna pruriens* (negra and ceniza), tropical alfalfa, *Cajanus cajan*, *Vigna radiata* (mung bean), *V. unguiculata* (cowpea), *Crotalaria juncea*, *Indigofera endecaphylla*, *Phaseolus trinervius*, and *Ipomea batatas* (sweet potato) which have rapid canopy coverage to suppress the establishment of weeds. Melon (*Colocynthis citrullus* L.) planted at a density of 5,000 plants/ha suppressed weed growth of *Commelina diffusa* for five months, enhancing establishment and yield of melon in Nigeria [49]. Use of vigorous healthy planting material and close spacing of the crop may also be used. It has been shown that spacings of 1.2 x 1.2 m (6,944 plants/ha) and 1.5 x 1.2 m (4,444 plants/ha) gave high yields and "natural" control of these weeds [8, 66].

Field studies conducted in St. Vincent and the Grenadines in 2003/2004 compared several treatments including 3 cover crops in suppressing *Commelina diffusa* weed infestations in banana at 63 days after application (DAA) [30]. The cover crops included *Arachis pintoi* (wild peanuts) which was sown by seed and stem cuttings, 16 cm apart, *Mucuna pruriens* (velvet beans) drilled 30 cm apart and *Desmodium heterocarpon* var *ovalifolium* (CIAT 13651) broadcast at a rate of 5 kg/ha. Best results were obtained from *Desmodium heterocarpon* (86.7%) followed by *Arachis pintoi* (52.1%) and *Mucuna pruriens* (43.3%). *Desmodium heterocarpon* was also found to be competitive to *C. diffusa* significantly suppressing its growth in Farmer Participatory Research trials also conducted in St. Vincent in 2005/2006 [30].

Mulching is another viable option for management of the weed. Mulching with rice straw, cut bush, grass, coffee hulls, water hyacinth or even the dead or senescent banana leaves, pruned suckers and old stems could significantly suppress weed growth. Black plastic mulch also provides good weed control as it stifles weed seed growth and development when light penetration is reduced. There are no reports of work done on the use of these mulches for suppression of Commelina species. In field studies in St. Vincent and the Grenadines in 2003/2004 three dead mulches were compared using senescent banana leaves (traditional practice of farmers) applied to a depth of 3-5 cm, coffee hulls applied to a depth of 3-5 cm and black plastic polyethylene tarp at 1.0 mils thickness [30]. Results indicate a 94.5% and 95.6% suppression of weeds including *C. diffusa* with coffee hulls and banana mulch treatments respectively and 100% suppression with black plastic mulch.

9. Mechanical management

Commelina diffusa is particularly difficult to control by cultivation, partly because broken pieces of the stem readily take root and underground stems with pale, reduced leaves and flowers are often produced [32]. The plant is easy to rake up, roll up or hand pull and very small infestations can be dug out. It can be bagged and well baked in the sun, however, follow – up work is essential as any small fragment of the stem remaining will regrow and needs to be removed and destroyed off - site. Mechanical control using the weed whacker may also contribute the spread of stem cuttings in addition to damaging the banana root system as much of the plant lies within the top 15 cm of the soil [30].

To investigate the effect of cutting and depth on the regeneration potential of *C. diffusa* greenhouse studies were conducted in 2004/2005 (Isaac et al. unpublished data 2005) using three cutting types: tip cuttings (2 nodes, 2 leaves), 2 node pieces only and 1 node, 1 leaf piece buried at depths including 0 (control), 2.5, 5.0 and 7.0 cm to demonstrate emergence patterns. These cuttings were intended to simulate cuttings made from a weed whacker and the practice of burying the weed. Regeneration was observed from all cuttings from 0 – 5.0 cm depths but no growth was observed at 7.0 cm. *C. diffusa* dry matter (DM) was highest at surface level (0cm - control) for all cuttings and reduced with increased depth. Results indicate that for effective management of *C. diffusa* by cutting, nodes must be reduced to less than half with no leaves which may starve the plants' photosynthetic ability and hence suppress regeneration. Burial should be up to 5.0 cm to ensure that there is no emergence of the weed. Similar studies [5] indicated that cuttings buried deeper than 2 cm failed to regenerate.

Research has shown that soil solarization, a hydrothermal process of heating moist soil, can successfully disinfect soil pests and control weeds [1, 4, 15, 56]. Soil solarization by covering with plastic sheeting for 6 weeks in the warmer months will weaken the plant. After removing the plastic any regrowth can be dug out or sprayed, however, this method will not be effective in full shade. Solarization can be used alone or in combination with other chemicals or biological agents as the framework for an IPM programme for soilborne pests in open fields. In field trials in St. Vincent, soil solarization using clear polyethylene plastic at 0.5 mils under Fairtrade banana plants showed variable suppression of C. diffusa as the weed emerged under the clear plastic showing chlorotic and suppressed growth symptoms, resuming its full growth potential after removal of the plastic covering 2 months after application (Isaac et al. unpublished data 2005). Seed germination of C. benghalensis was found to increase by soil solarization in studies conducted in Brazil [43].

10. Organic management

Attempts have also been made to find organic treatments for control of Commelina species in banana in St. Vincent and the Grenadines [30]. DTE corn weed blocker (corn gluten meal) pre-emergent weed blocker and slow release fertilizer (9-1-0) which controls emerging weeds was applied at a rate of 10 kg/ha. Burnout® (concentrated vinegar and acetic acid) (20%), urea (20%), and fertilizer solution (20%) were also used to evaluate their efficacy on the control of Commelina species and other weed species. All treatments showed varying levels of control for up to 3 weeks. Best results were obtained from Burnout® which caused phytotoxic damage on the leaves of actively growing plants offering 43% control. This was followed by urea (41%), fertilizer solution (34%) and corn weed blocker (20%). Urea, fertilizer and corn weed blocker treatments resulted in the general stunting of plants in addition to the burning of leaves. However, stems and roots remained intact. Similar results using treatments high in nitrogen were obtained in Russia [59] where seed production of C. benghalensis and stunted growth under artificial dense competition in cereals resulted. These results indicate that there is no evidence that this Commelina species competes for nitrogen. In fact the species does not pose any threat in competing for nutrients with banana. Repeat applications of these treatments are therefore necessary for the effective management of Commelina species in organic farming systems.

Studies conducted in Brazil in soybean-wheat rotations under no-tillage conditions showed reductions in the seedbank of C. benghalensis in areas infested with Brachiaria plantaginea [73]. Analysis of the soluble fraction of B. plantaginea indicated a predominance of aconitic acid (AA) among the aliphatic acids and ferulic acid (FA) among the phenolic acids. Laboratory bioassays using C. benghalensis were carried out to evaluate phytotoxic effects of pure organic acid solutions and dilute extracts of B. plantaginea on seeds germination, root development and fungal germination and AA and FA solutions and the extract of B. plantaginea extract reduced germination and root length of *C. benghalensis* [73]. Both AA and FA have the potential for use as bio-herbicides.

11. Biological management

There have not been many reports on biological control of Commelina species. *Commelina diffusa* is grazed by small ruminants, pigs and cows. Because this species is very fleshy and has a high moisture content, it is difficult to use it as fodder for domestic stock [27]. However, recent research has indicated that *C. diffusa* compared well with many commonly used fodder crops and could contribute as a protein source for ruminants on smallholder farms [30]. There have also been reports of foraging of this weed by *Gallus domesticus* (chickens) [30].

There are no reports of promising insect candidates for biological control reported on Commelina spp. in the USA [63, 64]. In Korea and China there have been reports of *Lema concinnpennis* and *Lema scutellaris* (Coleoptera: Chrysomelidae) two leaf-feeding species on *C. communis* [86]. *Noelema sexpunctata* (Coleoptera: Chrysomelidae) another leaf-feeding species was also reported on *C. communis* [45].

In Central Virginia, USA, *Pycnodees medius* (Hemiptera: Miridae) was found to cause tissue necrosis on *C. communis* [33]. Various insects were also screened for their potential as biocontrol agents of weeds in rice and it was found that *Necrobis ruficollis* (blue beetle), *Rhaphidopalpa africana* (yellow beetle), *Conocephalus* sp., *Tetragrnathidae* spp. and *Paracinema tricolor* (grasshopper) were promising [45]. Feeding and nymphal development (up to 3rd and 4th instar) of *Cornop aquaticaum* (grasshopper) were reported on *C. africana* L., and *Murdannia africana* (Vahl.) [25]. It was also observed that *Rhaphidopalpa africana* beetles fed more than the others on the weed, *C. benghalensis* L. [25].

There are records of agromyzid leaf miners which may be promising sources of candidate biological control agents [75]. *Liriomyza commelinae* (Diptera: Agromyzidae), a leaf-miner, was however reported on *C. diffusa* in Jamaica [20, 61]. *Commelina diffusa* is the main food plant of *L. commelinae*, however, it is susceptible to predation by the formicid: *Crematogaster brevispinosa* as well as competition and exposure to the sun (high temperatures) which causes high mortality [20].

There are prospects for the management of invasive alien weeds in Latin America using co-evolved fungal pathogens in selected species from the genera Commelina [14]. Pathogens recorded in the native range of Commelina species include: *Cercospora benghalensis* Chidd., *Cylindrosporium kilimandscharium* Allesch. (Hyphomycete), *Kordyana celebensis* Gaum, (Exobasidiales: Brachybasidiaceae), *Phakopsora tecta* H.S. Jacks and Holw (Uredinales: Phakopsoraceae), *Septoria commelinae* Canonaco (Coelomycete), *Uromyces commelinae* Cooke (Uredinales: Pucciniaceae), *Phoma herbarum* [14, 23, 76]. These mycobiota would appear to be good potential agents for classical biological control (CBC) [14]. Although some of the most promising (e.g. the rusts *Phakopsora tecta* and *Uromyces commelinae*) are already present in the New World, they are restricted to certain regions and could be redistributed [14]. The uredinal state of a rust was found widespread on *C. diffusa* in Hawaii [22] sometimes causing death of parts above ground. Studies aimed at identifying mycoherbicidal biocontrol agents have been conducted in Brazil on three endemic pathogens of *C. benghalensis* which were: a bacterium (Erwinia sp.) and two fungi (*Corynespora cassiicola* and *Cercospora* sp.) [38, 39].

12. Conclusion and recommendations

The Commelina species are very persistent, noxious weeds which must be managed using an integrated approach to weed management. Weed management strategies that are narrowly focused will ultimately cause shifts in weed populations to species that no longer respond to the strategy resulting in adapted species, tolerant species or herbicide-resistant biotypes [51], which is the case with Commelina species in cropping systems. The integrated approach should utilize alternative strategies such as those mentioned in this paper including the most practical options, cultural and mechanical not negating the judicious use of herbicides. Such combinations should provide significant management levels of Commelina species for both conventional as well as organic growers using a pesticide free production PFP approach. Utilization of the useful benefits of Commelina species after uprooting will also serve to check the heavy use of herbicides in cropping systems.

The integrated approach must begin very early as once an infestation is really entrenched it presents several difficulties because of the pernicious growth habit of this weed. Successful management of *C. benghalensis* will require a multi-component approach including an effective herbicide that provides soil residual activity [80]. Recent studies on the management of Commelina species have, however, still focused primarily on effective herbicides and herbicide mixtures for their control despite hard evidence of the development of herbicide-resistant biotypes. Additionally, the adoption within recent years of GM crops particularly herbicide – resistant crops presents serious issues involving their negative ecological impact as already there are reports of Commelina species prominence in some agroecosystems due to simple and significant selection pressure brought to bear by these herbicide – resistant crops and the concomitant use of the herbicide [52].

The best way to control Commelina species for small holders in developing countries would be by implementing an integrated approach that embraces a variety of options which should be attuned to the individual farmer's agronomic and socio – economic conditions (soil type, climate, costs, local practices and preferences). For example, in banana growing areas in the Windward Islands, the growth of the weed can be suppressed by a single application of a herbicide or weed whacking very early before extensive spread of the weed followed by planting a competitive cover crop like *Desmodium heterocarpon* that would not only prevent re-invasion but improve soil fertility.

Future research in developing effective management strategies for *Commelina benghalensis* should:

• Develop an accurate predictive model for *C. benghalensis* germination

- Evaluate the seedbank longevity of *C. benghalensis*

- Determine the primary dispersal mechanism(s)

- Characterize the environmental limits of *C. benghalensis* in the U.S.A. [80].

Surely this list can be expanded to include other Commelina species such as *C. diffusa* which is definitely a problematic weed in the cropping systems in the Windward Islands. The research direction should also:

- Determine threshold levels of *C. diffusa* in crops such as banana

- Evaluate the allelopathic potential of Commelina species by extracting hydro alcoholic compounds which could be used as a possible bioherbicide in controlling other problem weeds

- Screen for mycobiota with good potential for CBC such as the rust species Uromyces commilinae which has been identified in several Caribbean Islands.

- Determine the reasons for reduced seed production of *C. diffusa* species found under banana fields in the Windward Islands as compared to higher seed numbers (both aerial and underground) of *C. benghalensis* species in the USA.

Author details

Wendy-Ann Isaac[1], Zongjun Gao[2] and Mei Li[2]

1 Department of Food Production, Faculty of Food and Agriculture, The University of the West Indies, St. Augustine, Trinidad

2 Institute of Plant Protection, Shandong Academy of Agricultural Sciences, Jinan, China

References

[1] Abu-irmaileh, B. E. Weed control in vegetables by soil solarization, In FAO Plant Production and Protection Paper (Amman, Jordon), (1991).

[2] Baker, C. A, & Zettler, F. W. Viruses infecting wild and cultivated species of the Commelinaceae. Plant Disease (1988). , 72(6), 513-518.

[3] Barnes, J. Managing hairy wandering Jew. Queensland Government, Department of Primary Industries Publication #QL03056; (2003).

[4] Benjamin, A, & Rubin, B. Soil solarization as a means of weed control. Phytoparasitica (1982).

[5] Budd, G. D, Thomas, P. E. L, & Allison, J. C. S. Vegetative regeneration depth of ger-
     mination and seed dormancy in *Commelina benghalensis* L. Rhodesian Journal of Agri-
     cultural Research (1979). , 17, 151-153.

[6] Burns, J. H. A comparison of invasive and non-invasive dayflowers (Commelinaceae)
     across experimental nutrient and water gradients. Diversity and Distributions
     (2004). , 10, 387-397.

[7] CABICrop Protection Compendium, Global Module, (2002). edition. Wallingford,
     UK.   CAB   International.   Available   from: http://www.cabi/compendia/cpc/
     index.htmaccessed 15 January 2007)

[8] Chako, E. K, & Reddy, A. Effect of planting distance and intercropping with cowpea
     on weed growth in banana. In: Proceedings of the 8th Asian-Pacific Weed Science So-
     ciety Conference. (1981). , 137-141.

[9] Chivinge, O. A, & Kawisi, M. The effect of node number on the regeneration of wan-
     dering Jew (*Commelina benghalensis* L.). Zimbabwe Journal of Agricultural Research
     (1989). , 27(2), 131-138.

[10] Carmona, A. Flex (fomasafen) and cobra (lactofen): Two products with potential for
     broadleaf weed control for leguminous covers in oil palm plantations. ASD Oil Palm
     Papers (1991). , 4, 1-5.

[11] Dias, T. C. S, Alves, P. L. C. A, & Lemes, L. N. Interference periods of *Commelina ben-
     ghalensis* after coffee establishment. Planta daninha (2005). , 23(3), 398-404.

[12] Duke, J. A, & Ayensu, E. S. Medicinal plants of China. Reference Publications, Inc.
     (1985).

[13] Edmunds, J. E. Association of *Rotylenchulus reniformis* with 'Robusta' banana and
     Commelina sp. roots in the Windward Islands. Tropical Agriculture, Trinidad
     (1971). , 1971(48), 1-55.

[14] Ellison, C. A, & Barreto, R. W. Prospects for the management of invasive alien weeds
     using co-evolved fungal pathogens: a Latin American Perspective, Biological Inva-
     sions (2004). , 6(1), 23-45.

[15] Elmore, C, & Heefketh, K. A. Soil solarization an integrated approach to weed con-
     trol. Proceeding of the 35th Annual California Weed Conference, 143. Department of
     Botany, California University, Davis CA 95616, USA. (1983).

[16] Explore biodiversity and the wild classroomCommelinaceae: (Spiderwort family).
     (2002). http://www.explorebiodiversity.com/Plants/Commelinaceae.htmaccessed 10
     February 2006)

[17] Fish, L. Commelinaceae. In O.A. Leistner (ed.) Seed Plants of Southern Africa. Strelit-
     za. National Botanical Institute, Pretoria. (2000). , 591-593.

[18] Faden, R. B. The misconstrued and rare species of Commelina (Commelinaceae) in the eastern United States. Annals of Missouri Botanical Gardens (1993). , 80, 208-218.

[19] Fournet, J, & Hammerton, J. L. Weeds of the Lesser Antilles. Institute of National Research Agronomy, Paris, France. (1991).

[20] Freeman, B. E, & Smith, D. C. Variation of density-dependence with spatial scale in the leaf-mining fly *Liromyza commelinae* (Diptera: Agromyzidae). Ecological Entomology (1990). , 15, 265-274.

[21] Gao ZongJunLi Mei and Gao XingXiang. Laboratory toxicity of 20 herbicides against Bengal dayflower (*Commolina bengalensis* L.). (Abstract) VI International Weed Science Congress, Hangzhou, China, 17- 22 June, 2012. Published by the International Weed Science Society, (2012). , 140.

[22] Gardener, D. E. Rust on *Commelina diffusa* in Hawaii. Plant Disease (1981). , 65(8), 690-691.

[23] Gu ZuMin, Ji MingShan, Li XingHai and Qi ZhiQiu. Effects of environmental factors on effectiveness of *Phoma herbarum* strain SYAU-06 against *Commelina communis. Chinese* Journal of Biological Control (2009). in Chinese with English abstract)., 25(4), 355-358.

[24] Hammerton, J. L. Weed Problems and Weed Control in the Commonwealth Caribbean, Tropical Pest Management (1981). , 27(3), 379-387.

[25] Hill, M. P, & Oberholzer, I. G. Host specificity of grasshopper, *Cornops aquaticum*, a natural enemy of water hyacinth. In Proceedings of the X International Symposium on Biological Control of Weeds. Spencer, N.R. (Ed.) 4- 14 July 1999, Montana State University, Bozeman, Montana, USA. , 349-356.

[26] Hilton, H. W. Herbicide Tolerant Strains of Weeds. Honolulu, HI: Hawaiian Sugar Planters Association Annual Rep. (1957). , 69-72.

[27] Holm, L. G, Pluknett, D. L, Pancho, J. V, & Herberger, J. P. The World's Worst Weeds: Distribution and Biology. The University Press of Hawaii, Honolulu. (1977).

[28] Holm, L. G, Doll, J, Holm, E, Pancho, J. V, & Herberger, J. P. World Weeds, Natural Histories and Distribution. JohnWiley and Sons, New York, (1997).

[29] Inserra, R. N, Dunn, R. A, Mcsorley, R, Langdon, K. R, & Richmer, A. Y. Weed hosts of *Rotylenchulus reniformis* in ornamental nurseries of Southern Florida. Nematology Circular (Gainesville) (1989).

[30] Isaac, W. A. I, Brathwaite, R. A. I, Cohen, J. E, & Bekele, I. Effects of alternative weed management strategies on *Commelina diffusa* Burm. infestations in Fairtrade banana (*Musa* spp.) in St. Vincent and the Grenadines. Crop Protection (2007a). , 26-1219.

[31] Isaac, W. A. I, & Brathwaite, R. A. I. *Commelina* species- Review of the weed status of the genus and possibilities for alternative weed management in the tropics. AgroThesis(2007b). , 5(1), 3-18.

[32] Ivens, G. W. East African Weeds and their Control. Oxford University Press, Nairobi, Kenya. (1964).

[33] Johnson, S. R. *Commelina communis* (Commelinacea) as host to *Pycnoderes medius* knight (Hemiptera: Miridae) in central Virginia, USA. The Entomologist (1997). , 116, 205-206.

[34] Kennedy, R. A, Eastburn, J. L, & Jensen, K. G. C. C4 evolution of intermediate characteristics. American Journal of Botany (1980). , 67, 1207-17.

[35] Khan, R. M, Kumar, S, & Reddy, P. P. Role of plant parasitic nematodes and fungi in guava wilt. Pest Management in Horticultural Ecosystems (2001). , 7(2), 152-161.

[36] Liu Bo, Li HaiYan, Li Feng and Wang XianFeng.Dayflower control efficacy of a fomesafen plus imazethapyr plus clomazone 18% EC package mixture in soybeans. Agrochemicals (2006). in Chinese with English abstract)., 45(9), 636-638.

[37] Li Wei Dong, Wang Guang Xiang and Zhang Ge Chuan.Control effect of fomesafen 250 g/L SL on broadleaf weeds in soybean field. Journal of Anhui Agricultural Sciences (2011). in Chinese with English abstract)., 39(18), 10934-10935.

[38] Lustosa, D. C, Oliveira, J. R, & Barreto, R. W. Ocorrência de uma bacteriose em *Commelina benghalensis* (L) no Brasil, In 33 Congresso Brasileiro de Fitopatologia, 2000, Belém. Fitopatologia Brasileira (2000). , 25, 324-324.

[39] Lustosa, D. C, & Barreto, R. W. Primeiro relato de Cercospora commelinicola Chupp em Commelina benghalensis L. no Brasil. In: 34 Congresso Brasileiro de Fitopatologia, (2001). São Pedro. Fitopatologia Brasileira 2001; , 26, 364-364.

[40] Lu Xing Tao, Zhang Tian Tian, Zhang Yong, Kong Fan Hua, Ma Wei Yong, Ma Shi Zhong and Zhang Cheng Ling.Weed control efficacy and potao safety of rimsulfuron. Agrochemicals (2011). in Chinese with English abstract)., 50(11), 845-847.

[41] Ma Hong, Guan ChengHong and Tao Bo.The tolerance to imazethapyr in different leaf stages of dayflower (*Commelina communis* L.). Acta Phytophylacica Sinica (2009). in Chinese with English abstract)., 36(5), 450-454.

[42] Ma Hong, Guan ChengHong and Tao Bo.Tolerance to imazethapyr and physiological difference of dayflower (*Commelina communis* L.) at different leaf stages. Chinese journal of oil crop sciences (2010). in Chinese with English abstract)., 32(1), 136-138.

[43] Marenco, R. A, & Castro-lustosa, D. Soil solarization for weed control in carrot. Pesquisa Agropecuária Brasileira, Brasilia (2000). , 35(10), 2025-2032.

[44] Mead, F. W. Bureau of nematology: detections of special interest. Triology Technical Report (1990). , 29(1), 3-6.

[45] Morton, T. C, & Vencl, F. V. Larval beetles from a defence from recycled host-plant chemicals discharged as fecal wastes. Journal of Chemical Ecology (1998). , 24, 765-785.

[46] Myint, A. Common weeds of Guyana. Demerara, Guyana : National Agricultural Research Institute. (1994).

[47] National American Plant Protection Organization (NAPPO)Pest facts sheet- *Commelina benghalensis* L. June (2003). pp.

[48] Nkwiine, C, Tumuhairwe, J. K, Gumisiriza, C, & Tumuhairwe, F. K. Agrobiodiversity of banana (*Musa* spp.) production in Bushwere, Mbarara district, Uganda, In Agricultural biodiversity in smallholder farms of East Africa. F. Kaihura and M. Stocking (Eds.) United Nations University Press. (2003).

[49] Obiefuna, J. C. Biological weed control in plantains (*Musa* AAB) with Egusi melon (*Colocynthis citrullus* L.). Biological Agriculture and Horticulture (1989). , 6, 221-227.

[50] Oppong, F. K, Osei- Bonsu, K, Amoah, F. M, & Opoku- Ameyaw, K. Evaluation of Basta (glufosinate ammonium) for weed control in Coffee. Journal of the Ghana Science Association (1998). , 1(1), 60-68.

[51] Owen, M. D. K. The value of alternative strategies for weed management. (Abstract) III international Weed Science Congress, Foz do Iguassu, Brazil, 6- 11 June, (2000). Published by the International Weed Science Society, Oregon, U.S.A. , 50.

[52] Owen, M. D. K, & Zelaya, I. A. Herbicide- resistant crops and weed resistance to herbicides. Paper presented at the Symposium 'Herbicide- resistant crops from biotechnology: current and future status' held by the Agrochemicals Division of the American Chemical Society at the 227th National Meeting, Anaheim, CA, March, (2004). , 29-30.

[53] Pancho, J. V. Seed sizes and production capabilities of common weed species in the rice fields of the Philippines. Philippine Agriculturist (1964). , 48, 307-316.

[54] Prostko, E. P, Culpepper, A. S, Webster, T. M, & Flanders, J. T. Tropical Spiderwort identification and control in Georgia field crops. Circ. 884. University of Georgia College of Agriculture and Environmental Science / Coop. Ext. Ser. Bull., Tifton. (2005). http://pubs.caes.uga.edu/caes-pubs/pubs/PDF/c884.pdfaccessed 20 September 2006).

[55] Queneherve, P, Chabrier, C, Auwerkerken, A, Topart, P, Martiny, B, & Martie-luce, S. Status of weeds as reservoirs of plant parasitic nematodes in banana fields in Martinique. Crop Protection (2006). , 25, 860-867.

[56] Ragone, D, & Wilson, J. E. Control of weeds, nematodes and soil born pathogens by soil solarization. Alafua Agricultural Bulletin, 14. University of Hawaii, USA., (1988). , 13, 13-20.

[57] Robinson, A. F, Inserra, R. N, Caswell-chen, E. P, Vovlas, N, & Troccoli, A. *Rotylenchulus* species : identification, distribution, host ranges and crop plant resistance. Nematropica (1997). , 15, 165-170.

[58] Rodriguez, M. G, Sanchez, L, & Rodriguez, M. E. Plant parasitic nematodes associated with coffee (*Coffea arabica*) in Cajalbana, Cuba. Revista de Proteccion Vegetal (2000). , 15(1), 38-42.

[59] Shcherbakova, J. A. The effect of sowing depth of agricultural crops and fertilizers on the growth and development of *Commelina communis*. Sibirskii Vestnik Sel'skokhozyaistvennoi Nauki, (1974). , 6, 33-37.

[60] Singh, N. D. Studies on selected hosts of *Rotylenchulus reniformis* and its pathogenicity to soybean (*Glycine max*). Nematropica(1975). , 5(2), 46-51.

[61] Smith, D. Impact of natural enemies on the leaf mining fly Liriomyza commelinace, In Proceedings of the Interamerican Society for Tropical Horticulture, (1990). abstract only)., 34, 101-104.

[62] Stamps, R. H. Prodiamine suppresses spreading dayflower (*Commelina diffusa*) facilitating hand-weeding in leatherleaf fern (*Rumohra adiantiformis*) ground beds. Journal of Environmental Horticulture(1993). , 11(2), 93-95.

[63] Standish, R. J. Prospects for biological control of *Tradescania fluminensis* Vell. (Commelinaceae). Doc Science Internal Series 9, New Zealand Department of Conservation. (2001). http://www.doc.govt.nzaccessed 15 March 2006).

[64] Standish, R. J. Experimenting with methods to control Tradescantia fluminensis an invasive weed of native forest remnants in New Zealand. New Zealand Journal of Ecology (2002). , 26(2), 161-170.

[65] Sun YiHui, Zhang RongBao, Zhou JinHui, Xiao Gang, Sun YanHui and Guan HongDan.Herbicidal effects and safety of Imazethapyr AS for weeds control in soybean field. Pesticide Science and Administration (2008). in Chinese with English abstract)., 29(1), 27-29.

[66] Terry, P. J. Weed Management in Bananas and Plantains. In Weed Management for Developing Countries. Labrada, R., J.C. Caseley and C. Parker (eds.) FAO Plant Production & Protection Paper 120. (1996). , 11-315.

[67] Tian Jing and Zhao ChangShanCloransulam-methyl 84% WG controlling destructive weeds in soybean field. Agrochemicals (2009). in Chinese with English abstract)., 48(5), 376-378.

[68] Tonzani, R, Cardoso, G. V, Zonta, E, Merry, W, Parraga, M. S, & Pereira, M. G. Effects of four weed extracts on lettuce. (Abstract) III International Weed Science Congress, Foz do Iguassu, Brazil, 6-11 June, (2000). Published by the International Weed Sicence Society, Oregon, U.S.A. 2000. , 34.

[69] Tuffi Santos, L.D., Meira, R.M.S.A., and Santos, I.C.Effect of glyphosate on the mor-
      pho-anatomy of leaves and stems of *C. diffusa* and *C. benghalensis*. Planta daninha
      (2002)., 22(1), 101-107.

[70] Urich, R, Coronel, I, Silva, D, Cuberos, M, & Wulff, R. D. Intraspecific variability in
      Commelina erecta: response to phosphorus addition. Canadian Journal of Botany
      (2003)., 81, 945-955.

[71] Valverde, R. A. Brome mosaic virus isolates naturally infecting *Commelina diffusa* and
      Commelina communis. Plant Disease (1983).

[72] Van Rijin, P. J. Weed Management in the humid and sub- tropics. Royal Tropical In-
      stitute, Amsterdam, The Netherlands. (2000).

[73] Voll, E, Franchini, J. C, Cruz, R. T, Gazziero, D. L. P, Brighenti, A. M, & Adegas, F. S.
      Chemical interactions of *Brachiara plantaginea* with *Commelina benghalensis* and *Acan-
      thospermum hispidium* in soybean cropping systems. Journal of Chemical Ecology
      (2004).

[74] Walker, S. R, & Evenson, J. P. Biology of *Commelina benghalensis* L. in south-eastern
      Queensland. 1. Growth, development and seed production. Weed Research UK
      (1985a)., 25(4), 239-244.

[75] Walker, S. R, & Evenson, J. P. Biology of *Commelina benghalensis* L. in south-eastern
      Queensland. 2. Seed dormancy, germination and emergence. Weed Research UK.
      (1985b)., 25(4), 245-250.

[76] Waterhouse, D. F. Editor). Biological control of weeds: Southeast Asian prospects.
      Canberra, Australia; Australian Center for International Agricultural Research
      (ACIAR). (1994).

[77] Webster, T. M. Weed survey- southern states: broadleaf crops subsection. Proceed-
      ings of the Southern Weed Science Society (2001)., 54, 244-259.

[78] Webster, T. M. and MacDonald, G.E. A survey of weeds in various crops in Georgia.
      Weed Technology (2001)., 15, 771-790.

[79] Webster, T. M, Burton, M. G, Culpepper, A. S, York, A. C, & Prostko, E. P. Tropical
      Spiderwort (*Commelina benghalensis*): A tropical invader threatens agroecosystems of
      the Southern United States. Weed Technology (2005)., 19(3), 501-508.

[80] Webster, T. M, Burton, M. G, Culpepper, A. S, Flanders, J. T, Grey, T. L, & York, A. C.
      Tropical Spiderwort (*Commelina benghalensis* L.) Control and Emergence Patterns in
      Preemergence Herbicide Systems. Journal of Cotton Science (2006a)., 10, 68-75.

[81] Webster, T, Grey, T, Burton, M, Flanders, J, & Culpepper, A. Tropical Spiderwort
      (*Commelina benghalensis*): the worst weed in cotton? In Proceedings of the 2006 Belt-
      wide Cotton Conference, January 3-6, (2006). San Antonio, Texas) 2006b., 2181-2183.

[82] Webster, T, Flanders, J, & Culpepper, A. Critical period of tropical spiderwort (*Commelina benghalensis*) control in cotton. Weed Science Society of America Abstracts (2006). c , 80.

[83] WeedScience org. Group O/4 resistant spreading dayflower (*Commelina diffusa*), USA: Hawaii. (2005). http://www.weedscience.org/Case/Case.asp?ResistID=394accessed 15 March 2007)

[84] Wilson, A. K. Commelinacea- review of the distribution, biology and control of the important weeds belonging to this family. Tropical Pest Management (1981). , 27(3), 405-418.

[85] Yang YuTing, Lin ChangFu, Geng HeLi, Sun BaoXiang and William H. Ahrens.Studies on the action of bromoxynil and 2,4-D butyl ester herbicides combinations against dayflower (Commelina communis L.). Pesticides (2001). in Chinese with English abstract)., 40(7), 37-38.

[86] Zhang XiuRong, Ma Shu, Ying, Dai BingLi, X.R. Zhang, S.Y. Ma and B.L. Dai.Monophagy of *Lema scutellaris* on *Commelina communis*. Acta Entomologica Sinica (1996). , 39, 281-285.

[87] Zimmerman, A. De nematoden der koffiewortels. Deel I Mededeel's. Lands Plantentium (Buitenzorg) (1898).

Chapter 9

Herbicides — A Double Edged Sword

Mona H. El-Hadary and Gyuhwa Chung

Additional information is available at the end of the chapter

1. Introduction

Weeds represent a global agronomic problem that threatens the productivity of cultivated crops. Weeds compete with cultivated crops for the available moisture, nutrients and light. Consequently, weeds significantly reduce either crop yield or quality. Control of weeds is essential to maintaining the production of economic crops. Weed control may be achieved either through manual eradication or herbicide application. Balanced usage of herbicides should be considered in controlling weeds. Low concentrations of herbicides may act as growth regulators for the main crop metabolism [1]. However, in some cases, herbicides may affect the main crop adversely by interfering with its essential biochemical processes such as respiration, photosynthesis, protein metabolism, and hydrolytic enzyme activity [1].

Herbicide interference with the morphology, physiology and biochemical pathways of treated plants varies according to the characteristic actions of the herbicide and depends upon the degree of tolerance or susceptibility of the crop plant species. Environmental factors and soil conditions affecting plant growth, as well as herbicide formulation, herbicide degradation and application method would significantly influence the effects of herbicides on treated plants. Once an herbicide reaches the site of action in the treated plants, the biochemical processes are affected. Herbicides differ in their site of action and may have more than one site of action. As the herbicide concentration increases in plant tissue, additional sites of action may become involved. The effect of herbicides on growth, productivity and different metabolic activities has been studied extensively in many investigations such as in **El-Hadary** [1].

1.1. A word from the authors

Authors intended to give some examples for commercial herbicides that were applied in agronomic systems within the past fifty years. These examples include those herbicides which may now be internationally prohibited but are still used in the developing and underdeveloping countries due to their low price and the little information available about them. References have been included that cover a long era of research concerning herbicide application in order to include those prohibited herbicides. Also, references were included that focus on research that was conducted in under- and developing countries.

1.2. Herbicides

This chapter will discuss different herbicide groups, classification, selectivity, interference with metabolic processes and hazardous action upon crop plants. Also, the relation to naturally occurring phenomena, such allelopathy and future prospects of genetic engineering in the production of plant herbicides themselves, will be mentioned.

2. Classification of herbicides (Broad lines)

There are different broad lines upon which herbicides could be classified:

2.1. Application timing

Time of application of an herbicide is so critical for getting satisfying results. Herbicides application is achieved either pre-emergence or post-emergence of the weed seedlings. Pre-emergence involves herbicide application prior to seed germination while post-emergence means application after seed germination and active growth. Moreover, post-directed application refers to targeting the treatment to a particular portion of the plant once emerged and growing.

2.2. Application method

Herbicides may be applied either as a foliar spray or a soil treatment. The application method may take either the broadcast pattern through treatment of the entire area or the spot pattern through specified area treatment.

2.3. Chemical groups

The chemical group to which an herbicide belongs indicates its mode of action. A good classification and description for herbicides is provided by "Compendium of Pesticide Common Names" at the web site of *http*://www.alanwood.net/pesticides/class_herbicides.html.

2.4. Mode of action

Herbicides poisonous action goes either by contact or systematically. Herbicides can be classified according to their mode of action into two categories; non-selective herbicides and selective herbicides. Non-selective herbicides are characterized by having a general poisonous effect to the plant cells while selective herbicides can recognize the plant which they affect and kill it by interference with its principle biochemical processes.

3. Selectivity of herbicides

Selectivity of herbicides for eradicating weeds can be achieved through employing some factors related to:

3.1. Biochemical differences

Based on the biochemical differences between weeds and crops, or even weeds between each other, selectivity can be achieved. There is a great diversity of types of weeds usually growing in one crop. When employing an herbicide based on biochemical differences, the crop plant would possess a defense mechanism that is usually absent in most of the competing weed species. Consequently, the herbicide would react with the biochemical metabolism of the weeds without any fatal interference on main crop metabolism.

3.2. Morphological differences

The selectivity which depends upon morphological differences is characteristic for post-emergence herbicides. Dicotyledonous plants have leaves spread out and exposed meristematic tissue, so that the toxin is directed to the growing point situated at the center of a rosette. While upright leaves of monocotyledonous plants enable plants to form a sheath around the meristem that protects it from receiving the herbicidal spray (Figure 1) [1]. Therefore, such morphological differences can be recruited to work with monocotyledon crops against dicotyledon weeds.

3.3. Chronological selectivity

Chronological selectivity utilizes the time period necessary for growing both weeds and crop plants. In other words, it depends upon the fact that some weeds are shallower rooted and grow more rapidly than the crop plants. In consequence, many of the potentially more competitive weeds that emerge before the crop can be sprayed by a foliage spray. The time of application of the herbicide is important for chronological selectivity to be successful. That means if the non-selective herbicides are applied too early, many of the germinating weed seedlings will escape and break through the soil surface; however, the crop may be damaged if those herbicides are applied too late (Figure 1) [1].

3.4. Positional selectivity

Positional selectivity is based upon the localization of weeds on the soil surface related to the main plant crop position. If seeds, tubers, etc., of the crops are large compared with those of the weeds, they become sown or placed quite deeply in the soil compared with the more shallow competitive weed seeds. Consequently, positional selectivity can often be achieved by spraying the soil surface with soil acting herbicides. These herbicides are able to destroy weed seeds growing in the top few millimeters of the soil, whereas the large seeds of the crop are protected by the fact that they are sown deeper in the soil. Bacteria and other microorganisms attack and inactivate most herbicides when used at economic concentrations so the potential hazard to the crop is reduced (Figure 1) [1].

3.5. Placement selectivity

Placement selectivity can be achieved for non-selective substances when it is possible to direct a foliar spray in such a way that it makes contact only with the leaves of weeds and not the crop [2].

3.6. Genetic engineering

If the mode of action of an herbicide is known and the target proves to be a protein, genetic engineering may well allow the crop gene coding for that protein to be isolated. It is then possible to alter that crop gene so that it is less affected by the herbicide [2].This will be discussed in detail at the end of the chapter.

4. Herbicide interference with physiological and biochemical processes and plant response

Mode of action of herbicides can lead to various physiological and biochemical effects on both growth and development of the emerging seedlings as well as the established plants. These physiological and biochemical effects are followed by various types of visual injury symptoms on susceptible plants. The incidental damage extent depends on the selectivity of the herbicide as well as the applied concentration. The herbicide application is always recommended at a certain dose termed as recommended dose (R), above which, a great damage to the crop plant may be obtained. Overdoses threaten not only the crop plant but also the environment and human health. Some herbicides in lower than recommended doses may act as growth regulators for crop plants [1,2].

Even recommended doses may have undesired effects upon the crop. The undesired effects might occur in the form of chlorosis, defoliation, necrosis, morphological aberrations, growth stimulation, cupping of leaves, marginal leaf burn, delayed emergence, germination failure, etc. These injury symptoms may appear on any part of the plant.

The various physiological and biochemical processes affected by herbicides are grouped under five broad categories including: respiration, mitochondrial activities, photosynthesis, protein synthesis, nucleic acid metabolism, and hydrolytic enzyme activities. Most herbicides can affect at least one or all of these processes. The following discusses their effect on various biochemical processes.

4.1. Respiration and mitochondrial activities

Cellular respiration that takes place in mitochondria involves the synthesis of ATP and the transport of electrons and protons from respiratory substances to oxygen. Herbicides affect

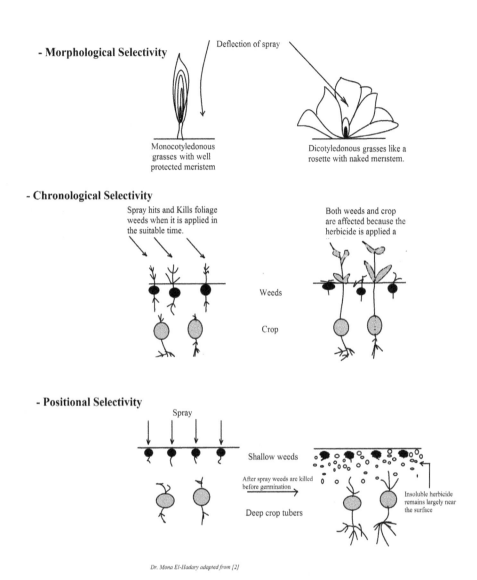

- **Morphological Selectivity**

Deflection of spray

Monocotyledonous
grasses with well
protected meristem

Dicotyledonous grasses like a
rosette with naked meristem.

- **Chronological Selectivity**

Spray hits and Kills foliage
weeds when it is applied in
the suitable time.

Both weeds and crop
are affected because the
herbicide is applied a

Weeds

Crop

- **Positional Selectivity**

Spray

Shallow weeds

After spray weeds are killed
before germination

Deep crop tubers

Insoluble herbicide
remains largely near
the surface

Dr. Mona El-Hadary adapted from [2]

Figure 1. Factors Exploitable to Achieve Selectivity of Herbicides [1] as adapted from [2].

the mitochondrial activities by uncoupling the reaction responsible for ATP synthesis or interfering with electron transport and energy transfer. Uncouplers act on the membranes of the mitochondria in which phosphorylation takes place. Electrons leak through the membranes so that the charges that they normally separate are lost. As a result, energy is not accumulated for ATP synthesis [3].

4.2. Photosynthesis

Pigment content and photosynthetic activity are affected by herbicidal applications. The mode of action of herbicides on the photosynthesis process depends on the chemical group to which the herbicide belongs [3]. Herbicides affect chloroplast organization and pigment formation especially chlorophyll which is the principle absorbing pigment. Chlorophyll bleaching is a potent inhibitor for photosynthetic electron transport and CO_2 fixation.

Herbicides affect photosynthetic activity via different ways including photosynthetic pigments. The primary site of action is located at photosystem II (PSII) since they cause blocking of the Hill reaction. The oxygen evolution step is inhibited by interfering with the reducing side rather than the oxidizing side of PSII [4]. The inhibition of electron transfer through PSII causes a block in the whole transport chain as the inhibition of the noncyclic photophosphorylation and ATP synthesis. Consequently, the production of NADP is blocked and the function of the protective carotenoid system is prevented [5]. Urea herbicides inhibit both noncyclic and cyclic electron transport by forming a complex with oxidized form of an unknown component located in the electron transfer pathway close to PSII. This component also takes part in cyclic electron transport.

The photosystem I (PSI) also could be reduced by some herbicides but it requires much higher concentrations of the herbicide than that required for the inhibition of PSII. Since PSII precedes PSI and the former is blocked completely at concentrations which do not affect PSI.

In a study conducted by **El-Hadary** [1], it was observed that photosynthetic activity measured in wheat chloroplasts (variety Giza 163) was greatly reduced throughout the growth by using Brominal as an example for bromphenol herbicides but lower concentrations (1/4R, 1/2R and R) increased the activity. Pigment content represented as chlorophyll, a/b ratio and carotenoids showed a similar results [1]. In the same study, sulfonyl-urea herbicides such as Granstar were examined. It was observed that low Granstar concentrations stimulated the photolytic activity of chloroplasts while high concentrations reduced it. However, Granstar reduced a/b ratios throughout the growth stages, except a slight increase at the fruiting stage with 1/2R. Carotenoids were decreased only with high Granstar concentrations [1].

4.3. Protein and nucleic acid metabolism

Protein synthesis takes place mainly in three stages involving initiation, elongation and termination of the polypeptide chain. Blocking any one of these stages by the herbicide will cause inhibition of protein and nucleic acid synthesis. The herbicides that inhibit photosynthesis and ATP formation can lead to inhibition of protein synthesis as a secondary effect. The damage that is caused by an herbicide is governed by its chemical group. There are numerous studies that investigate effects of the herbicidal chemical groups upon protein and nucleic acid metabolism [2].

For instance, sulfonyl-urea herbicides block the biosynthesis of the branched chain amino acids in higher plants [6,7]. Aliphatic herbicides like Dalapon cause degradation of protein to ammonium compounds as detected in *Setaria lutescens* and sugar beets [8]. While acetamide herbicides such as propachlor, alachlor and prynaclor inhibited the protein content and RNA synthesis as reported in barley [9-12]. Also, metalachlor inhibited protein synthesis in barley [13]. RNA and protein synthesis in tomato were found to be inhibited by propanil [14].

Benzoic and phenylacetic herbicides had variable effects on protein. For example, chloramben had no effect on RNA and protein synthesis on susceptible species [15]. On the other hand, it was suggested that foliar-applications of dicamba increased RNA and protein levels in susceptible plants by removal of histone from the DNA template [16]

Carbamate herbicide groups include a large number of herbicides such as asulam, barban, chlorpropham, propham, desmedipham and phenmedipham. [17]. Barban was found to inhibit protein synthesis and the degree of inhibition was related to the susceptibility of the plant species. For example, barban increased nucleotide content of wild oat shoots associated with disruption of RNA and protein synthesis. Chlorproham and propham inhibited amino acid incorporation into protein and induced a reduction in protein synthesis [18]. DNA, RNA and protein synthesis are also inhibited at high concentrations (10-3 M) of propham [19].

Fluridone, paraquat, perfluidone and propanil treatments were found to reduce soluble protein levels in soybean [20]. Paraquat and diquat readily act on proteins, modifying their structure and function (e.g.lysozome) since they interact with dibasic and dicarboxylic amino acids like ornithine and glutamic acid [21].

Oxadiazon at high doses inhibited protein synthesis in soybean while RNA and DNA synthesis were less sensitive to oxadiazon [22]. Combination of 2,4-D and glufosinate had an additive effect on protein synthesis in both sorghum and soybean [22]. On the other hand, sethoxydim, R- 25788 [N, N dichloroacetamide] or R- 28725 at low doses did not inhibit protein or RNA synthesis in cells of both sorghum and soybean but sethoxydim significantly inhibited DNA synthesis while R-25788 stimulated it [23]. Thus, the combined effects of sethoxydim and the two Safeners (R- 25788 and R- 28725) on protein and RNA synthesis were additive while on DNA synthesis they were antagonistic.

The application of haloxyfop to *Zea mays* and soybean cell suspension, increased ^{14}C labeled free amino acids level and incorporation of ^{14}C leucine as a precursor revealed that haloxyfop did not inhibit protein synthesis [24].

Napropamide reduced DNA synthesis, RNA root cells of *Pea* and protein [25]. The inhibitory effect of napropamide on the mitotic cycle resulted from an inhibition in the synthesis of cell cycle specific protein. In contrast, 0.5 R, 1R and 1.5 R of metribuzin stimulated total and protein-N accumulation in soybean. Consequently, protein content was increased while RNA and DNA levels decreased [26]. Protein content of soybean yield was reported to be increased by application of 100 ppm GA$_3$ (gibbrellic acid) and 2g/L Librel separately or together [27].

Metoxuron had a remarkable inhibition on the total protein biosynthesis, while bromoxynil accelerated the biosynthesis of low molecular proteins (water-soluble proteins) and inhibited the biosynthesis of high molecular proteins (sodium hydroxid soluble proteins) in wheat (*Triticum aestivum*, var. Sakha 69) [28]. Bromoxynil at low doses (0.4 and 0.8 kg / fed) enhanced protein content and RNA synthesis in wheat plants after 30 to 60 days from foliar spraying [29].

Nitrogen in wheat grains, consequently protein, was found to be increased by treating wheat plants with Brominal at the 2-leaf stage [30]. Different bromoxynil levels increased the protein percentages in wheat grains [31]. The foliar spray with bromoxynil increased significantly the protein content in wheat grains [32]. Application of bromoxynil at the full recommended rate significantly increased grain nitrogen and proteins in both wheat and barley. The increase was evaluated by multiplying grain nitrogen by 5.7 as a factor in both wheat and barley [33]. Protein content in wheat vegetation (Giza 163) was significantly increased at the vegetative stage and flowering stage while decreased at the fruiting stage as a response to either low or high Brominal treatments [1]. In contrast, the protein content of wheat root was reduced. Also, protein profiling of grains is greatly altered with an induction for 19kDa and 25kDa but an inhibition for 66kDa, 100kDa and 110kDa was obtained [1].

The action of urea herbicides on protein and nucleic acid metabolism has been reported by many researchers. Although fluometron can cause an increase in the low molecular weight fraction of DNA, RNA and protein synthesis [34], diuron and monuron inhibited the same parameters as reported [35]. However, the monomethylated derivative of isouron [N-[5-(1,1-dimethy ethyl-3-iso) (azol]-N-methylurea] suppressed the protein synthesis in soybean[36].

Sulfonylurea herbicides were found to inhibit branched chain amino acids valine, leucine and isoleucine (e.g. Granstar; DPX- L 5300; tribenuron) [6, 7]. Aflon (urea herbicide), when sprayed at 1/2 R and R doses on *Phaseolus vulgaris*, induced a DNA increase in both shoot and root while RNA content was increased in shoot only [37]. Moreover, RNA content of roots was mostly decreased in response to R and 2R aflon treatments but increased as a result of the 1/2 R application [37]. Protein content of the wheat shoot system was increased with all Granstar concentrations at the vegetative stage and with low concentrations (1/2R and R) at both flowering and fruiting stages. In contrast, protein levels were decreased with 5/2R at the flowering stage and with 3/2R and 2R- and 5/2R at the fruiting stage [1]. Granstar treatments reduced the contents of root proteins at the vegetative stage and flowering stages but increased it at the fruiting stage. Protein profiling of grain proteins exerted an induction for 19kDa and 25kDa and complete suppression for 66kDa, 100kDa and 110KDa [1].

4.4. Hydrolytic enzyme activities

Enzymes of plants were affected greatly by herbicide treatments and their effect differs according to the chemical group to which the herbicide belongs. The following examples represent some effects of herbicides on the enzyme activities of some plant species.

One of the major metabolic processes that take place during seed germination is the production of hydrolytic enzymes such as α-, β-amylases that degrade stored carbohydrates into simple sugars. The production of hydrolytic enzymes requires the synthesis and presence of proteins, polyribosomes and nucleic acids. Thus, an effect of the herbicide on protein formation as mentioned above, would affect the synthesis of the hydrolytic enzymes [1, 3]. **El-Hadary** [1] reported that use of either Brominal or Granstar at different levels below and above the recommended rate induced stimulation for amylolytic enzyme activity (α and β-amylase); however, an incidence of a slight reduction in β-amylase activity was observed with 2R and higher doses of Granstar [1].

Dalapon, which is an aliphatic herbicide, did not affect the activity of hydrolytic enzymes like protease, α- amylase and dipeptidase in barley seeds [38]. Acetamides such as alachlor, propachlor and prynachlor which all were applied at pre-emergence caused an inhibition for seed germination in barley by reducing the synthesis of α-amylase enzyme [39].

It was reported that propaclor inhibited the gibberellic acid (GA$_3$) induced production of α-amylase in barley seeds [40]. Similarly, alachlor, propachlor and prynachlor were found to inhibit α-amylase as well as protease synthesis in barley seeds [41, 42]. It was suggested that these herbicides may act as repressors for gene action preventing the normal expression of the hormonal effect of GA$_3$ through the synthesis of DNA-dependent RNA. This was confirmed when higher levels of GA$_3$ overcame alachlor inhibition by removing the repressor effect [42]. In addition, the effect of these acetamide herbicides on α- amylase and protease was suggested to be secondary and these herbicides possibly act on the biosynthetic reactions (like protein synthesis) required for the synthesis of these hydrolytic enzymes.

Chloroamben and dicamba, which belong to the benzoic and phenylacetic acid herbicide groups, were found to inhibit GA$_3$-induced α-amylase synthesis and the development of amylase activity in barley seeds [40, 43]. This agrees with effect of trifluarlin, as an example for dinitroanilines, which was found to inhibit the *de novo* synthesis of hydrolytic enzymes such as protease [44] and dipeptidase in squash cotyledons [45], phytase in barley seedlings, squash cotyledons and maize embryos [39], and α-amylase in barley seeds [40].

Nitriles such as bromoxynil and ioxynil also inhibited proteolytic and amylolytic enzyme activities [46, 45]. Also, thiocarbamate herbicides were found to inhibit GA$_3$- induced α-amylase synthesis in susceptible weeds [17]. Acifluorfon was found to stimulate the activity of chalcone synthase, phenylalanine ammonia lyase and isoflavone 7-0- glucosy transferase which are responsible for the accumulation of isoflavonoids in soybean leaves [47].

The increase of galactonolactone oxidase was reported in common beans as a result of acifluorfen application; this enzyme is responsible for lipid peroxidation. Acifluorfen was found to increase the activity of galactonactone reductase, which prevented further oxidation of lipids [48]. Other herbicides, alachlor and glyphosate, were observed to inhibit 5- enolpyr-uvyl shikimate-3-phosphate (EPSP) synthase enzyme. This enzyme is responsible for the synthesis of all cinnamate derivatives (intermediates in flavonoids biosynthesis pathway) leading to reduced flavonoid synthesis in higher plants [49].

Sulfonylureas herbicides act by inhibiting acetolactate synthase enzymes, thereby blocking the biosynthesis of the branched chain amino acids in higher plants [7]. According to **Gronwald** [50], carbomothioate herbicides inhibited one or more acyl- CoA elongase enzymes which catalyze the condensation of malonyl CoA with fatty acid acyl-CoA substrates to form a very long chain fatty acid, used in the synthesis of surface lipids.

The effects of triazine, urea and nitroaniline herbicides on amylase and acid proteolytic activities of wheat grain cultivars, Salwa, Grana and Liwilla were studied by **Wybieralshi and Wybieralska** [51]. The studied herbicides were found to inhibit amylase activity in Salwa and Liwilla, but increased it in Grana. Acid proteolytic activity in Liwilla and Salwa was reduced especially by Igran 80 (terbutryn) and Dicuran 60 (Chlorotoluron), while the activity in Grana was not affected. In contrast, amylase, dehydrogenase, cellulase and xylanase activities were increased by application of the herbicides Pyramin (chloridazon), Ro-neet (cycloate) and Venzar (lenacil) when applied on the soil with 5% (w/w) addition of wheat straw [52]. Other studies suggested that application of SAN 9789 (norflurazon) as a metabolic inhibitor to *Sinapis alba* seedlings destroyed the chloroplasts but had no effect on α-amylase activity. This is due to the fact that α-amylase is a cytosolic enzyme [53].

The levels of leaf β-amylase and starch debranching enzyme in pea seedlings were found to slightly decrease in response to norflurazon-treatment [54]. However, inhibitors of chloroplastic functions, i.e.; diuron (DCMU), atrazine, tentoxin, paclobutrazol and San 9785 (4 - chloro-5-(dimethylamino)-2-phenyl-3 (2H)- pyridazinone) caused either no or only slight increases in α-amylase activity. In contrast were the inhibitors of plastidic protein synthesis lincomycin and chloramphenicol that cause an increase in α-amylase activity in pea seedlings. It is concluded that there was an inverse relationship between α-amylase activity and chlorophyll concentration in pea petals and stems [55]. Similarly an inhibition of α-amylase induction in barley seeds was reported [56]. Also, Li found that juglone decreased the content of total soluble protein and α-amylase activity induced by gibberellin by 74% and 78% in the aleuron cells of barley. It was concluded that juglone may be a metabolic inhibitor which prevents many (if not all) physiological and biochemical processes involving SH-groups in compounds such as amino acids, peptides and enzymes [57].

The activities of α-and β-amylases of castor bean and maize Giza 2 seedlings and adult plants supplemented with low concentration (0.5-2.5 μg/g) of metribuzin either alone or in combination with NaCl at 50 μg/g were increased significantly [58] but higher metribuzin concentration (5-10μg) had an opposite response. Application of 1.5-4.5kg/ha thiobencarb and butachlor six days after transplanting of 30-day-old rice seedlings affected the enzyme activities of the seedlings whether they were grown alone or with the competitive barnyard grass [59]. Moreover, both herbicides reduced α-amylase activity by increasing the concentration but a sharp increase in α-amylase activity was noted at 96h post-treatment in both species. In addition, protease (proteinase) activity was maximized after post-treatment at both 48h.and 24h in rice and grass, respectively.

Butachor (1000-3000 g/ha) and oxyfluorfen (100-300g/ha) effect on α-amylase activity and chlorophyll content in 46 rice cultivars was dependent on the degree of tolerance of each cultivar [60]. It was concluded that rice cultivars ADT-37, ASD-16 and ASD-18 were highly tolerant to butachor, whereas ADT-36, ADT-38 and PY-3 were highly susceptible. However, tolerance to oxyfluorfen was high in ASD-18 and AS-18696, while IR-50 was highly susceptible [60].

4.5. Lipid synthesis and oxidation

Substituted ureas, uracils, triazine, benzonitriles and bipyridyls markedly accelerated the photo-oxidations (lipids- per-oxidation) but peroxidation was completely prevented by NADH or NADPH [5]. Lipid peroxidation in higher plants (Duranta and Cassia) was induced by oxyfluorfen [61] but the peroxidative cell damage is controlled by antioxidative systems such as vitamins "C" and "E".

Lipid peroxidation and galactonlactone oxidase increased in response to the treatment of *Phaseolus vulgaris* leaves with acifluorfen [48] and the activity of glutathione reductase also increased to prevent further oxidation. Gronowald studies on herbicides concluded that the carbothioates group impaired the synthesis of surface lipids (waxes, cutin, and subrin) by inhibiting acyl- CoA elongases while chloroacetamide herbicides inhibited *de novo* fatty acid biosynthesis. Similarly, pyridazinones herbicides decreased the degree of unsaturation of plastidic galactolipids while aryloxyphenoxy pypropionic acid and cyclohexanedione herbicides inhibited *de novo* fatty acid synthesis. The target site for all these classes is the enzyme acetyl-CoA carboxylase [50].

The total lipid content as well as *gluco*-and *phospho*-lipid content of maize seedlings markedly decreased by application of perfluidone while in sunflower cotyledons total lipids were not affected but glycolipids increased at the expense of phospholipids [62]. Also, a decrease in lipid synthesis in soybean by Isouron was reported [36] but an increase in seed oil of soybean was obtained by 0.5R or 1R metribuzin application [26].

4.6. Carbohydrate content

Carbohydrate content is one of the most affected parameters in response to herbicide application. **Yakout** [28] demonstrated that treating wheat (*Triticum aestivum* var. Sakha 69) with metoxuron showed a slight reduction in the available carbohydrates with relatively no change in sucrose content while bromoxynil showed an increase in different carbohydrate levels. Also, the total reducing substances (may include sugars, phenolic substances, ascorbic acid, organic acid, etc.) were increased for both treatments [28].

Inhibition of the accumulated reducing sugars, sucrose and polysaccharides, in soybean leaves was observed in response to 1R and 1.5R metribuzin application and, consequently, seed carbohydrate content decreased with increasing metribuzin concentration [26]. Terbytryn herbicide was found to decrease starch content and increase sugar content in pre-emergence and post emergence applications [63]. On the other hand, bromoxynil was reported to significantly increase soluble and total carbohydrates at low doses while a higher dose (1.2 kg/Fed) inhibited their synthetic rate in wheat plants [29]. Similarly, the results of **El-Hadary** [1] found that *mono-, di- and poly*-saccharides and, consequently, total carbohydrates were increased with low doses but decreased with high doses of either Brominal or Granstar [1]. The incidental increase with low concentrations was attributed to that some herbicides act as growth regulators in low doses.

Urea herbicides including afalon-S at low doses of 1/2R and R increased the soluble and insoluble sugar contents of shoots at different stages of growth and development of *Phaseolus vulgaris* while a reverse situation was obtained in the case of a 2R application. The root tissue treated with various concentrations suffered from an obvious decrease in the content of the different carbohydrate fraction relative to those of the control [37].

The content of reducing sugars and sucrose of *Ricinus communis* cultivar Balada and maize cultivar Giza 2 seedlings and adult plants supplemented with low concentrations (0.5-2.5μg/g) of metribuzin either alone or in combination with NaCl at 50μg/g were increased significantly but decreased in response to higher concentrations (5-10μg). On the other hand, polysaccharide content of R. *communis* and maize seedlings as well as adult plants were significantly decreased in response to low concentrations of metribuzin and increased significantly at higher concentrations either alone or in combination with NaCl. Total carbohydrate content detected in R. *communis* treated with metribuzin were greater than those detected in presence of herbicide and NaCl combination [58].

Thiobencarb and butaclor herbicides when applied at 1.5-4.5 kg/ha after transplanting 30 days old rice seedlings and barnyard grass grown alone or with rice were found to have no effect either on total carbohydrate or starch and reducing sugars in rice and grass [59].

4.7. Plant growth response and yield

Plant growth and yield are greatly affected by herbicidal applications depending on the age, tolerance, dose and the active chemical group of the herbicide. The author in a previous work pointed that Brominal application on wheat induced an increase in the number of grains per spike with 1/4 R. 1/2R and R while higher doses caused a significant reduction [1]. Also, grain yield showed a detectable reduction in monosaccharides, disaccharides, polysaccharides and, consequently, total carbohydrate levels with all Brominal concentrations [1].

The percentage of germination and seedling growth of barley was decreased greatly by applications of bromoxynil [64]. But the same herbicide in different concentrations encouraged wheat growth [31]. Also, growth parameters such as plant height, weight and leaf area of wheat plants at 75 days after sowing were increased significantly by foliar application of bromoxynil at rate of 1.0 L/Fed [32, 65]. Moreover, a good seedling establishment of wheat was obtained by combinations of bromoxynil and fenoxaprop [66]. Low metribuzin concentrations (0.5-2.5μg/g) either alone or in combination with NaCl (50 μg/g) caused an increase in different growth parameters such as leaf area, length of shoot and root, water content and dry matter accumulation in both *Ricinus communis* cultivars, and maize cultivars Giza 2 throughout the different growth stages [58]. In contrast, the higher metribuzin concentration (5-10μg) affected the same parameters oppositely [58].

Productivity of the plant is affected in terms of 100 grains weight in response to herbicides treatment. The yield of wheat grains (var.Sakha 69) increased by bromoxynil application [28]. A dose of 1.5 kg/ha of bromoxynil brought an increase in weight of 100 grains [30,67]. The highest yield was obtained when one liter/fed bromoxynil was applied at the third-leaf stage [68]. The number of wheat grains/ear and grain yield were increased at a low dose (0.8kg/ fed.) of bromoxynil [29,69] while a higher dose of the same herbicide (1.2 kg/ fed) reduced the yield of wheat varieties; i.e. Sakha 69, Giza 157 and Giza 160 [29]. On the other hand, it was noticed that higher doses of bromoxynil resulted in a marked increase in both yield and grains/ear when crops were poorly developed at the time of spraying [70]. However, the application of 2.5, 3.0 liter bromoxynil /ha at the third-leaf and flowering stages on wheat significantly decreased the grain yield [71] as well as the number of spikes per plant, main spike length, weight of 100 grains and straw per plant [32].

Herbicidal effects may be varied when they are applied in combination. For example, a marked increase was observed in the grain yield, ears/plant and number of ears in barley by using a combination of bromoxynil, ioxnil and mercoprop [72]. An increase of about 20% was recorded in grain wheat yield when oxitril 4, which is a combination of oxitril and bromoxynil, was used at 130g/liter and applied at rates of 1.5,4 and 5 liters/ha [73]. In winter wheat a marked increase in yield was mentioned in response to half rate applications of various commercial herbicides (active ingredients bromoxynil, ioxynil, mocoprop, cyanazine, fluroxypyr, metasulfuron-methyl, and clopyralid) [74].

Urea herbicide such as Granstar (metasulfuron- methyl 75% water dispersible granules) was found to suppress the growth rate of wheat and barley by about 20% while weeds were completely destroyed[75]. Its application with a dose of 20-40 g/ha in 200-500 liter/ha prior to planting resulted in 50% suppression [76]. The author in a previous work applied Granstar at a dose of 0.5R, 1R, 1.5R and 2R on wheat at 40-days old and reported an increase in grains no./spike [1]. However, a great decrease in monosaccharides, disaccharides, polysaccharides and, consequently, total carbohydrate levels was obtained in wheat grains with both low and high Granstar concentrations [1]. Also, chlorsulfuron was mentioned to reduce both the third leaf growth rate and shoot dry weight of wheat seedlings but not the root dry weight [76].

The urea herbicide metoxuron was reported to decrease wheat grain yield (var. Sakha 69) [28]. It was found that 100-seed weight of soybean was decreased by using metribuzin at rates of 0.5R, 1R and 1.5R [26]. Wheat yield was markedly increased by using tribenuron at a rate of 0-125g [77]. However, sulfonylurea herbicides, Chisel [Chlorosulfuron+thifensulfuron -methyl] and Granstar, significantly increased the productive tillering in some wheat varieties [78]. Application of trifluralin alone in the spring followed by some post herbicides resulted in a reduction in vegetative growth, shoot dry weight and wheat grain yield [79]. An application of 0.126 mM perfluidon herbicide was reported not only to decrease both fresh and dry weight but also shoot length of maize seedlings [62].

5. Hazardous action of herbicides in the agricultural environment and human health

Although the benefits gained from herbicides usage in weeds control, herbisides have undesired effects on man health and environment. Their residues remain in the soil for many years, affecting crops, water canals, grazing animals and human health and even the pollution of air.

Herbicides and pesticides have been suspected by the "National Cancer Research Institute" as a probable cause of certain cancers especially cancers of the brain, prostate, stomach and lip, as well as leukemia, skin melanomas and Hodgkin's lymphoma [80]. They also cause reproductive problems as well as infertility and nervous system diseases. The National Academy of Sciences reported that infants and children, because of their developing physiology, are more susceptible to the negative effects of herbicides and pesticides in comparison to adults. Herbicides may cause human poisoning since they affect humans through three mechanisms of entry: ingestion, inhalation and dermal absorption. In under-developed countries, the least expensive pesticides are utilized due the inability of farmers to purchase more expensive, safer products. As a byproduct of pesticide use, farmers and their families are affected daily with health problems directly resulting from pesticide exposure [81]. Herbicide toxicity and risks are not only limited by their direct use but can also present risks indirectly. Indirect risks are represented by herbicidal traces that remain in the edible plants themselves as well as the residues in the soil that may remain for a number of years before it can be degraded. Moreover, the leakage of these herbicides and their residues in water canals, vaporization and sublimation in air may be poisonous to the surrounding living organisms.

6. Natural herbicides

Allelopathy phenomenon serves the agricultural community so much. The following section discusses the related concepts to allelopathy and recruiting it as natural herbicides for weed management to be an alternative or to minimize conventional herbicide use.

6.1. Allelopathy term

Allelopathy is a natural biological phenomenon of interference among organisms in such a way that an organism produces one or more biochemicals that influence the growth, survival, and reproduction of other organisms. Allelopathy is the favorable or adverse effect of one plant on another due to direct or indirect release of chemicals from live or dead plants (including microorganisms).

6.2. Allelochemical term

Allelochemicals, or allelochemics, are a subset of low molecular weight secondary metabolites such as alkaloids, phenolics, flavonoids, terpenoids, and glucosinolates which are produced during growth and development but are not used by the allelopathic plant [82]. Allelochemicals may have beneficial (positive allelopathy) or detrimental (negative allelopathy) effects on the target organisms. Allelochemicals with negative allelopathic effects contribute in plant defense against herbivory. Also, allelochemicals could be recruited in weed management as alternatives to herbicides.

Allelochemicals are listed as six classes [83] that possess actual or potential phytotoxicity. The classes are namely alkaloids, benzoxazinones, cinnamic acid derivatives, cyanogenic compounds, ethylene and other seed germination stimulants, and flavonoids which have been isolated from over 30 families of terrestrial and aquatic plants. Like synthetic herbicides, there is no common mode of action or physiological target site for all allelochemicals.

6.3. Allelochemical occurrence

Allelochemics are present in different parts of the plant; leaves, flowers, fruits, stems, bark, roots, rhizomes, seeds and pollen. They may be released from plants into the environment through volatilization, leaching, root exudation, and decomposition of plant residues. Rainfall causes the leaching of allelopathic substances from leaves which fall to the ground during period of stress, leading to inhibition of growth and germination of crop plants [84, 85].

6.4. Allelochemical classification and biosynthesis

According to the different structures and properties of allelochemicals, they can be classified into the following categories: water-soluble organic acids, straight-chain alcohols, aliphatic aldehydes, and ketones; simple unsaturated lactones; long-chain fatty acids and polyacety-lenes; quinines (benzoquinone, anthraquinone and complex quinines); phenolics; cinnamic acid and its derivatives; coumarins; flavonoids; tannins; steroids and terpenoids (sesquiter-pene lactones, diterpenes, and triterpenoids) [86]. The biosynthetic pathways of the major allelopathic substances are shown in Figure 2 [87].

6.5. Allelochemical interference and biological activity

The allelochemical interference implies their interference with each other as well the interference with other surrounding plants. Several chemicals can be released together and may exert toxicities in an additive or synergistic manner. Allelopathic interferences often result from the mixing action of several different compounds. Allelopathic plant extracts can effectively control weeds since mixtures of allelopathic water extracts are more effective than the application of single-plant extract. Combined application of allelopathic extracts and reduced herbicide dose (up to half the standard dose) give as much weed control as the standard herbicide dose in several field crops. Lower doses of herbicides may help to reduce the development of herbicide resistance in weed ecotypes [88]. Allelopathy thus offers an attractive environmentally friendly alternative to pesticides in agricultural pest management [88].

Response of the receiver plants to allelochemicals is not only concentration dependent but also controlled by the biochemical pathway in the receiver plant. Generally, low concentrations of allelochemicals are stimulatory while it is inhibitory with higher concentrations [89]. Allelo-chemical concentrations in the producer plant may also vary over time and in the plant tissue produced. Foliar and leaf litter leachates of Eucalyptus species, for example, are more toxic than bark leachates to some food crops. Typically, allelochemical concentration in field situations is below the required inhibitory level that can affect sensitive plants.

Receiver plant response to antagonistic allelochemicals is detected as certain signs on growth and development of the plants that are exposed to allelochemicals. The effect includes the inhibition or retardation of germination rate; seeds darkness and swelling; root or radicle reduction, curling of the root axis, lack of root hairs; increased number of seminal roots, swelling or necrosis of root tips; shoot or coleoptile extension; discolouration, reduced dry weight accumulation; and lowered reproductive capacity. These morphological effects may be secondary for primary events due to interference with different biochemical pathways of the receiver plant [90].

Biological activity of allelochemicals could be increased by some modifications so the end product could be more active, selective, or persistent. This is attributed to the potential

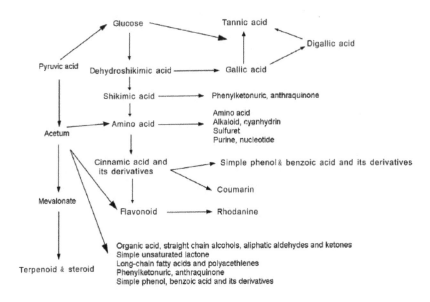

Figure 2. The Biosynthetic Pathways of the Major Allelopathic Substances [87]

phytotoxicity of alkaloids, benzoxazinones, cinnamic acid derivatives, cyanogenic compounds, ethylene and other seed germination stimulants, and flavonoids that always represent the secondary products of allelopathic plants. Biodegradable natural plant products rarely contain halogenated atoms and possess structural diversity and complexity, constituting one such class of chemicals and these can act directly as herbicides or may provide lead structures for herbicidal discovery [91]. Selection of allelopathic plants is a good and commonly used approach for identification of plants with biologically active natural products [91].

Different crops such as beet (*Beta vulgaris* L.), lupin (*Lupinus lutens* L.), maize (*Zea mays* L.), wheat (*Triticum aestivum* L.), oats (*Avena sativa* L.) and barley (*Hordeum vulgare* L.) are known to have an allelopathic effect on other crops (Rice, 1984b). For instance, some wheat cultivars were found to significantly inhibit both germination and radicle growth of annual ryegrass. The allelopathic potential of wheat cultivars was positively correlated with their allelochemical (total phenolics) content [92]. However, different allelopathic compounds of some crops important in weed management are presented in Table 1 [93].

6.6. Allelopathic plants impact

There are some examples of plants that act as natural herbicides, such as black walnut, sunflowers, sagebrush and spotted knapweed. An herbicidal chemical called catechin was extracted from the roots of spotted knapweed and can be synthesized on a larger scale and applied to a number of other invasive plants due to selectivity. Another popular species with

Crops	Scientific name	Allelochemicals
Rice	Oryza sativa L.	Phenolic acids
Wheat	Triticum aestivumL.	Hydroxamic acids
Cucumber	Cucumis sativus L.	Benzoic and Cinnamic acids
Black mustard	Brassica nigra L.	Allyl isothiocyanate
Buck wheat	Fagopyrium esculentum L.	Fatty acids
Clovers and	*Trifolium spp.*	Isoflavonoids and Phenolics
Sweet clover	*Melilotus spp.*	Phenolics
Oats	Avena sativa L	Phenolic acids and Scopoletin
Cereals	-	Hydroxamic acids
Sudangrass		Phenolic acids and Dhurrin
Sorghum	Sorghum bicolor L.	Sorgoleone

Table 1. Allelochemicals of Some Important Crops

natural herbicide abilities is the black walnut tree whose leaf extraction is often used in commercially-produced natural herbicides [94].

Other natural pre-emergent herbicides are used to control weed growth such the natural herbicide corn gluten meal. Corn gluten meal was originally developed as a medium for growing fungus, but its inhibitory effect upon the germination of weeds and grasses was detected. A cover crop of rye could work as a natural herbicide between soybean crops [94].

Herbicidal effects have been identified and quantified for more than twenty allelochemicals in *Vulpia* residues. Those present in large quantities possessed low biological activities, while those present in small quantities possessed strong inhibitory activities. Interference between different allelochemicals controls the overall phytotoxicity of *Vulpia* residues which varies according to the individual chemical structure and occurred quantity. This interference provides a pattern for suggested artificial combinations of these allelochemicals prepared in aqueous solution. Biological tests for different combinations of *Vulpia* extracts demonstrated the existence of strong synergistic effects among the identified allelochemics. Moreover, exploration of the composition of a cluster of allelochemicals, which are simple in structure, possess various biological activities and few barriers to synthesis and production; this provides an alternative option for developing new herbicides from individual plant allelochemicals [94].

Selective activity of tree allelochemicals on crops and other plants has also been reported. For example, *Leucaena leucocephala*, the miracle tree promoted for revegetation, soil and water conservation and animal improvements in India, also contains a toxic, non-protein amino acid in leaves and foliage that inhibits the growth of other trees but not its own seedlings. *Leucaena* species have also been shown to reduce the yield of wheat but increase the yield of rice. Leachates of the chaste tree or box elder can retard the growth of pangolagrass but stimulate growth of bluestem, another pasture grass. Examples that are shown in Table 2 represent some allelopathic plants and their impact as reported in published research [95].

6.7. Allelochemical modes of action

Allelochemical action goes mainly through affecting photosynthesis, respiration cell division, enzymes function and activity, endogenous hormones and protein synthesis. This suggests allelochemical action on the molecular level and gene expression [86]. Some phenolics such as ferulic acid and cinnamic acid can inhibit protein synthesis or amino acid transport and the subsequent growth of treated plants. This is attributed to the ability of all phenolics to reduce integrity of DNA and RNA [86]. A series of physiological and biochemical changes in plants induced by phenolic compounds are shown in Figure 3 [87].

Allelopathic Plant	Impact
- Rows of black walnut interplanted with corn in an alley cropping system	- Reduced corn yield attributed to production of juglone, an allelopathic compound from black walnut, found 4.25 meters from trees
- Rows of Leucaena interplanted with crops in an alley cropping system	- Reduced the yield of wheat and tumeric but increased the yield of maize and rice
- Lantana, a perennial woody weed pest in Florida citrus	- Lantana roots and shoots incorporated into soil reduced germination and growth of milkweed vine, another weed
- Sour orange, a widely used citrus rootstock in the past, now avoided because of susceptibility to citrus tristeza virus	- Leaf extracts and volatile compounds inhibited seed germination and root growth of pigweed, bermudagrass, and lambsquarters
- Red maple, swamp chestnut oak, sweet bay, and red cedar	- Preliminary reports indicate that wood extracts inhibit lettuce seed as much as or more than black walnut extracts
- Eucalyptus and neem trees	- A spatial allelopathic relationship if wheat was grown within 5 m
- Chaste tree or box elder	- Leachates retarded the growth of pangolagrass, a pasture grass but stimulated the growth of bluestem, another grass species
- Mango	- Dried mango leaf powder completely inhibited sprouting of purple nutsedge tubers.
- Tree of Heaven	- Ailanthone, isolated from the Tree of Heaven, has been reported to possess non-selecitve post-emergence herbicial activity similar to glyphosate and paraquat
- Rye and wheat	- Allelopathic suppression of weeds when used as cover crops or when crop residues are retained as mulch.
- Broccoli	- Broccoli residue interferes with growth of other cruciferous crops that follow

Table 2. Examples of Allelopathy from Published Research.

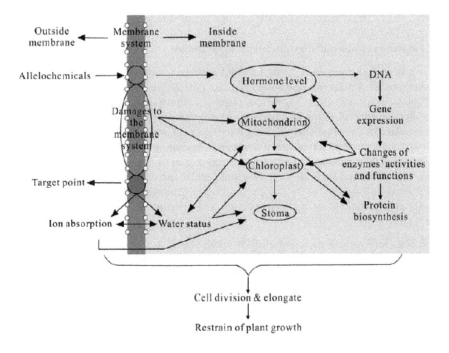

Figure 3. Mechanism of Allelochemicals [87].

6.8. Strategies of allelopathic plants application as natural herbicides

The strategy of allelochemical application is based on their antagonistic or synergistic action. Antagonistic properties of allelopathic plants are utilized in companion cropping system. Growing a companion plant which is selectively allelopathic against certain weeds and does not interfere appreciably with crop growth can greatly reduce weed establishment [96].

The interaction of weeds with crops may be positive; for instance, controlled densities of wild mustard (*Brassica campestris* L.) were interplanted with broccoli (*Brassica oleracea* var. Premium crop), crop yield increased by as much as 50% compared with broccoli planted alone [97].

Allelochemicals may be utilized as stimulators to weed seed germination before sowing the main crops, so that the germinated weeds could be eradicated easily. *Striga asiatica* is a good example for this case since it grows as a parasite to cereal grains in the southeastern United States. *Striga* normally germinates in response to compounds released from its host plants [98]. A germination stimulant, a p-benzoquinone compound from a natural host (sorghum) for *Striga* was identified. This stimulatory compound is used to induce germination of *Striga* and eradicate it before cropping its host. Ethylene was found to be a very effective germination stimulant. Also, ethylene stimulates *Striga* to germinate in the absence of a host [99] since its use as a gas at about 1.5kg/ha has been used effectively via a soil injection to trigger "suicidal" germination of *Striga* and to deplete the numbers of dormant seeds in soil [100].

6.9. Limitation of using allelopathic plants as herbicides

Recruiting allelopathy in weed management is limited by both the allelopathic plant itself and the environment. Production, release and phytotoxicity of allelochemicals are altered by biotic and abiotic soil factors [101, 102] such as plant age, temperature, light and soil conditions, microflora, nutritional status, and herbicide treatments. Toxicity of allelochemicals may be either cleared or increased after releasing into the soil by action of microbes [103] since the toxicity is influenced by soil texture. For instance, amounts of water-soluble phenolics in *P. lanceolata* leaf leachate amended soil varied depending on the soil textural classes if it is clay, sandy-loam, sand, or silty-loam [104]. Some allelopathic agents are active only under hot and dry climates as they work in the vapor phase such as monoterpenes because the high vapor density of the essential oils may penetrate into soil, affecting adversely the under growing plants [105].

High costs for synthesizing many allelochemicals stands as a limiting factor for utilizing allelochemicals. Also, the hazardous action of allelochemicals on human beings limits their use. They may be toxic [91] carcinogenic [106] or even cause thyroid, liver and kidney diseases in monogastric animals [107].

Allelopathic potentiality of some plants is influenced either by the availability or deficiency of nutrient. The deficiency of nutrients favors the production of secondary metabolites. For example in aerobic P-deficient soil, rice roots excrete organic anions, particularly citrate, to solubilize and enhance phosphorus uptake [108]. Some allelochemicals affect the growth of the plant itself, i.e., autotoxic effect as some derivatives of benzoic and cinnamic acids from the root exudates of cucumber since it inhibits root antioxidant enzymes and leaf photosynthesis, transpiration and stomatal conductance in cucumber [109].

Natural herbicides sound attractive as alternatives for herbicides but their application is still surrounded with much concern since they affect humans and environmental equilibrium. The agricultural community cannot discard the use of synthetic herbicides completely at the present time but their use can be reduced up to a certain extent by utilizing allelopathic potentiality as an alternative weed management strategy for crop production.

7. Future prospects for rationalization of herbicide usage by molecular biology

Rationalization of herbicidal use targets mainly the production of plants which are herbicidal themselves by recruiting allelopathic characters. Allelopathy is considered a genetically influenced factor [91]. Allelopathic characteristics are more likely to evolve in competitive populations such as in wild types [110]. Therefore, it is possible to enhance weed suppressive potential of crop cultivars or to transfer allelopathic characteristics from wild types or unrelated plants into commercial crop cultivars through conventional plant breeding methods or other genetic recombination strategies. There are two methods for creating herbicidal plant crops that have been suggested; regulation of gene expression related to alleochemicales biosynthesis; or insertion of genes to produce allelochemicals that are not found in the crop [88].

7.1. Gene insertion

The allelopathic phenomenon as mentioned before refers to the ability of some plant species to suppress other species by releasing allelochemicals, which are not toxic to the originating plant but toxic to surrounding vegetation. Breeding allelopathic cultivars by molecular approaches are more complicated than developing an herbicide-resistant crop. Genetic engineering of allelochemicals bases on their overexpression as valuable secondary metabolites in plants [111]. Most secondary metabolites being used as allelochemicals are products of a multi-gene system might which have to be developed and transformed into the specific crop to produce allelochemicals [112, 113].

Gene insertion targets the change of the recent biochemical pathways into another one which is able to produce new allelochemicals through the insertion of transgenes. Although there is great difficulty to satisfy this approach, it represents the promising molecular approaches available for application in the near future. Various reviews in this trend and reference book on molecular biology of weed control [112, 113] were conducted.

7.2. Regulation of gene expression related to allelochemicals

Regulation of gene expression by a biologist first requires accurate identification of the target allelochemical(s), to determine enzymes and the genes encoding them. Accordingly, a specific promoter can be inserted into crop plants to enhance allelochemical production. Allelochemicals are conditionally expressed by biotic and abiotic factors since some metabolites having allelopathic potential might be newly synthesized or highly elevated in rice plants by UV irradiation [114]. For instance, there is a differential response to UV or other environmental stresses among rice cultivars. The phenylpropanoid pathway intermediates of several allelopathic rice cultivars have the highest content of p-coumaric acid. The latter is a key reaction in the biosynthesis of a large number of phenolic compounds in higher plants. Phenolic compounds are derived from cinnamic acid by the catalysis of 4-hydroxylase (CA4H) enzyme. The activity of CA4H was measured to determine its response to UV irradiation in rice leaves of different varieties. *Kouketsumochi* showed induction for CA4H activity by UV after 24 h of UV irradiation for 20 min while the rice cultivar AUS 196 showed no response. The increase in CA4H enzyme activity as a required enzyme in conversion of cinammic acid into p-coumaric acid suggested a role for CA4H gene in the elevation of the allelopathic function in rice plants [114].

Responsiveness to environmental stresses and plant-plant interaction may be conferred by a specific promoter. A promoter which its induction is responsive to an elicitor can be used to regulate genes that are responsible for coding allelochemicals. The expression of phytoalexins and pathogenesis related genes in plants were reported in response to UV treatment and other plant defense inducers [115, 116]. UV was found to stimulate phytoalexine production in pepper. The effective motifs response to UV light was determined in tobacco by examining the expression of GUS activity of plants transformed with the constructs of various CASC (*Capsicum annuum* sesqiterpene cyclase) promoters fused into GUS gene [115]. This was followed by UV irradiation of the transgenic plants to assure the induction of the CASC promoters through examining GUS activity of the transgenic plants. The levels of GUS activity for transgenic plants with pBI121-KF1 and pBI121-KF6 were significantly elevated by UV-irradiation and had a two-to-threefold increase approximately over the untreated-transgenic plants. In contrast, GUS expression in the transgenic plants with pBI121-CaMV 35S was not changed by UV, and in the other constructs had only a very small increase [117]. The CASC promoters of both KF-1 and KF-6 were suggested to contain cis-acting elements capable of conferring quantitative expression patterns that were exclusively associated with UV irradiation. The regulation of genes associated with allelopathy could be achieved by developing a specific promoter responsive to plant-weed competition or environmental stresses. The CASC promoters of KF-1 and KF-6 obtained may be specific to UV. Thus, this promoter can be used for the overexpression of specific promoters constructed to allelochemical-producing genes [116]. To regulate the CA4H gene in the phenylpropanoid pathway, specific promoters, the CASC-KF1 and KF6, were fused to CA4H gene. The gene constructs were introduced into the binary plant expression vector pIG121-HMR with reverse primer harbouring *Bam*HI site and forward primer harbouring *Hind*III site as illustrated in Figure 4 [118].

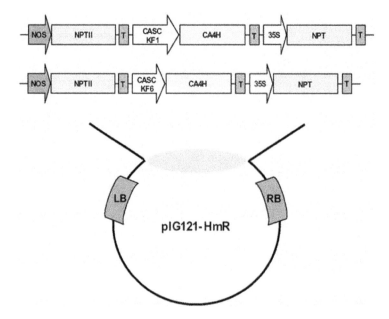

Figure 4. The Gene Cassette with Specific Promoters Responsive to UV Irradiation in pIG121-HmR [117].

8. Conclusion

Herbicides are widely used in agricultural communities on a large scale for eradicating weeds. Herbicides function by affecting different biochemical processes in weeds. Herbicides in low doses act as growth regulators for the main crop but high doses may cause crop damage. However, uncontrolled herbicide use can cause hazardous effects not only upon the main crop but also human health and the surrounding environment [80, 81]. Moreover, heavy doses of herbicides create the problem of herbicide resistance development in weeds. There is an urgent need to identify natural alternatives that can meet the demands of agrosystems without affecting the surrounding environment. Hence, the idea of recruiting the allelopathic phenomenon of some plants in inhibiting the growth of weed vegetation has been investigated. Allelopathy cannot cancel the use of herbicides completely but can minimize it. Allelopathic plant use has limitations in the application because of the potential toxicity. Thus, molecular biology can aid the agricultural community by engineering crops to be herbicides themselves through gene insertion and regulation depending on well-defined allelopathic genes or promoters, respectively. Even with well-characterized allelopathic genes, it might be very difficult to transfer genes into crops.

Author details

Mona H. El-Hadary[1,2*] and Gyuhwa Chung[3]

*Address all correspondence to: drmona3000@yahoo.com

1 Department of Molecular Biology, Genetic Engineering and Biotechnology Research Institute (GEBRI) Minufiya University, Egypt

2 Department of Botany, Faculty of Science, Damanhour University, Egypt

3 Department of Biotechnology, Chonnam National University, Korea

References

[1] El-Hadary, M. H. Effect of Brominal and Granstar Herbicides on Growth and Some Metabolic Activities in Wheat. MSc. thesis. Faculty of Science Tanta University Egypt; (1988).

[2] Hassall, K. A, Editors- Ebert, E, & Kayser, H. Staub T- Book Review: The Biochemistry and Uses of Pesticides. Structure, Metabolism, Mode of Action and Uses in Crop Protection. (2nd Edition). VCH Verlagsgesellschaft mb: Germany; (2003). DOI: 10.1002/anie.199202422.

[3] Rao, . . Principles of Weed Science (2nd ed.). India Science Publishers ISBN 1-5808-069: 1983; 125-160.

[4] Radosevich, S. R, Steinbak, K. E, & Arntsen, C. G. Effect of Photosystem II Inhibitors on Thylkaloid Membranes of Two common Groundsel (*Senecio vulgaris*) Biotypes. Journal of Weed Science (1979). , 27-216.

[5] Giannopolitis, C. N, & Ayers, G. S. Enhancement of Chloroplsat Photooxidations with Photosynthesis-Inhibiting Herbicides and Protection with NADH or NADPH. Journal of Weed Science (1978). , 26 440.

[6] Brown, H. M. Mode of Action, Crop Selectivity and Soil Relations of the Sulfonylurea Herbicides. Journal of Pesticide Science (1990). , 29(3), 263-281.

[7] Moberg, W. K, & Cross, B. Herbicides Inhibiting Branched-Chain Amino Acid Bio-synthesis. Journal of Pesticide Science (1990). , 29(3), 241-246.

[8] Anderson, R. N, Linck, A. J, & Behrens, R. Absorption, Translocation, and Fate of Da-lapon in Sugar Beets and Yellow Foxtail. Journal of Weeds (1962). , 10-1.

[9] Duke, W. B. An Investigation of the Mode of Action of Chloro-N-isopropylacetani-lide. PhD. Thesis. Illinois Urbana University USA; (1967). , 2.

[10] Duke WB; Slife FWHanson JB. Studies on Mode of action of chloro-N-isopropylaceta-nilide. Abstr. Weed Science Society of America (1967). , 2.

[11] Duke, W. B, Slife, F. W, Hanson, J. B, & Butler, H. S. An Investigation on the Mecha-nism of Propachlor. Journal of Weed Science (1975). , 23-142.

[12] Rao, V. S, & Duke, W. B. The Effects of Acetanilide Herbicides on Polysome and Pro-tein Formation. Abstr. Weed Science Society of America (1974).

[13] Deal LM; Reeves JT ; Larkins BA, Hess FD. Use of an *in vitro* Protein Synthesizing System to Test the Mode of Action of Chloroacetamides. Journal of Weed Science (1980). , 28-334.

[14] Hofstra, G, & Switzer, C. M. The Phytotoxicity of Propainil. Journal of Weed Science (1968). , 16-23.

[15] Gruenhagen, R. D, & Moreland, D. E. Effect of Herbicides on ATP Levels in Excised Soybean Hypocotyls. Journal of Weed Science (1971). , 19-319.

[16] Arnold, W. F, & Nalewaja, J. D. Effect of Dicamba on RNA and Protein. Journal of Weed Science (1971). , 301-305.

[17] Mann JD; Jordan LSDay BE. The Effects of Carbamate Herbicides on Polymer Syn-thesis. Deeds (1965a). , 13-63.

[18] Mann JD; Jordan LSDay, B.E. A survey of Herbicides for their Effect upon Protein Synthesis. Journal of Plant Physiology (1965b). , 40-840.

[19] Rost, T. L, & Bayer, D. E. Cell Cycle Population kinetics of Pea Root Tip Meristems Treated with Propham. Journal of Weed Science (1976). , 24-81.

[20] Hoagland, R. E, & Duke, S. O. Relationships Between Phenylalanine Ammonia-Lyase Activity and Physiological Response of Soybean (*Glycine max*) Seedlings to Herbicides. Journal of Weed Science (1983). , 31(6), 845-852.

[21] Szogyi, M, Cserhati, T, & Szigeti, Z. Action of Paraquat and Diquat on proteins and Phospholipids. Journal of Pesticide Biochemistry and Physiology (1989). , 34(3), 240-245.

[22] Hatzios, K. K. Effects of Glufosinate on the Metabolism of Sorghum as Influenced by 2,4-D Asparagine and Glutamine. In: Proceedings of the Southern Weed Sci. Society, 38th Annual Meeting; , 480.

[23] Hatzios, K. K, & Moon, P. A. Combined Effects of Sethoxydim and Chloroacetamide Safeners on the Metabolism of Sorghum Protoplasts and Soybean Cells. In: Proceedings of the Southern Weed Science Society, 38th Annual Meeting; , 462.

[24] Cho HY; Widholm JWSlife EW. Effects of Haloxyfop on Corn (*Zea mays*) and Soybean (*Glycine max*) Cell Suspension Cultures. Journal of Weed Science (1986). , 34(4), 496-501.

[25] Ditomaso, J. M, Rost, T. L, & Ashton, F. M. The Comparative Cell Cycle and Effects of Herbicide Napropamide on Root tip Meristems. Journal of Pesticide Biochemistry and Physiology (1988). , 31(2), 166-174.

[26] Gabr, M. A, & Shakeeb, M. A. Metabolic Changes Associated with growth of Soybean as Affected by Pre-emergence Application of Metribuzin. Canadian Journal of Botany (1988). , 66(12), 2380-2384.

[27] Salem, S. M. Effect of Some Growth Regulators and Micronutrients on Growth and Productivity of Soybean Plants. Bulletin of Faculty of Agriculture Cairo University Egypt (1989). , 40(1), 213-224.

[28] Yakout, G. A, & Soliman, E. L-S. h. a. r. a. k. y A. S. FS. The Effect of Bromoxynil and Mell oxuron Herbicides on Wheat Leaves. Alexandria Science Exchange Egypt (1987). , 8(4), 1-15.

[29] Fathi, S. F, & Shaban, A. Response of Some Wheat Cultivars to Bromoxynil. Zagazig Journal of Agriculture Research Egypt (1991). , 18(3), 729-738.

[30] Morsy, M. A, Zaitoon, M. I, Hanna, L. H, & Ibrahim, I. Z. Effect of Brominal on Yield Components and Uptake of Some Plant Nutrients in Wheat. Annals of Agricultural Science Faculty of Agricultural Science Mashtohor, Zagazig Univesty Egypt.

[31] EL-Desoky IR. The Infleunce of Some Herbicide Mixtures on Wheat and Associated Weeds. PhD thesis. Faculty of Agriculture Cairo University Egypt, 1990.

[32] Shehzad, M. A, Nadeem, M. A, & Iqba, M. Weed Control and Yield Attributes Against Post-emergence Herbicides Application in Wheat Crop. Global Advanced Research Journal of Agricultural Science Punjab, Pakistan (2012). , 1(1), 007-016.

[33] Grundy, A. C, Botman, N. D, & Williams, F. R. Effects of Herbicide and Nitrogen Fertilizer Application on Grain Yield and Quality of Wheat and Barley. Journal of Agricultural Science (1996). , 126(4), 379-385.

[34] Ali-zade, M. A, & Ismailove, A. A. The Use of Herbicides in Cotton Field and their Effect on the Nucleic Acid Content of Cotten Leaves. Weed Abst., (1979). , 28-145.

[35] Ashton, M. Y. de Villiers OT; Glenn RK, Duke W. B. Localization of Metabolic Sites of Action of Hexbicides. Pesticides. Journal of Biochemistry and Physiology (1977). , 7-122.

[36] Henry, W. T, & Hatzioz, K. K. Comparative Effects of Three Urea Herbicidal Derivatives on the Metabolism of Enzymatically Isolated Soybean Leaf Cells. Journal of Weed Researches (1987). , 27-23.

[37] El-Shafey, A. S. EL-Akkad SS. Effect of soil treatment with the herbicide Afalon-S on Certain Physiological Aspects in *Phaseolus vulgaris*. Journal of Desert Researches (1992). , 42(2), 15-18.

[38] Lotlikar, P. D, Remmert, L. F, & Freed, V. H. Effect of D and other Herbicides on Oxidetive Phosphorylation on Mitochondria from Cabbage. Journal of Weed Science (1968). , 2, 4.

[39] Penner, D. Herbicide and inorganic Phosphate on Phytase in Seedlings. Journal of Weed Science (1970). , 18-301.

[40] Moreland, D. E, Malhotra, S. S, Gruenhayen, R. D, & Shokrahii, E. H. Effects of Herbicides on RNA and Protein Synthesis. Journal of Weed Science (1969). , 17-556.

[41] Rao, . . Mechanism of action of acetanilide herbicides. PhD thesis. Cornell University Ithaca New York USA; 1974 p116.

[42] Rao, V. S, & Duke, W. B. Effect of Alachlor, Propachlor and Prynchlor on GAInduced Production of Protease and α-Amylase. Journal of Weed Science (1976). , 3.

[43] Penner, D. Herbicidal Influence on Amylase in Barley and Squash Seedlings. Journal of Weed Science (1968). , 16-519.

[44] Ashton, F. M. Relationship between Light and Toxicity Symptoms Caused by Atrazine and Monuron. In: Weeds (1965). , 13-164.

[45] Tsay, R, & Ashton, F. M. Effect of Several Herbicides on Dipeptidase Activity of Squash Cotyledons. Journal of Weed Science (1971). , 19-682.

[46] Penner, D, & Ashton, F. M. Influnce of Dichlobenil, Endothal and Bromoxynil on Kinin Control of Protolytic Activity. Journal of Weed Science (1968). , 16-323.

[47] Casio, E. G, Weissenbock, G, & Moclure, J. W. Acifluorfen-Induced Isoflavonoids and Enzymes of their Biosynthesis in Mature Soybean Leaves. Whole Leaf and Mesophyll Responses. Journal of Plant Physiology (1985). , 78(1), 14-19.

[48] Schmidt, A, & Kunert, K. J. Lipid Peroxidation in Higher Plants, the Role of Gluta-thione Reductase. Journal of Plant Physiology (1986). , 82(3), 700-702.

[49] Lydon, J, & Ducke, S. O. Pesticide Effects on Secondary Metabolism of Higher Plants. Journal of Pesticide Science (1989). , 25(4), 361-373.

[50] Gronwald, J. W. Lipid Biosynthesis Inhibitors. Journal of Weed Science (1991). , 39(3), 435-449.

[51] Wybieralshi, J, & Wybieralska, A. Enzyme Activity of Wheat Grain Treated with Herbicides Journal of Chemosphere (1988). , 17(1), 159-163.

[52] Pietr, S. J, & Jablonska, E. The Effect of Action of Herbicides on Some Chemical Pa-rameters and the Enzymatic Activity of soils. Polish Journal of Soil Science (1987). , 169(2), 17-23.

[53] Manga, V. A, & Sharma, R. Lack of Functionan Interrelationship between β-amylase Photoregulation and Chloroplast Development in Mustard (*Sinapis alba* L.) Cotyle-dons. Journal of Plant and Cell Physiology (1990). , 31(2), 167-172.

[54] Saeed, M. Regulation of Amylolytic Enzymes in the Photosynthetic Tissues of Pea (*Pisum sativum* L.). Dissertation Abstracts International, Science and Engineer (1990).

[55] Saeed, M, & Duke, S. H. Chloroplastic Regulation of Apoplastic α-amylase Activity in Pea Seedlings. Journal of Plant Physiology (1990). , 93(1), 131-140.

[56] Yoshikawa, H, Fujimolto, E, & Doi, K. Synthesis and Biolgical Activtiy of Benzalde-hyde O-alkyloximes as Abscisic Acid Mimics. Journal of Bioscience, Biotechnology and Biochemistry (1992).

[57] Li, H. H, Nishimura, H, Hasegawa, K, & Mizutani, J. Some Physiological Effects and the Possible Mechanism of Action of Juglone in plants. Journal of Weed research-To-kyo (1993). , 38(3), 214-222.

[58] Hasaneen MNAEL-Saht HM, Bassyoni FM. Growth, Carbohydrates and Associated Invertase and Amylase. Journal of Biologia Plantarum (1994). , 36(3), 451-459.

[59] Kumar, J, & Prakash, J. Effect of Thiobencarb and Butachlor on Photosynthesis, Car-bohydrate Content, Amylase and Protease Activity in Rice (*Oryza sativa*) and Barn-yard Grass (*Echinochloa crus galli*), Indian Journal of. Agricultural Science (1994). , 64(1), 9-14.

[60] Kathiresan, R. M, Gurusamy, A, Brown, H, Cussans, G. W, Devine, M. D, & Duke, S. O. Fernandez, Quintanilla C, Helweg A, Labrada RE, Landes M, Kudsk P, Streibig JC. Herbicide Tolerance in Rice cultivars. In: Proceedings of the Second International Weed Control Congress, Copenhagen, Denmark, 25-28 June (1996). , 1(4), 955-962.

[61] Finckh, B. F, Kunert, K, & Vitamin, C. and "E" an Antioxidative System Against Herbicide-Induced Lipid Peroxidation in Higher Plants. Journal of Agriculture and Food
Chemistry (1985). , 33(4), 574-577.

[62] Valadon LRGKates M. Effect of Perfluidon on Metabolism of Lipids in Maize (*Zea
mays* L.) and Sunflowe (*Helanthus annuus* L.). Journal of Plant-Growth Regulation
(1984). , 1984(3), 2-111.

[63] Bansal, G. L, & Sharma, V. K. Effect of Terbutryn on Maize (*Zea mays*) and Watergrass (*Echinochloa colonum*). Changes in Chlorophyll Content and Carbohydrates. Indian Journal of Weed Science (1989).

[64] Abdou, R. F, & Ahamed, S. A. Cytological and Developmental Effects of Four Herbicides on Barely. Journal of Rachis (1989). , 8(2), 14-16.

[65] El-Bagouri, I. H, Wassif, M. M, Kadi, M. A, & Sabet, S. A. Response of Barley to Foliar
Application of Some Micro Nutrients Under the Conditions of Saline Water Irrigation and Highly Calcareous Soil. Desert Intuitional Bulletin A.R.E. (1983). , 14-1.

[66] Malik, N. Meadow bromegrass and crested wheat grass forage yield response to herbicides applied during establishment. Bibliographic Citation Journal production Agric. (1991). , 4(4), 508-515.

[67] Majid, A, Hussein, M. R, & Mkhtar, M. A. Studies on Chemical Weed Control in
Wheat. Pakistan Journal of Agricultural Research (1983). , 21(4), 167-171.

[68] Gonzalez, M. J, & Ferrandez, G. A. Early Weed Control in Wheat. Revista de los
CREA (1987). , 124-5.

[69] Ashraf, M. Y, & Bahig, N. A. Response of Wheat (*Triticum aestivum* L.) to herbieidal
Wheat Control. Biblographic citation Nucleus- Karahi. (1989).

[70] Fogelfors, H. Different Herbicide Doses in Barley Studies of the Actual Requirement.
Swedish Crop Protection Conference. Weeds and Weed Control (1991). , 32-53.

[71] Montazeri, M, & Saber, H. K. Response of Golestan Wheat Cultivar to D and Bromoxynil at Different Growth Stages. Journal of Revista de los CREA Seed and Plant
(1992). , 2, 4.

[72] Botman, N. D. Effects of Herbicide Use, Fungicide Use and Position in the Field on
the Yield and Yield Components of Spring Barley. Journal of Agricultural Science
(1992). , 118(1), 17-28.

[73] Hallgern, E. Effects of Some Herbicides or Mixtures of Herbicides on Annual Dicots
as a Whole and on Grain Yield at Different Doses, Development Stages and Weed
Densities. Vaxtodling, Institutionen for Vaxtodling, Sveriges Lantbruk, suniversitet,
(1993).

[74] Grundy, A. C, Botman, N. D, & Williams, F. R. Effects of Herbicide and Nitrogen Fertilizer Application on Grain Yield and Quality of Wheat and Barley. Journal of Agricultural Science (1996). a(4) 379-385.

[75] SpiridonovYu. YA, Raskin MS, Samus MV, Grishakova OM, Shestakov VG, Yakovets VI, kirillova NA. Effectiveness of Preparations of Sulfonylurea Derivatives in Weed Control Communication 3. The Effectiveness of Granstar in Sowings of Cereal Crops. Journal of Agrokhimiya, (1990). , 8-116.

[76] Dong, B, Rengel, Z, & Graham, R. D. Effects of Herbicides Chlorsulfuron on Growth and Nutrient Uptake Parameters on Wheat Genotypes Differing in Zn-efficiency. Journal of Plant and Soil (1995).

[77] Stewart, V. R, & Keener, T. K. Evaluation of Four Sulfonylurea Herbicides for Broad Leaved Weed Control in Winridge Winter Wheat. Proceedings of the Western Society of Weed Science (1989).

[78] Drozd, D. Reaction of Spring Wheat Varieties to New Generation Herbicides (Chisel and Granstar). Biuletyn Instytutu Hodowli- I- Ak limatyzacji. Roslin (1995). , 194-199.

[79] Clay, S. A, Gaffney, J. F, & Wrage, L. I. Spring Wheat Cultivar Responses to Trifluralin and Post-emergence Herbicides. Journal of Weed Technology (1995). , 9(2), 352-355.

[80] eHow living healthy. The Effects of Herbicides and Pesticides on Humans., by Flint D. http://www.ehow.com/facts_5636303_effects-herbicides-pesticides-humans.html

[81] Kato, M. Elyanne Ratcliffe MPH, Rohrer WH. Agricultural Pesticide Exposure and its Negative Health Effects. Children Global Medicine. www.dghonline.org. http://www.globalmedicine.nl/index.php/global-medicine-1/agricultural-pesticide-exposure.

[82] Rice, E. L. Allelopathy." (2nd ed.) Academic Press: New York; , 421.

[83] Putnam, A. R. Weed Tech. (1988). , 2-510.

[84] Rice, E. L. Allelopathy. Academic Press: New York; (1974 3). p.

[85] Mann, J. Secondary Metabolism (2nd edi.). Clarendon Press: Oxford;(1987). p.

[86] Li, Z. H, Wang, Q, Ruan, X, Pan, C. D, & Jiang, D. A. Phenolics and Plant Allelopathy. Journal of Molecules (2010). 1420-3049Available at www.mdpi.com/journal/molecules-doi:10.3390/molecules15128933., 15-8933.

[87] Wang, Q, Ruan, X, Li, Z. H, & Pan, C. D. Autotoxicity of Plants and Research of Coniferous Forest Autotoxicity. Sci. Sil. Sin. (2006). , 43-134.

[88] FarooqJabran M, Cheema K, Wahid ZA, Siddique A, Kadambot HM The Role of Allelopathy in Agricultural. Journal of Pest Management Science; (2011). Available at http://onlinelibrary.wiley.com/doi/10.1002/ps.2091/abstract., 2011(67), 5-493.

[89] Lovett, J. V. Allelochemicals, Mycotoxins and Insect Pheromones and Allomones. In: Phytochemical Ecology. Chou CH and Waller GR (ed.).Taipei: ROC; (1989). , 49-67.

[90] Rice, E. L. Botany Review. (1979). , 45-15.

[91] Duke, S. O, Dayan, F. E, Romagni, J. G, & Rimando, A. M. Natural Products as Sources of Herbicides: Current Status and Future Trends. Journal of Weed Research (2000). , 40-99.

[92] Wu, H, Pratley, J, Lemerle, D, Haig, T, & Verbeek, B. Differential Allelopathic Potential among Wheat Accesions to Annual Ryegrass. In: DL Michalk Dl, Pratley JE (eds.) NSW 2650: Proceedings of the 9th Australian Agronomy Conference of the Australian Society of Agronomy: "Agronomy, growing a greener future?", NSW July 1998, Charles Sturt University, Wagga Wagga; (1998). Avalible from http://www.regional.org.au., 2650, 20-23.

[93] Bhadoria PBSAllelopathy: A Natural Way towards Weed Management. American Journal of Experimental Agriculture (2011). , 1(1), 7-20.

[94] An, M, Pratley, J. E, & Haig, T. Allelopathy: From Concept to Reality. In: DL Michalk Dl, Pratley JE (eds.) NSW 2650: Proceedings of the 9th Australian Agronomy Conference of the Australian Society of Agronomy: "Agronomy, growing a greener future?", NSW July 1998, Charles Sturt University, Wagga Wagga; (1998). Avalible from http://www.regional.org.au., 2650, 20-23.

[95] Ferguson, J. J, & Rathinasabapathi, B. Allelopathy: How Plants Suppress Other Plants. Publication series of Horticultural Sciences Department, Florida Cooperative Extension Service, Institute of Food and Agricultural Sciences, University of Florida. HS944 document; July 2003. Reviewed May (2009). EDIS Web site at http://edis.ifas.ufl.edu/hs186.

[96] Putnam, A. R, & Duke, W. B. Annual Reviewof Phytopathology. (1978). , 1978, 16-413.

[97] Jimenez-osornio, J. J, & Gliessman, S. R. In "Allelochemicals: Role in Agriculture and Forestry". Waller GR (ed.). American Chemical Society Washington DC (1987). , 262-274.

[98] Matusova, R, & Rani, K. Verstappen FWA, Franssen MCR, Beale MH,. Bouwmeester HJ. The Strigolactone Germination Stimulants of the Plant-Parasitic *Striga* and *Orobanche* spp. are Derived From the Carotenoid Pathway. Journal of Plant Physiology (2005). , 139(2), 920-934.

[99] Egley, G. H, & Dale, J. E. Ethylene, 2-Cloroethylphosphonic acid, and witched germination. Proceeding 23rd Annual Meeting Southern Weed Science Society; (1970). , 372.

[100] Eplee, R. E. Ethylene: a switched seed stimulant. Weed Science (1975). , 23-433.

[101] Huang, P. M, Wang, M. C, & Wang, M. K. Catalytic Transformation of Phenolic Compounds in the Soil. In Inderjit, et al. (ed.) Principles and Practices in Plant Ecology: Allelochemical interactions. CRC Press: Boca Raton FL; , 1999-287.

[102] InderjitCheng, H.H., Nishimura, H. Plant phenolics and terpenoids: Transformation, Degradation, and Potential for Allelopathic Interactions. In S. Inderjit, et al. (ed.): Principles and practices in Plant Ecology: Allelochemical interactions. CRC Press: Boca Raton FL; , 1999-255.

[103] InderjitAllelopathy Symposium. Soil Environmental Effects on Allelochemical Activity. Journal of Agronomy (2001). , 93-79.

[104] InderjitDakshini KMM. Allelopathic Effect of *Pluchea lanceolata* (Asteraceae) on Characteristics of Four Soils and Tomato and Mustard Growth. American Journal of Botany. (1994). , 81, 799-804.

[105] Koitabashi, R, Suzuki, T, Kawazu, T, Sakai, A, Kuroiwa, H, & Kuroiwa, T. Cineole Inhibits Roots Growth and DNA Synthesis in the Root Apical Meristem of *Brassica campestris* L. Journal of Plant. Research (1997). , 110, 1-6.

[106] InderjitBhowmik PC. The Importance of Allelochemicals in Weed Invasiveness and the Natural Suppression. In: Inderjit, Mallik, A.U. (ed.), Chemical Ecology of Plant: Allelopathy of Aquatic and Terrestrial Ecosystems. Birkhauser Verlag AG: Basal; , 187-192.

[107] Van Etten, C. H, & Tookey, H. L. In: CRC Handbook of Naturally Occurring Food Toxicant. Rechcighl M. Jr (ed.) CRC Press: Boca Raton; , 1983-15.

[108] Kirk GJDSantos EE, Santos MB. Phosphate Solubilization By Organic Anion Excretion from Rice Growing in Aerobic Soil: Rates of Excretion and Decomposition, Effects on Rhizosphere pH and Effects on Phosphate Solubility and Uptake. Journal of New Phytopathology (1999). , 142, 185-200.

[109] Yu, J. Q, & Matsui, Y. Phytotoxic Substances in the Root exudates of *Cucumis sativus* L. Journal of Chemistry and Ecololgy (1994). , 20-21.

[110] Putnam, A. R. editor.Tang CS- The Science of Allelopathy. John Wiley and Sons: New York; (1986 3). p.

[111] Canel, C. From genes to Phytochemicals: the Genomics Approach to the Characterization and Utilization of Plant Secondary Metabolism. Journal of Acta Horticulturae. (1999). , 500-51.

[112] Gressel, J. Molecular Biology of Weed Control. Taylor and Francis Publishers, London. Hahlbrock, K. and D. Scheel 1989. Physiology and Molecular Biology of Phenylpropanoid Metabolism. Annual Review of Plant Physiology Plant Molecular Biology (2002). , 40-347.

[113] Fei, F. H, Chun, L. C, Ze, Z, Zeng, H. L, Fang, Y. D, Cheng, L, & Zhong, H. X. The UDP-glucosyltransferase multigene family in *Bombyx mori*. Journal of Bio Med Central Genomics (2008). doi:10.1186/1471-2164-9-563.

[114] Kim, H. Y, Shin, H. Y, Sohn, D. S, Lee, I. J, Kim, K. U, Lee, S. C, Jeong, H. J, & Cho, M. S. Enzyme Activities and Compounds Rrelated to Self-Defense in UV-Challenged Leaves of Rice. *Korean* Journal of Crop Science (2000). , 46(1), 22-28.

[115] Back, K, He, S, Kim, K. U, & Shin, D. H. Cloning and Bacterial Expression of Sesquiterpene Cyclase, a Key Branch Point Enzyme for the Synthesis of Sesquiterpenoid Phytoalexin Capsidiol in UV-Challenged Leaves of *Capsicum annuum*. Journal of Plant Cell Physiology (1998). , 39(9), 899-904.

[116] El-Hadary, M. H. Molecular Studies on Some Pathogenesis-Related (PRs) Proteins in Tomato Plants. PhD thesis. Genetic Engineering and Biotechnology Research Institute (GEBRI) Minufiya University, Egypt; (2007).

[117] Shin, D. H, Kim, K. U, Sohn, D. S, Kang, S. U, Kim, H. Y, Lee, I. J, & Kim, M. Y. Regulation of Gene Expression Related to Allelopathy. *In:* Kim KU, Shin DH (eds.) *Proc. of the Inernational. Workshop in Rice Allelopathy.* Kyungpook National University, Taegu, Korea, August (2000). Institute of Agricultural Science and Technology, Kyungpook National University, Taegu 2000;109-124, 17-19.

[118] Kim, K. U, & Shin, D. H. The importance of allelopathy in breeding new cultivars. Agriculture and Consumer Protection. FAO Corporate Document Repository-Weed Management for Developing Countries (Addendum 1). Available at http://www.fao.org/docrep/006/Y5031E/y5031e0f.htm.

Permissions

The contributors of this book come from diverse backgrounds, making this book a truly international effort. This book will bring forth new frontiers with its revolutionizing research information and detailed analysis of the nascent developments around the world.

We would like to thank Andrew J. Price and Jessica A. Kelton, for lending their expertise to make the book truly unique. They have played a crucial role in the development of this book. Without their invaluable contribution this book wouldn't have been possible. They have made vital efforts to compile up to date information on the varied aspects of this subject to make this book a valuable addition to the collection of many professionals and students.

This book was conceptualized with the vision of imparting up-to-date information and advanced data in this field. To ensure the same, a matchless editorial board was set up. Every individual on the board went through rigorous rounds of assessment to prove their worth. After which they invested a large part of their time researching and compiling the most relevant data for our readers. Conferences and sessions were held from time to time between the editorial board and the contributing authors to present the data in the most comprehensible form. The editorial team has worked tirelessly to provide valuable and valid information to help people across the globe.

Every chapter published in this book has been scrutinized by our experts. Their significance has been extensively debated. The topics covered herein carry significant findings which will fuel the growth of the discipline. They may even be implemented as practical applications or may be referred to as a beginning point for another development. Chapters in this book were first published by InTech; hereby published with permission under the Creative Commons Attribution License or equivalent.

The editorial board has been involved in producing this book since its inception. They have spent rigorous hours researching and exploring the diverse topics which have resulted in the successful publishing of this book. They have passed on their knowledge of decades through this book. To expedite this challenging task, the publisher supported the team at every step. A small team of assistant editors was also appointed to further simplify the editing procedure and attain best results for the readers.

Our editorial team has been hand-picked from every corner of the world. Their multi-ethnicity adds dynamic inputs to the discussions which result in innovative outcomes. These outcomes are then further discussed with the researchers and contributors who give their valuable feedback and opinion regarding the same. The feedback is then collaborated with the researches and they are edited in a comprehensive manner to aid the understanding of the subject.

Apart from the editorial board, the designing team has also invested a significant amount of their time in understanding the subject and creating the most relevant covers. They scrutinized every image to scout for the most suitable representation of the subject and create an appropriate cover for the book.

The publishing team has been involved in this book since its early stages. They were actively engaged in every process, be it collecting the data, connecting with the contributors or procuring relevant information. The team has been an ardent support to the editorial, designing and production team. Their endless efforts to recruit the best for this project, has resulted in the accomplishment of this book. They are a veteran in the field of academics and their pool of knowledge is as vast as their experience in printing. Their expertise and guidance has proved useful at every step. Their uncompromising quality standards have made this book an exceptional effort. Their encouragement from time to time has been an inspiration for everyone.

The publisher and the editorial board hope that this book will prove to be a valuable piece of knowledge for researchers, students, practitioners and scholars across the globe.

List of Contributors

Jamal R. Qasem
Department of Plant Protection, Faculty of Agriculture, University of Jordan, Amman, Jordan

Maria Aparecida Marin-Morales, Bruna de Campos Ventura-Camargo and Márcia Miyuki Hoshina
Department of Biology, Institute of Biosciences, São Paulo State University (UNESP), SP, Brazil

Valdemar Luiz Tornisielo, Rafael Grossi Botelho, Paulo Alexandre de Toledo Alves, Eloana Janice Bonfleur and Sergio Henrique Monteiro
Laboratory of Ecotoxicology, Center for Nuclear Energy in Agriculture, University of São Paulo, Piracicaba, SP, Brazil

Thaís C. C. Fernandes, Marcos A. Pizano and Maria A. Marin-Morales
Universidade Estadual Paulista, IB-Campus de Rio Claro, Rio Claro/SP, Brazil

Dorota Soltys and Urszula Krasuska
Laboratory of Biotechnology, Plant Breeding and Acclimatization Institute - National Research Institute, Mlochow, Poland

Renata Bogatek and Agnieszka Gniazdowska
Department of Plant Physiology, Warsaw University of Life Sciences – SGGW, Warsaw, Poland

Andrew J. Price
United States Department of Agriculture, Agricultural Research Service, National Soil Dynamics Laboratory, Auburn, Alabama, USA

Jessica A. Kelton
Auburn University, Auburn, Alabama, USA

Istvan Jablonkai
Institute of Organic Chemistry, Research Centre for Natural Sciences, Hungarian Academy of Sciences, Budapest, Hungary

Wendy-Ann Isaac
Department of Food Production, Faculty of Food and Agriculture, The University of the West Indies, St. Augustine, Trinidad

Zongjun Gao and Mei Li
Institute of Plant Protection, Shandong Academy of Agricultural Sciences, Jinan, China

Mona H. El-Hadary
Department of Molecular Biology, Genetic Engineering and Biotechnology Research Institute (GEBRI), Minufiya University, Egypt
Department of Botany, Faculty of Science, Damanhour University, Egypt

Gyuhwa Chung
Department of Biotechnology, Chonnam National University, Korea